1292

FUNDAMENTALS OF NETWORK ANALYSIS

FUNDAMENTALS OF NETWORK ANALYSIS

G. H. Hostetter

CALIFORNIA STATE UNIVERSITY, LONG BEACH

HARPER & ROW, PUBLISHERS, New York
Cambridge, Hagerstown, Philadelphia, San Francisco,
London, Mexico City, São Paulo, Sydney

1817

To Donna

Sponsoring Editor: Charlie Dresser
Project Editor: Penelope Schmukler
Production Manager: Marion Palen
Compositor: Syntax International Pte. Ltd.
Printer and Binder: The Maple Press Company
Art Studio: J & R Technical Services Inc.

FUNDAMENTALS OF NETWORK ANALYSIS

Library of Congress Cataloging in Publication Data

Hostetter, G. H.   1939-
    Fundamentals of network analysis.

    Includes index.
    1. Electric network analysis.   I. Title.
TK454.2.H67   621.319′2   79-9085
ISBN 0-06-042909-7

CONTENTS

CHAPTER TWO Source-Resistor Network Solutions Using Equivalent Circuits

CHAPTER THREE Systematic Simultaneous Equations

CHAPTER FOUR   Source-Resistor Network Properties

Part Two   Inductive and Capacitive Networks

CHAPTER FIVE   Network Differential Equations

Preface

In recent years there have been two trends that are of great significance and concern because they signal major changes in professional technical education.

The first trend involves the changing backgrounds of students entering engineering and scientific programs. In times past it was highly probable that entering students would already have considerable technical background—in military service, as radio amateurs, or as home experimenters. Today, such backgrounds are much less common.

This is not to say that our students are less motivated or less creative. Probably they are more so. But their preparation is generally quite different than in the past, in that a large initial measure of underlying perspective can no longer be assumed. It is thus increasingly necessary and important to start at the beginning and to allow time at first for assimilation of basic concepts.

The second trend is the increasing curricular pressures to concentrate more and more material into a limited amount of time. Some of the greatest pressures have been sustained by the more traditional subjects such as network analysis. Because of the ever wider breadth of topics spanned by electrical engineering, the luxury of several courses required of all students is no longer desirable for many programs. The concepts and skills once taught in two or three courses may now be condensed into a single course. For such a course to capture the student's interest and imagination, an extraordinary degree of perspective and organization is required, together with a style that encourages logical thinking.

Educational Objectives

One does not invest the tremendous amount of time and effort necessary to produce a beginning text of this nature without feeling very strongly that major improvements can be made over previous books. In particular, I believe that it is possible to present the essentials of traditional electrical engineering network analysis in a single course, and that it can be delivered in a manner that is compatible with present student capabilities and needs. Of course, additional topics in networks, beyond this basic core, should be available in other courses for more specialized needs.

Because it is largely self-contained, this text is also quite suitable for non-majors, or for programs in which the various branches of engineering are not differentiated at this level.

This first course in network analysis develops a thorough background in three fundamental areas:

1. Source-resistor networks
2. Switched networks
3. Sinusoidally driven networks

It is designed to give students the following basic skills:

1. Comfort with every aspect of source-resistor network analysis, including equivalent circuits and solutions of networks involving controlled sources.
2. A solid background in the solution of switched first-order networks and an acquaintance with switched RLC networks.
3. A high degree of capability and confidence with sinusoidal network response.

Organization and Pace

The presentation is not linear, with a fixed number of pages to be absorbed per week from beginning to end. At the start of their study most students need a great deal of reassurance and attention to detail. The development is thus gentle and deliberate in the early chapters, with explanations becoming more concise and thought provoking as the reader's sophistication increases.

Suggested scheduling of the various chapters over a 14-week semester and over a 10-week quarter are as follows:

Semester schedule			Quarter schedule	
Week	Chapter		Week	Chapter
1	1		1	1, 2
2	2		2	3
3	3		3	4
4	4		4	5
5	5		5–6	6, 7
6–7	6		7–8	8
8	7		9	9
9–10	8		10	10
11–12	9			
13–14	10			

Acknowledgments

The development and writing of this material has involved three institutions. I first taught courses in networks and began writing course notes at the University of Washington. Much of my viewpoint toward networks and education was shaped there. The manuscript took form over several years of writing, rewriting, and student feedback at California State University, Long Beach. The final assembly and proofs were completed during a sabbatical leave spent at the University of California, Irvine.

I am greatly indebted to my colleagues, especially G. H. Cain, T. Jordanides, H. John Lane, R. T. Stefani, and M. E. Valdez, who class tested the manuscript and offered encouragement in addition to many helpful comments and suggestions.

Professor Larry P. Huelsman of the University of Arizona reviewed the manuscript in detail. His suggestions were greatly appreciated and contributed substantially to the final result.

Special thanks are due to Cynthia Klepadlo who supervised typing of the manuscript and drew the original figures.

G. H. HOSTETTER

Then said a teacher, Speak to us of Teaching.
And he said:
No man can reveal to you aught but that which already lies half asleep in the dawning of your knowledge.

Kahlil Gibran
The Prophet *

* Quoted with the kind permission of the publisher, Alfred Knopf, New York (1961).

INTRODUCTION
To the Student

This book was conceived and written with a burning desire to provide a solid foundation of electrical engineering material, proceeding in a logical, organized, and interesting manner. It is a book of *concepts* and *analysis*, a way of thinking.

Teaching and Learning

There is a common idea to the effect that teaching consists of a transfer of knowledge from one mind to another by a hazy process resembling osmosis or the spread of germs. To the contrary, learning is a highly individual experience. A good teacher and a good textbook may allow you to proceed more rapidly than would otherwise be practical, by organizing and pacing the material. You however, are the one who does the learning; *you* are the one who achieves the understanding.

One of the goals of a professional education is for you eventually to reach the point where you can make good progress in a subject on your own, without classes and teachers. Many of you will one day explore brand new subjects before there are teachers or textbooks to teach them.

Experience and confidence

The understanding you are seeking does not involve much that can be mentally recorded and then played back for the next examination. What you will need beyond reasoning ability is experience and confidence, and this is most easily obtained by a steady effort which includes solving a lot of problems.

Work toward *knowing*, by reason, that you have the correct answers to problems. The reinforcement of checking answers with the book can be very helpful, but do not waste your time misusing them by approaching problems in a trial and error fashion.

Effective studying

Too many students spend countless hours of agony imagining that they are studying when they are largely just punishing themselves. Here are some principles that may increase the effectiveness of your studying:

1. Study the subject regularly.
2. Give the subject your *full* attention.
3. Make every effort to eliminate interruptions.
4. Test your own understanding frequently.
5. Reinforce your learning by periodic review. Written summaries, perhaps including brief examples, are very helpful.
6. Measure your level of understanding by how well you could explain the subject to a colleague for the first time, answering any questions that might arise. Explaining a subject to yourself builds your capability for conscious reasoning.

Nonmajors

If your primary interest is in a different field, relax. We begin from the beginning. Here is a good opportunity to exercise your mind and to test your analytical skills. It is also a good chance to extend your horizons a bit; major innovation and invention generally come from minds that range far, minds that are able to apply knowledge in one area to problems in another area.

The previous remarks about studying are especially important to you if you wish to maximize the effectiveness of the time you invest.

After some brief preliminaries, we begin with ideas about electric current and voltage.

FUNDAMENTALS
OF NETWORK
ANALYSIS

Yes! the apparatus of which I speak, and which will doubtless astonish you, is only an assemblage of a number of good conductors of different sorts arranged in a certain way. Thirty, forty, sixty pieces or more of copper, or better of silver, each in contact with a piece of tin, or what is much better, of zinc and an equal number of layers of water or some other liquid which is a better conductor than pure water, such as salt-water or lye and so forth . . . .

Alessandro Volta
From a communication to the Royal
Society, Milan, 1800.

It is this state of electricity in a series of electromotive and conductive substances which I will briefly call electric current; and as I will continuously be compelled to talk about two opposed directions according to those in which the two electricities move, I propose that each time the subject comes up and in order to avoid repetition, I will describe the direction of the electric current by referring to that of the positive electricity.

Andre-Marie Ampere
From *Memoires sur l'Electrodynamique*
French Academy of Sciences, Paris, 1820

The amount of current in a galvanic chain is directly proportional to the sum of all tensions and inversely proportional to the total reduced length of the chain.

Georg Simon Ohm
From *The Galvanic Chain,*
Mathematically Treated,
Berlin, 1827

The first part of this book is devoted entirely to the analysis of networks containing just sources and resistors. This emphasis is with very good reason: All of the techniques for the solution of source-resistor networks apply to the solution of networks in general.

A chapter is devoted to network solution by equivalent circuits because this method is simple, offers considerable insight, and is widely used in practical applications such as electronic design.

PART ONE
Source-Resistor Networks

To the Instructor

These first four chapters are concerned exclusively with networks consisting of sources and resistors. It is doubtful that knowledge of networks in general can ever surpass the level of one's understanding of their algebraic properties, so a very thorough foundation in these fundamentals is established.

It is the author's experience that, without a careful tour through these basics, it is common for students to experience considerable difficulty with advanced material because of a lack of familiarity with elementary details.

Reference senses. Particular attention is given from the start to the routine incorporation of correct algebraic signs into Kirchhoff's laws, and voltage-current and power relationships. Students who have been allowed, even encouraged, to gloss over the matter of reference senses at this early stage are often later plagued by having written network equations which are only within a few algebraic signs of being correct. Most important though, this carelessness detracts from the solid, assured, reasoning approach to problems we want so much to foster in our students.

Equivalent circuits. Although the *theory* of networks is served by the conversion of network problems to mathematical problems, a great deal of understanding and insight is provided by equivalent circuit solutions. And equivalent circuits is the usual network solution method used in fields such as power and electronics. A step-by-step approach is emphasized, keeping track of which signals are the same in the equivalent networks at each stage.

Network equations. Before proceeding to develop systematic equation-writing methods, linear algebraic equations are discussed briefly and Cramer's rule is summarized.

Nodal equations are then introduced, followed by a complementary presentation of mesh equations. Voltage sources are accommodated in nodal equations by considering the source currents to be unknown; in return, node-to-node voltages are known. Current sources, similarly, are routinely incorporated into mesh equations.

Controlled sources are simply written into the equations in terms of the controlling signal. Then the controlling signal is related, if necessary, to the nodal or mesh variables.

Since the systematic nodal equations are a general method of network solution, and probably the easiest to apply for nonplanar networks, the discussion of mesh equations for nonplanar networks, while complete, is brief.

Network properties. Because of its importance to electronic design, an early section of Chapter Four is devoted to the subject of finding equivalent resistances, particularly where controlled sources are involved. Another section covers superposition of sources and of source components.

The Thévenin and Norton equivalents are derived using Cramer's rule and systematic network equations. Tests to find the source functions and resistance are then presented, including cases in which the two-terminal network includes controlled sources.

With an eye toward future concerns, material on maximum power transfer, transfer ratios, and delta-wye transformation follows.

CHAPTER ONE
Fundamental Concepts and Methods

1.1 Introduction

This first chapter contains a good deal of terminology, together with the two fundamental laws governing the behavior of electrical networks: Kirchhoff's current law and Kirchhoff's voltage law.

All of the basic relationships for electrical networks could easily be printed on half a page, but there is much more to know than this. To communicate with fellow engineers and scientists, it is necessary to be familiar with many technical terms. Systematic and efficient solution methods are necessary, methods that insure timely answers to broad classes of problems. One needs to learn, without spending a lifetime rediscovering them, a number of important basic network properties. And a great deal of experience in solving problems is helpful in gaining the insight that is important for innovation and invention.

When you complete this chapter, you should know

1. What electric current and voltage are.
2. Kirchhoff's current and voltage laws and how to apply them.
3. What a network diagram is and the meanings of the terms element, conductor, node, and loop.

4. How electrical power flow is related to an element's voltage and current.
5. What voltage sources, current sources, and resistors are.
6. How electrical power flow in a resistor is related to the resistor current and to the resistor voltage.
7. How to solve networks, quickly and easily, in situations where all elements are in parallel or where all elements are in series.
8. The current divider rule and the voltage divider rule and how to use them.

The International System of units (Systéme International d'Unités) or "SI units" is used exclusively in network analysis. Apart from stating the appropriate, consistent unit when a quantity is introduced, units will not be continually restated in this text.

1.2 Electric Current and Reference Direction

1.2.1 Current Flow in Conductors

An electric current is a flow of electric charge. In networks one is interested in charge flow in wires or *conductors*. Often, but not always, these conductors are made of a metal such as copper.

The outer electrons of each molecule of a metallic conductor are, in a sense, pooled and shared by neighboring molecules, giving rise to a "sea" of relatively free electrons throughout the material. The electron sea is uniformly distributed within the conductor; the attraction of the ionized molecules for nearby electrons and the repulsion between neighboring electrons tend to equalize the density of electrons in the sea.

Although the free electrons may mill about within the conductor, very few of them ever have sufficient energy to break free of the conductor surface unless the conductor temperature is very high or there is an extremely large electric force perpendicular to the conductor surface. The electron sea is thus confined to a conductor much like water in a pipe. Conductors are also commonly surrounded by an *insulating* material, one in which charge motion is all but completely arrested, to keep conductors (and their free electrons) from coming in unwanted contact with one another.

The sea of electrons in a metallic conductor may be made to flow by the application of externally produced electric forces. In a network, where the propagation time of any disturbance is negligibly small, the forces between the particles within the material act to cause equal currents across each cross section of the conductor. It makes sense then to speak of *the* current through a conductor in a network; the current is the same everywhere along a conductor.

Similar considerations apply to other conductors as well as the metallic ones. An ionized solution such as salt water, for example, is a conductor in which the charge carriers are both Na^{2+} and Cl^{2-} ions. The two types of ions may be made to flow in opposite directions, resulting in a net current.

1.2.2   Definition of Electric Current

Electric current is the equivalent rate of flow of positive charge in a given reference direction.

Network problems are only concerned with *equivalent* total charge flows, expressed in terms of either positive or negative charge. By convention, even though the mobile charges in a metallic conductor are the negative electrons, currents are described in terms of equivalent positive charge flow. The charge carriers in other devices (such as semiconductors) may be positive. And in others, both negative and positive charge motions contribute to the current.

Because there are two possible directions of equivalent positive charge flow in a conductor, a careful definition of current must involve a directional sense, a reference direction for the current.

1.2.3   Specifying Currents

The SI unit for current,

$$\frac{coulomb}{second} = ampere \text{ (symbol A)}$$

is named for Andre-Marie Ampere (1775–1836), who performed early experiments with magnetism and the magnetic effects of electric current.

The usual symbol for a current is the function of time $i(t)$. If several currents are to be indicated, the symbol i is subscripted to distinguish different currents.

The reference direction for a current may be indicated by an arrow beside the symbol for a conductor (a line) on a network diagram, as in Figure 1-1(a).

Figure 1-1   (a) Indicating the reference direction of a current
(b) Reversing the sense of a current reference direction

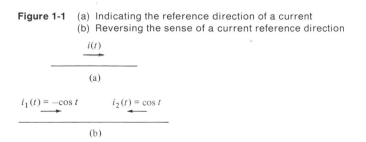

For constant currents it may be advantageous to choose reference directions, when it is convenient to do so, in the actual directions of equivalent positive charge flow. In more involved problems the actual directions of equivalent positive charge flow may not be obvious by inspection. Either direction may then be chosen (or assumed) as the reference direction. If the current with that reference direction turns out to be negative, then the actual direction of equivalent positive charge flow is counter to the direction of the arrow.

For currents that are time varying and change actual direction from time to time, it is impractical to change the senses of the reference arrows whenever the current changes direction. One reference direction is selected to define each current of interest; the current with that reference direction may be negative some times and positive others, a negative current meaning that the equivalent rate of positive charge flow is counter to the chosen reference direction.

Reversing the reference direction reverses the algebraic sign of the current, as in the example in Figure 1-1(b).

Definition of Electric Current

Electric current is the equivalent rate of flow of positive charge in a given reference direction.

The reference direction may be indicated by an arrow on a network diagram.

1.3   Kirchhoff's Current Law

In networks charge does not accumulate within a conductor. Thus the net rate of flow of charge into any conducting region is zero: The sum of the currents with reference directions entering a junction of conductors is equal to the sum of the current with reference directions leaving the junction. This relationship is known as *Kirchhoff's current law*. [Gustav Kirchhoff (1824–1887) was Professor of Physics, Heidelberg.]

The branching of a conductor may be represented by joined lines on a network diagram. Two examples of the application of Kirchhoff's current law to branching conductors are shown in Figure 1-2.

Given all but one of the junction currents, the remaining current is determined, as in Figure 1-3, where

$$i(t) = 6 \sin t - 10 + 5e^{-4t} - 3 \cos 2t$$

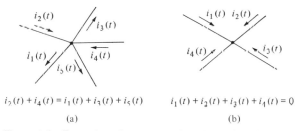

$$i_2(t) + i_4(t) = i_1(t) + i_3(t) + i_5(t)$$

(a)

$$i_1(t) + i_2(t) + i_3(t) + i_4(t) = 0$$

(b)

Figure 1-2 Examples of currents at junctions of conductors

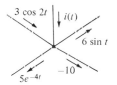

Figure 1-3 Determining one junction current from the others

D1-1

Find the current $I(t)$:

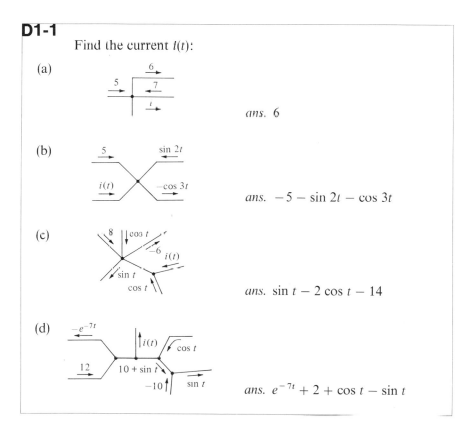

(a)

ans. 6

(b)

ans. $-5 - \sin 2t - \cos 3t$

(c)

ans. $\sin t - 2\cos t - 14$

(d)

ans. $e^{-7t} + 2 + \cos t - \sin t$

Kirchhoff's Current Law

The sum of the currents with reference directions entering a junction of conductors is equal to the sum of the currents with reference directions leaving the junction.

1.4 Voltage and Reference Polarity

1.4.1 Electric Potential in Networks

The electric potential, ϕ, at a point in a network is the potential energy per unit charge of a charge located at that point. The electric potential is generally a function of position in the network. There are conditions in nature for which an electric potential function does not exist. This is to say that it is possible that the energy of a charge depends not only where it is but how it got there. But such is not the case in that class of problems called "networks."

Electric potential in a network is analogous to gravitational potential. As in the gravitational case, the zero level of electric potential is arbitrary; only differences in potential energy are of physical significance. Voltage is a difference in electric potential. The voltage between two points is the energy per unit charge necessary to move the charge through the network from one point to the other.

The unit of voltage, the *volt*,

$$\frac{\text{newton-meter}}{\text{coulomb}} = \text{volt (symbol V)}$$

is in honor of Alessandro Volta (1748–1827), who invented the battery and whose early electrical experiments stimulated the interest of many other great minds of his time.

The usual symbol for a voltage is the function of time $v(t)$. If several voltages are to be indicated, the symbol v is usually subscripted to distinguish the different voltages.

1.4.2 Specifying Voltages

Voltage is a difference in electric potential between two points in a network. The points involved and the sense of the difference may be indicated on a diagram of the network with a bracket and a plus sign, such as that shown in Figure 1-4(a). The plus sign indicates which potential appears with a plus sign in the potential difference; the other potential, at the not-plus side of

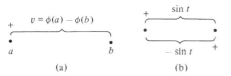

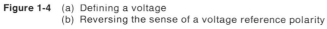

Figure 1-4 (a) Defining a voltage
(b) Reversing the sense of a voltage reference polarity

the bracket, is subtracted. This indication of the sense of the difference in potential is called the *voltage reference polarity*.

It is occasionally advantageous to select one of the two possible reference polarities for a voltage in preference to the other. For a constant voltage it is sensible to choose the reference so that the defined voltage is positive if the actual sense of the voltage is known. If the actual polarity of a voltage is not known, either polarity may be chosen to define it. If that voltage turns out to be negative, so be it; the potential at the point by the plus sign is smaller than the potential at the other point.

The most interesting voltages are time varying and change polarity from time to time. For these, one reference polarity is chosen, and the voltage so defined is sometimes positive, sometimes negative.

Plus signs are used for other purposes in connection with physical devices—for example, to identify the terminals of batteries, transformers, and meters. The plus sign on a voltage *reference polarity*, however, only indicates which potential is subtracted from which. It does *not* necessarily mean that that point nearest the plus sign is at a higher potential than the other point.

Reversing the reference polarity reverses the algebraic sign of the voltage. Figure 1-4(b) is an example.

Voltage

Voltage is a difference in electric potential between two points in a network.

The sense of the potential difference, the reference polarity of a voltage, may be indicated by a bracket and plus sign on a network diagram.

1.5 Kirchhoff's Voltage Law

The sum of the voltages around a loop, if the voltage reference polarities are chosen symmetrically, is zero. For example, for the four network voltages

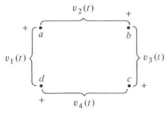

Figure 1-5 Example of voltages around a loop, with symmetric reference polarities

in Figure 1-5,

$$v_1 + v_2 + v_3 + v_4 = \phi(a) - \phi(d) + \phi(b) - \phi(a) + \phi(c) - \phi(b) + \phi(d) - \phi(c)$$
$$= 0$$

If the voltage reference polarities do not happen to be symmetric, plus to not-plus, plus to not-plus, etc., then the sum of the voltages with one reference polarity sense equals the sum of the voltages around the loop with the other reference polarity sense. Two examples of the application of Kirchhoff's voltage law are shown in Figure 1-6. The dots on the sketches represent points on conductors in a network.

Given all but one voltage in a loop, the remaining voltage may be found, as in Figure 1-7, for which

$$v(t) = 3 \cos t + e^{-4t} + 5 - 8 \sin 2t$$

Figure 1-6 Examples of voltages around a loop

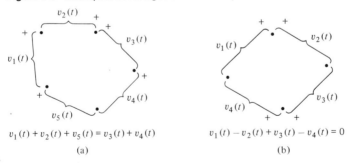

$$v_1(t) + v_2(t) + v_5(t) = v_3(t) + v_4(t)$$

(a)

$$v_1(t) - v_2(t) + v_3(t) - v_4(t) = 0$$

(b)

Figure 1-7 Determining one voltage from the others in a loop

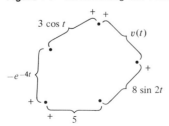

This result, the algebraic sum of the voltages around any closed loop in an electrical network is zero, is known as *Kirchhoff's voltage law.*

D1-2

Find the voltage $v(t)$:

(a)

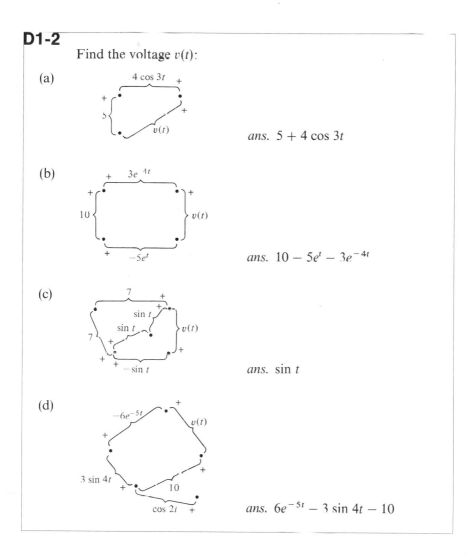

ans. $5 + 4\cos 3t$

(b)

ans. $10 - 5e^t - 3e^{-4t}$

(c)

ans. $\sin t$

(d)

ans. $6e^{-5t} - 3\sin 4t - 10$

Kirchhoff's Voltage Law

Around a closed loop in an electrical network the sum of voltages with one reference polarity sense equals the sum of the voltages around the loop with the other reference polarity sense.

1.6 Network Diagrams

1.6.1 Diagrams and Two-Terminal Elements

An electrical device terminating in two conductors within which charge does not accumulate is called a *two-terminal element*. The current entering one terminal is the same as the current leaving the other terminal. There are then only one current (with either reference direction) and only one voltage (with either reference polarity) to speak of in regard to a two-terminal element. The general symbol for an element is shown in Figure 1-8, together with a defined current through the element $i(t)$ and voltage across it $v(t)$.

An element might represent a battery, a light bulb, an electric heater, and so on. Or the element might be more complicated, representing a combination of simpler devices such as all those connected to an automobile's battery. The important characteristic is that the device has two terminals: All the components could be placed within a box with just two conductors coming out of that box.

An electrical network is a connection of elements. The structure of the network may be indicated by a drawing, a network diagram, which indicates the interconnection of elements. Examples are given in Figure 1-9. Specific types of elements are given more specific symbols than the "box" which is used to indicate an element in general.

Some physical devices, transistors, for example, are inherently three- or more-terminal devices. Fortunately they may be represented by an equivalent set of two-terminal elements.

Figure 1-8 Element with defined current and voltage

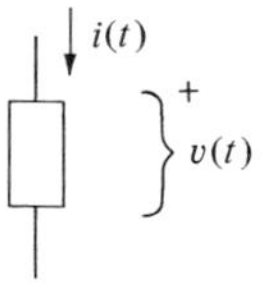

$$i(t)$$

$$+$$
$$v(t)$$

Figure 1-9 Element connections

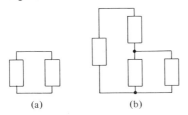

(a)   (b)

1.6.2 Ideal Conductors

An ideal conductor (Figure 1-10) is indicated by a line on a network diagram. It is used to show the interconnection of elements. The same electric current flows at any position along a length of conductor in a network.

Although a voltage, often small, may exist across a length of wire, the ideal conductor idealizes the wire to the point where there is zero voltage across its length. Figure 1-11 illustrates this. To model a wire for which the voltage between the ends of the wire is not negligible, the wire would be modeled by a more complicated element.

1.6.3 Nodes

A *node* is a complete set of conductor connections between elements. Each node is separated by one or more elements, and every two-terminal element connects between two nodes. In the example given in Figure 1-12, connections between conductors are indicated by dots, as is usual. The nodes are indicated by dashed lines.

Figure 1-10 Conductors joining elements in a network diagram

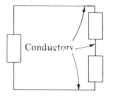

Figure 1-11 There is zero voltage across any length of conductor in a network diagram

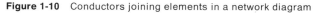

Figure 1-12 Example of network nodes

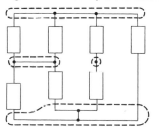

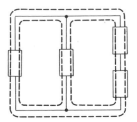

Figure 1-13 Example of network loops

1.6.4 Loops

A *loop* is a closed path through elements and conductors that does not cross itself. Figure 1-13 is an example in which the network loops are indicated by dashed lines.

A network that contains one or more loops is called a *circuit*. Every circuit is a network, but a network must contain at least one loop to be called a circuit.

D1-3

How many nodes and how many loops do each of the following networks have?

(a)

(b)

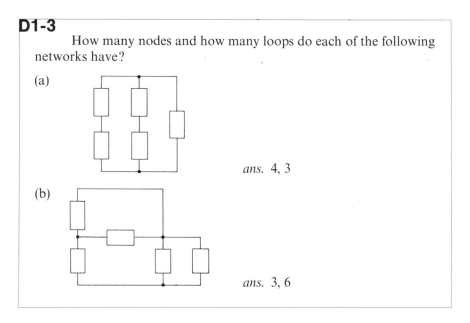

ans. 4, 3

ans. 3, 6

Network Diagrams

Network diagrams consist of elements, joined by conductors.

A node is a complete set of conductor connections between elements. A loop is a closed path through elements and conductors that does not cross itself.

1.7 Electrical Power Flow

1.7.1 Sink and Source Reference Relations

When the relationship between the voltage reference polarity and current reference direction for a two-terminal element is such that the current arrow has the sense entering the terminal nearest the plus sign, the voltage and current are said to be defined with the *sink* relation in the element. The opposite situation is called the *source* relation. The four possible reference orientations are shown in Figure 1-14.

The names *sink* and *source* arose from the fact that if v and i have the source relation and if v and i are both positive, the element is a source of electrical energy, that is, electrical energy is flowing out of the element. Similarly, if v and i have the sink relation and both are positive, the electrical energy flow is into the element; the element is a sink of electrical energy. The most interesting elements, however, accept energy at times and supply energy at other times, so it is best to treat the terms "sink" and "source" just as labels.

1.7.2 Power Flow Relations

A charge Q that moves from potential $\phi(a)$ to potential $\phi(b)$ in passing through an element, as in Figure 1-15, loses the energy

$$W = Q[\phi(a) - \phi(b)] = Qv$$

That is, the electrical energy W is transferred from the charge to the element. If W is negative, which would be the case if the charge flow or v were negative, electrical energy is transferred from the element to the charge.

Figure 1-14 Sink and source reference relations for an element
(a) Sink reference relation
(b) Source reference relation

(a)   (b)

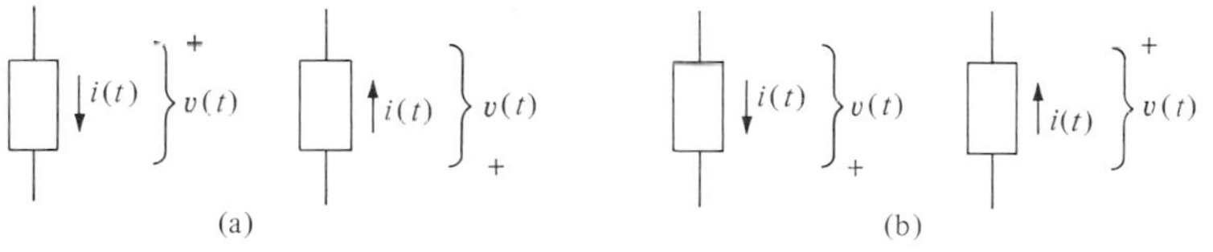

Figure 1-15 Charge flow through a potential difference

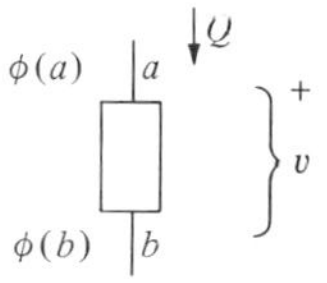

The rate of electrical energy flow, the electrical power flow, into an element is the rate of increase of W with time,

$$P_{into}(t) = \frac{dW}{dt} = \frac{dQ}{dt}\, v = iv$$

where i is the current in the sense from a to b, and v is the voltage $\phi(a) - \phi(b)$. The element voltage is not differentiated in this expression because, for all practical purposes, the effects of a time-varying current are propagated instantaneously in a network. The electrical power flow into a two-terminal element is then

$$P_{into}(t) = v(t)i(t)$$

where $v(t)$ and $i(t)$ have the sink reference relation. The SI unit of power is the *watt* (symbol **W**).

If $v(t)$ and $i(t)$ have the source reference relation, then

$$P_{into}(t) = -v(t)i(t)$$

as summarized in Figure 1-16. Negative power flow means that the actual flow of power is in the opposite direction.

Figure 1-16 Electrical power flow relations
(a) Source reference relation
(b) Sink reference relation

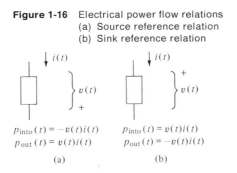

$p_{into}(t) = -v(t)i(t)$
$p_{out}(t) = v(t)i(t)$

$p_{into}(t) = v(t)i(t)$
$p_{out}(t) = -v(t)i(t)$

(a)   (b)

Figure 1-17 Examples of electrical power flow calculations

   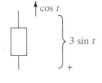 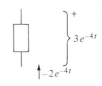

$p_{into}(t) = 10 \sin t$
$p_{out}(t) = -10 \sin t$

$p_{into}(t) = -50$
$p_{out}(t) = 50$

$p_{into}(t) = 3 \sin t \cos t$
$p_{out}(t) = -3 \sin t \cos t$

$p_{into}(t) = 6e^{-8t}$
$p_{out}(t) = -6e^{-8t}$

(a)   (b)   (c)   (d)

The electrical power flow out of an element is the negative of the flow into the element:

$$P_{out}(t) = -P_{into}(t)$$

Several examples of electrical power flow calculations are shown in Figure 1-17.

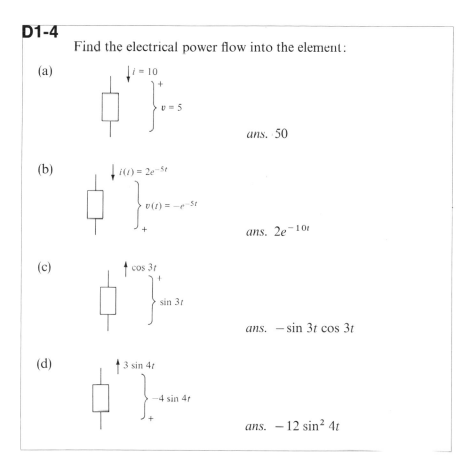

D1-4

Find the electrical power flow into the element:

(a)

$i = 10$

$v = 5$

ans. 50

(b)

$i(t) = 2e^{-5t}$

$v(t) = -e^{-5t}$

ans. $2e^{-10t}$

(c)

$\cos 3t$

$\sin 3t$

ans. $-\sin 3t \cos 3t$

(d)

$3 \sin 4t$

$-4 \sin 4t$

ans. $-12 \sin^2 4t$

Electrical Power Flow

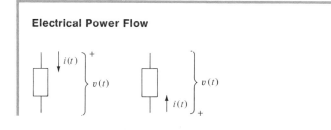

Sink reference relation:

$$P_{\text{into}}(t) = v(t)i(t)$$
$$P_{\text{out}}(t) = -v(t)i(t)$$

Source reference relation:

$$P_{\text{into}}(t) = -v(t)i(t)$$
$$P_{\text{out}}(t) = v(t)i(t)$$

1.8 Basic Source-Resistor Network Elements

1.8.1 Voltage Sources

An element that maintains a specified voltage between its terminals, no matter what the current through the element, is called a *voltage source*. The symbol for the voltage source is a circle, a terminal-to-terminal voltage function for the element, and a plus sign which indicates the polarity reference of the voltage. Figure 1-18(a) shows examples of voltage sources with specific source voltage functions.

If the voltage source function is stated in terms of some other network voltage or current, rather than explicitly as a function of time, the source is termed a *controlled* (or *dependent*) voltage source. The diamond-shaped symbol of Figure 1-18(b) is used then instead of the circle.

Figure 1-18   (a) Examples of fixed (or independent) voltage sources
(b) Examples of controlled (or dependent) voltage sources

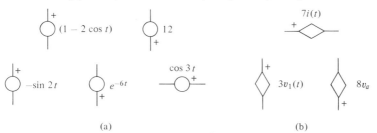

(a)   (b)

1.8.2   Current Sources

An element that maintains a specified current through its terminals, no matter what the voltage across the element, is called a current source. The symbol for a current source is a circle with an arrow indicating the source current reference direction. The current source function is indicated beside the symbol. Examples are given in Figure 1-19(a).

Controlled (or dependent) current sources, for which the source function is expressed in terms of another network voltage or current, are denoted by the diamond symbol, as in Figure 1-19(b).

1.8.3   Resistors

An element for which

$$v(t) = Ri(t)$$

when $v(t)$ and $i(t)$ have the sink reference is called a resistor. The symbol is a zig-zag line, and the constant of proportionality, the *resistance* of the element, is indicated beside the symbol, as in Figure 1-20.

If $v(t)$ and $i(t)$, instead, have the source reference (Figure 1-21),

$$v(t) = -Ri(t)$$

The proportional voltage-current relationship for a resistor is called *Ohm's law*, in honor of Georg Ohm, who is believed to have first suggested the proportional voltage-current relationship. The SI unit of resistance is

$$R = \frac{\text{voltage}}{\text{current}} = \text{ohm (symbol } \Omega)$$

Figure 1-19   (a) Examples of fixed (or independent) current sources
(b) Examples of controlled (or dependent) current sources

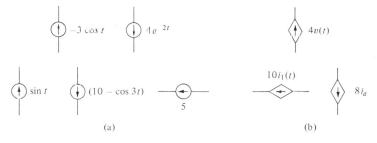

(a)   (b)

Figure 1-20   Resistor symbol and sink reference voltage-current relation

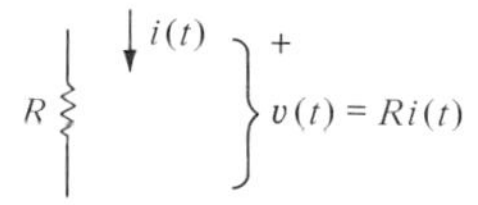

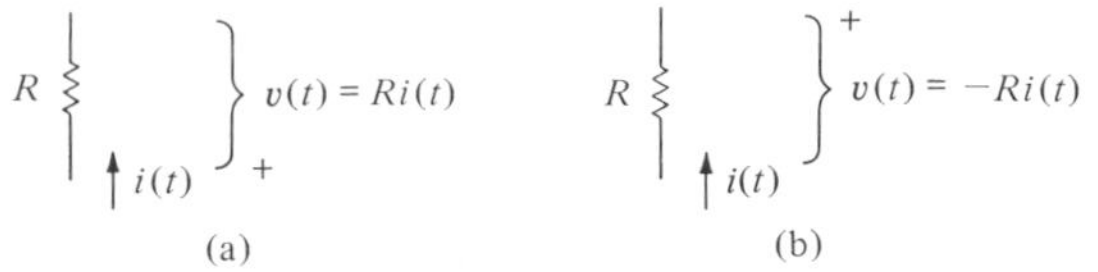

$v(t) = Ri(t)$

$v(t) = -Ri(t)$

(a)

(b)

Figure 1-21 Sink and source reference voltage-current relations for the resistor
(a) Sink reference relation
(b) Source reference relation

$i(t) = \dfrac{10}{4}$

$v(t) = -16 \sin 3t$

$i(t) = -\dfrac{7}{3} \cos t$

$v(t) = 50 - 5e^{-2t}$

(a)

(b)

(c)

(d)

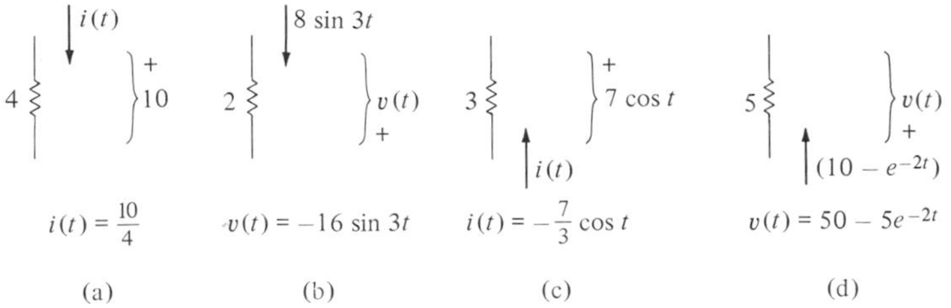

Figure 1-22 Example applications of Ohm's law

Physical devices called resistors are manufactured with a wide range of resistance, R, from a small fraction of an ohm to well over 100 million ohms. Resistors are marketed in many different sizes, the larger sizes capable of dissipating more heat power than the smaller sizes. The resistor is also a useful model of many other devices, such as a length of wire, a light bulb, and an electric heater.

Several examples of the application of Ohm's law are shown in Figure 1-22.

For a resistor of known resistance, if either the resistor voltage or the resistor current is known, the other is easily found via Ohm's law. The algebraic sign of the result is dependent on whether the defined element voltage and current have the sink or the source reference relation.

D1-5

Find $v(t)$:

(a)

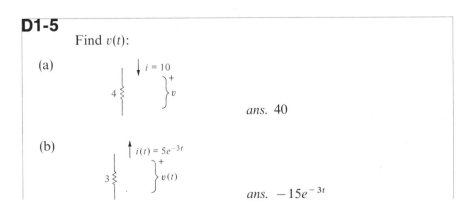

ans. 40

(b)

ans. $-15e^{-3t}$

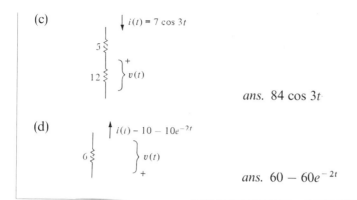

(c)

$i(t) = 7 \cos 3t$

$v(t)$

ans. 84 cos 3*t*

(d)

$i(t) = 10 - 10e^{-2t}$

$v(t)$

ans. $60 - 60e^{-2t}$

D1-6

Find $i(t)$:

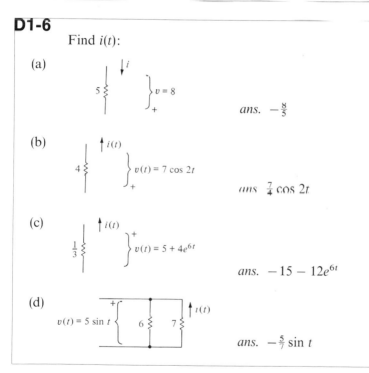

(a)

$v = 8$

ans. $-\frac{8}{5}$

(b)

$i(t)$

$v(t) = 7 \cos 2t$

ans. $\frac{7}{4} \cos 2t$

(c)

$i(t)$

$v(t) = 5 + 4e^{6t}$

ans. $-15 - 12e^{6t}$

(d)

$v(t) = 5 \sin t$ 6   7 $i(t)$

ans. $-\frac{5}{7} \sin t$

D1-7

Find the resistance of the resistor R:

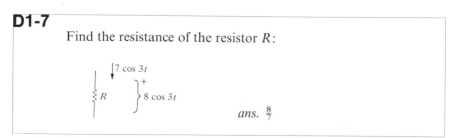

$7 \cos 3t$

R $8 \cos 3t$

ans. $\frac{8}{7}$

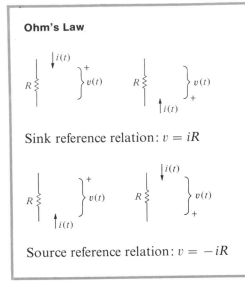

Ohm's Law

Sink reference relation: $v = iR$

Source reference relation: $v = -iR$

1.9 Power Relations for the Resistor

Using the voltage-current relation for the resistor, the power flow relations may be expressed in terms of the resistor voltage or the resistor current exclusively,

$$P_{\text{into}}(t) = i^2(t)R = \frac{v^2(t)}{R}$$

which hold for either the sink or the source reference since reversing the algebraic sign of v or i does not affect v^2 or i^2. Examples of resistor electrical power flow calculations are given in Figure 1-23.

Figure 1-23 Examples of electrical power flow calculations for resistors

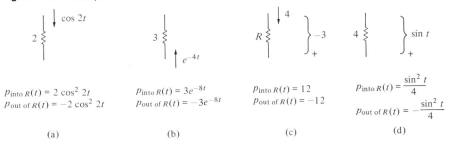

$p_{\text{into } R}(t) = 2\cos^2 2t$
$p_{\text{out of } R}(t) = -2\cos^2 2t$

$p_{\text{into } R}(t) = 3e^{-8t}$
$p_{\text{out of } R}(t) = -3e^{-8t}$

$p_{\text{into } R}(t) = 12$
$p_{\text{out of } R}(t) = -12$

$p_{\text{into } R}(t) = \dfrac{\sin^2 t}{4}$

$p_{\text{out of } R}(t) = -\dfrac{\sin^2 t}{4}$

(a)   (b)   (c)   (d)

For a resistor with positive resistance R, $P_{into\ R}$ is always nonnegative. This is to say that electrical power flow is always into the resistor; the resistor always dissipates electrical energy.

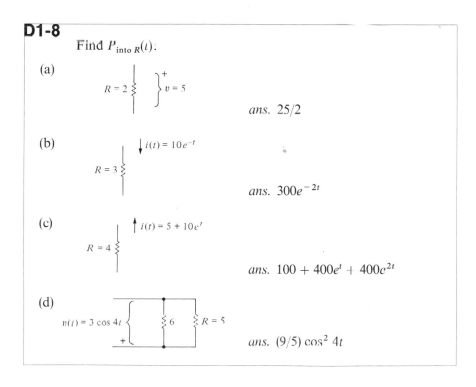

D1-8

Find $P_{into\ R}(t)$:

(a)

$R = 2$ $v = 5$

ans. $25/2$

(b)

$i(t) = 10e^{-t}$

$R = 3$

ans. $300e^{-2t}$

(c)

$i(t) = 5 + 10e^t$

$R = 4$

ans. $100 + 400e^t + 400e^{2t}$

(d)

$v(t) = 3 \cos 4t$ 6 $R = 5$

ans. $(9/5) \cos^2 4t$

Resistor Power Relations

$$P_{into\ R}(t) = i^2(t)R$$

$$P_{into\ R}(t) = \frac{v^2(t)}{R}$$

1.10   Solution of Parallel Networks

1.10.1   The Case of a Voltage Source between Nodes

Solving a network means finding any desired voltages and currents in the network. Consider first networks that have just two nodes, that is, networks where all of the elements are in *parallel*. If one of the parallel elements is

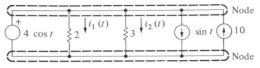

Figure 1-24 Parallel network with a voltage source between the nodes

a voltage source, then the only unknown element currents are resistor currents, which can be found by the application of the resistor voltage-current relations. For the network of Figure 1-24,

$$i_1(t) = \frac{4\cos t}{2}$$

$$i_2(t) = \frac{4\cos t}{3}$$

There cannot be two different voltage sources in parallel because there can be only one specific node-to-node voltage.

1.10.2 Nontrivial Two-Node Networks

A more interesting problem involves the parallel connection of just current sources and resistors. An example is shown in Figure 1-25. If the node-to-node voltage $v(t)$ were known, it would be a simple matter to find any other unknown signal in the network. So the fundamental problem here is to find the node-to-node voltage with one of the two possible reference polarities.

In terms of the voltage $v(t)$ indicated in the figure,

$$i_1(t) = \frac{v(t)}{2}$$

$$i_3(t) = -\frac{v(t)}{3}$$

$$i_4(t) = \frac{v(t)}{4}$$

Figure 1-25 Nontrivial parallel network example

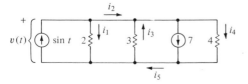

$$i_2(t) - \sin t - i_1(t) = \sin t - \frac{v(t)}{2}$$

$$i_5(t) = 7 + i_4(t) = 7 + \frac{v(t)}{4}$$

Alternatively,

$$i_2(t) = 7 + i_4(t) - i_3(t)$$

and

$$i_5(t) = \sin t + i_3(t) - i_1(t)$$

To find $v(t)$, apply Kirchhoff's current law at one of the two nodes, say, the top node. For the example of Figure 1-25,

$$\sin t - i_1 + i_3 - 7 - i_4 = 0$$

(If Kirchhoff's current law is applied at the bottom node, the same equation results, multiplied through by -1.) Now substitute for the resistor currents in terms of the node-to-node voltage $v(t)$:

$$\sin t - \frac{v(t)}{2} + \left(-\frac{v(t)}{3}\right) - 7 - \frac{v(t)}{4} = 0$$

Rearranging,

$$(\tfrac{1}{2} + \tfrac{1}{3} + \tfrac{1}{4})v(t) = \sin t - 7$$

which is easily solved for $v(t)$.

The corresponding equation for any such parallel network has the form

$$\begin{pmatrix} \text{sum of inverses of} \\ \text{resistances of resistors} \\ \text{connected between} \\ \text{the two nodes} \end{pmatrix} v(t) = \begin{pmatrix} \text{sum of current source functions for} \\ \text{current sources with reference direc-} \\ \text{tions entering the node closest to the} \\ + \text{ sign on the reference polarity for } v(t) \end{pmatrix}$$

$$- \begin{pmatrix} \text{sum of source currents of sources} \\ \text{with reference direction leaving the} \\ \text{node closest to the } + \text{ sign on the} \\ \text{reference polarity for } v(t) \end{pmatrix}$$

Figure 1-26 shows another example two-node network. Its solution is as follows:

$$(\tfrac{1}{2} + \tfrac{1}{4} + \tfrac{1}{6})v(t) = 3 \sin 5t + 10 - \cos t$$

$$v(t) = \tfrac{12}{11}(3 \sin 5t + 10 - \cos t)$$

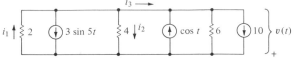

Figure 1-26 Another example parallel network

Given the node-to-node voltage, the other indicated signals are

$$i_1(t) = \frac{v(t)}{2} = \frac{6}{11}(3 \sin 5t + 10 - \cos t)$$

$$i_2(t) = -\frac{v(t)}{4} = -\frac{3}{11}(3 \sin 5t + 10 - \cos t)$$

$$i_3(t) = i_1(t) - 3 \sin 5t - i_2(t) = -\frac{6}{11} \sin 5t + \frac{90}{11} \cos t.$$

D1-9

Write and solve a two-node equation for the voltage $v(t)$, then find the other indicated signal using v.

When several answers are given, as they are in this set of drill problems, the voltages are given first, then the currents, in individual order of their subscripts.

(a)

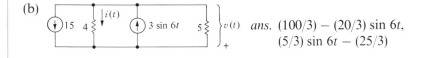

ans. 12, 6

(b)

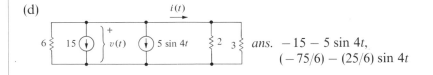

ans. $(100/3) - (20/3) \sin 6t,$
$(5/3) \sin 6t - (25/3)$

(c)

ans. $(10/7) \sin 3t - (40/7) \cos t,$
$(20/7) \cos t - (5/7) \sin 3t$

(d)

ans. $-15 - 5 \sin 4t,$
$(-75/6) - (25/6) \sin 4t$

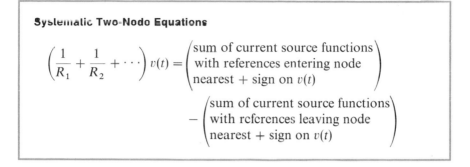

Systematic Two-Node Equations

$$\left(\frac{1}{R_1} + \frac{1}{R_2} + \cdots\right) v(t) = \begin{pmatrix}\text{sum of current source functions}\\ \text{with references entering node}\\ \text{nearest + sign on } v(t)\end{pmatrix}$$

$$-\begin{pmatrix}\text{sum of current source functions}\\ \text{with references leaving node}\\ \text{nearest + sign on } v(t)\end{pmatrix}$$

1.11 The Current Divider Rule

Situations are often encountered in which a known current flows through the parallel connection of two resistors. Part of this net current $i(t)$ flows through one resistor and the rest flows through the other resistor. Although this problem can be solved by writing a two-node equation every time it is encountered, it will save considerable time and effort to learn to write at a glance the fraction of $i(t)$ that flows through each of the two resistors. A general situation is pictured in Figure 1-27.

The two-node equation for $v(t)$ in Figure 1-27 is

$$\left(\frac{1}{R_1} + \frac{1}{R_2}\right) v(t) = i(t)$$

and $v(t)$ and $i_1(t)$ are related by

$$v(t) = i_1(t)R_1$$

Solving the above two equations for $i_1(t)$,

$$i_1(t) = \frac{R_1 R_2}{R_1(R_1 + R_2)} i(t) = \frac{R_2}{R_1 + R_2} i(t)$$

$$= \frac{(\text{opposite resistance})}{(\text{sum of the two resistances})} i(t)$$

Figure 1-27   Division of current between two resistors

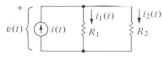

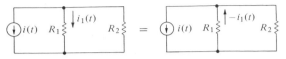

Figure 1-28 Current division with other reference senses

Similarly,

$$i_2(t) = \frac{\text{(opposite resistance)}}{\text{(sum of the two resistances)}} = \frac{R_1}{R_1 + R_2} i(t)$$

This relationship, the *current divider rule*, applies directly only when the reference directions of the various currents have the relative senses in the drawing of Figure 1-27. The results for other relative senses of the current reference directions follow easily. An example is shown in Figure 1-28, for which

$$-i_1 = \frac{R_2}{R_1 + R_2} i(t)$$

$$i_1 = \frac{-R_2}{R_1 + R_2} i(t)$$

D1-10

Use the current divider rule to find the current $i_1(t)$:

(a)

(b)

(c)

(d)

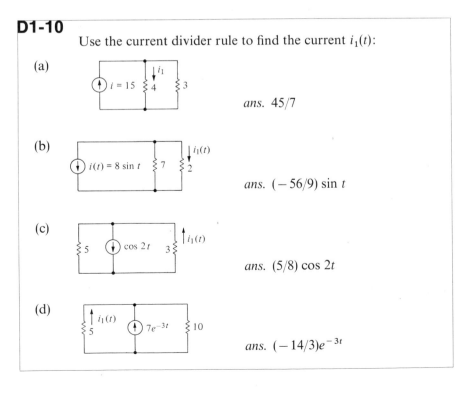

ans. 45/7

ans. $(-56/9) \sin t$

ans. $(5/8) \cos 2t$

ans. $(-14/3)e^{-3t}$

Current Dividor Rule

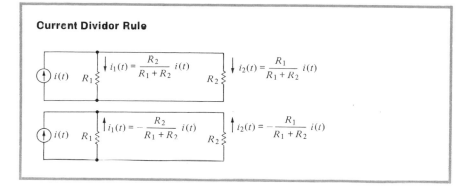

1.12   Solution of Series Networks

1.12.1   The Case of a Current Source in the Loop

A single-loop (or *single-mesh*) network is a *series* (end-to-end) connection of elements, as in the drawing of Figure 1-29. There cannot be two different current sources in the loop, otherwise one would be saying that the loop current is one function on the one hand and another function on the other.

If there is a current source in the loop, finding any other voltage in the network (there is only one current to speak of) is simple. For the example of Figure 1-29,

$$v_1(t) = 10e^{-6t}$$

$$v_2(t) = -15e^{-6t}$$

$$v_3(t) = v_1 - 10 = 10e^{-6t} - 10$$

$$v_4(t) = -10 + v_1(t) - v_2(t) + \sin 4t$$

$$= -10 + 10e^{-6t} - (-15e^{-6t}) + \sin 4t$$

$$= \sin 4t - 10 + 25e^{-6t}$$

1.12.2   Nontrivial Single-Loop Networks

A more interesting problem is the single-loop network with only voltage sources and resistors in the loop, as the one in Figure 1-30. If the loop

Figure 1-29   Series network with a current source in the loop

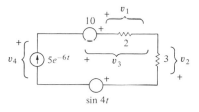

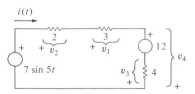

Figure 1-30 Nontrivial series network example

current $i(t)$ were known, it would be a simple matter to find any other signal in the network:

$$v_1(t) = 3i(t)$$
$$v_2(t) = 2i(t)$$
$$v_3(t) = -4i(t)$$
$$v_4(t) = v_3(t) - 12 = -4i(t) - 12$$

Applying Kirchhoff's voltage law around the loop gives

$$7 \sin 5t - v_2(t) - v_1(t) - 12 + v_3(t) = 0$$

Substituting for the resistor voltages in terms of the loop current $i(t)$,

$$7 \sin 5t - 2i(t) - 3i(t) - 12 + (-4i(t)) = 0$$
$$(2 + 3 + 4)i(t) = 7 \sin 5t - 12$$

Every network with a single loop and no current source in the loop has a loop equation of the form

$$\begin{pmatrix} \text{sum of resistances} \\ \text{around the loop} \end{pmatrix} i(t) = \begin{pmatrix} \text{sum of source functions of} \\ \text{sources that have the source} \\ \text{reference with the loop current } i(t) \end{pmatrix}$$
$$- \begin{pmatrix} \text{sum of source functions of} \\ \text{sources that have the sink} \\ \text{reference with the loop current } i(t) \end{pmatrix}$$

For the example of Figure 1-30,

$$i(t) = \tfrac{1}{9}(7 \sin 5t - 12)$$

and

$$v_1(t) = \frac{1}{3}(7 \sin 5t - 12)$$

$$v_2(t) = \frac{2}{9}(7 \sin 5t - 12)$$

$$v_3(t) = -\frac{4}{9}(7 \sin 5t - 12)$$

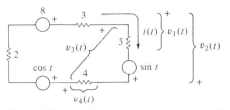

Figure 1-31 Another example series network

$$v_4(t) = -\frac{4}{9}(7 \sin 5t - 12) - 12 = -\frac{28}{9}\sin 5t - \frac{60}{9}$$

Figure 1-31 is another single-loop network example and is solved as follows:

$$(2 + 3 + 5 + 4)i(t) = \sin t - \cos t + 8$$

$$i(t) = \frac{1}{14}(\sin t - \cos t + 8)$$

Using $i(t)$, the other indicated network signals are as follows:

$$v_1(t) = \frac{5}{14}(\sin t - \cos t + 8)$$

$$v_2(t) = \sin t - \frac{5}{14}(\sin t - \cos t + 8)$$

$$= \frac{9}{14}\sin t + \frac{5}{14}\cos t - \frac{40}{14}$$

$$v_3(t) = v_1 - \sin t - v_4$$

$$= -\frac{5}{14}\sin t - \frac{9}{14}\cos t + \frac{72}{14}$$

$$v_4(t) = -\frac{4}{14}(\sin t - \cos t + 8)$$

D1-11

Write and solve a single-loop equation for the current $i(t)$, then use $i(t)$ to find the other indicated signal:

(a)

ans. 2, 6

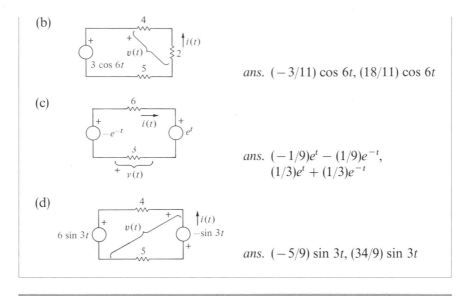

(b)

ans. $(-3/11)\cos 6t$, $(18/11)\cos 6t$

(c)

ans. $(-1/9)e^t - (1/9)e^{-t}$,
$(1/3)e^t + (1/3)e^{-t}$

(d)

ans. $(-5/9)\sin 3t$, $(34/9)\sin 3t$

Systematic Single-Loop Equations

$$(R_1 + R_2 + R_3 + \cdots)i(t) = \begin{pmatrix} \text{sum of voltage source functions} \\ \text{with source reference with } i(t) \end{pmatrix}$$
$$- \begin{pmatrix} \text{sum of voltage source functions} \\ \text{with sink reference with } i(t) \end{pmatrix}$$

1.13 The Voltage Divider Rule

Often there are situations in which it is desired to determine how a known voltage, $v(t)$, is distributed across two resistors in series. Some fraction of $v(t)$ is $v_1(t)$ and the remaining fraction of $v(t)$ is $v_2(t)$, as indicated in Figure 1-32. It should be carefully noted that there is only the direct connection between the two resistors in this situation; the same current flows through each.

Figure 1-32 Division of voltage across two resistors

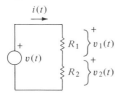

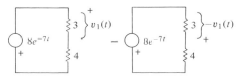

Figure 1-33 Voltage division with other reference polarities

This problem could be solved anew every time it is encountered but, better, it can be learned once and for all the fractions of $v(t)$ that are the voltages $v_1(t)$ and $v_2(t)$, thus saving considerable time and effort in the future.

The single-loop equation for the general network is

$$(R_1 + R_2)i(t) = v(t)$$

and v_1 is related to the loop current $i(t)$ by

$$v_1(t) = R_1 i(t)$$

Solving for v_1,

$$v_1(t) = \frac{R_1}{R_1 + R_2} v(t) = \frac{\text{(same resistor)}}{\text{(sum of the two resistors)}} v(t)$$

Similarly,

$$v_2(t) = \frac{\text{(same resistor)}}{\text{(sum of the two resistors)}} v(t)$$

$$= \frac{R_2}{R_1 + R_2} v(t)$$

These results, similar to those for the current divider rule, may be easily applied to situations in which the references have different senses than those for the basic problem. An example is shown in Figure 1-33, for which

$$-v_1 = \frac{3}{3+4} \cdot 8e^{-7t}$$

$$v_1(t) = -\frac{24}{7} e^{-7t}$$

D1-12

Use the voltage divider rule to find the voltage $v_1(t)$:

(a)

$$v = 10$$

$$\text{ans. } 20/3$$

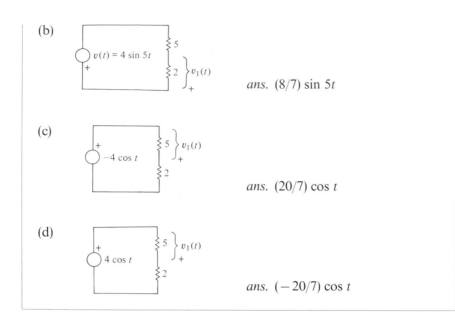

(b)

$v(t) = 4 \sin 5t$

ans. $(8/7) \sin 5t$

(c)

$-4 \cos t$

ans. $(20/7) \cos t$

(d)

$4 \cos t$

ans. $(-20/7) \cos t$

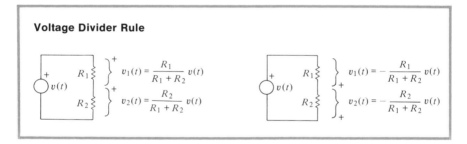

Voltage Divider Rule

$$v_1(t) = \frac{R_1}{R_1 + R_2} v(t)$$

$$v_2(t) = \frac{R_2}{R_1 + R_2} v(t)$$

$$v_1(t) = -\frac{R_1}{R_1 + R_2} v(t)$$

$$v_2(t) = -\frac{R_2}{R_1 + R_2} v(t)$$

Chapter One Problems

There are three types of problems at the end of each chapter. The *basic problems* consist of problems similar to the drill problems but which may require combining the techniques of several sections.

The *practical problems* involve practical numbers and SI units with multiplier prefixes, and are generally of the type encountered in engineering practice. Some of these problems introduce common situations in electrical measurement and instrumentation.

The *advanced problems* introduce topics and problems of a more theoretical nature.

Basic Problems

Kirchhoff's Laws

1.   Find the current $i(t)$:

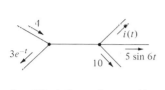

2.   Find the voltage $v(t)$:

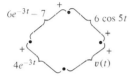

Resistance

3.   Find the voltage $v(t)$:

4.   Find the current $i(t)$:

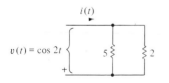

Parallel Networks

5.   Find the current $i(t)$:

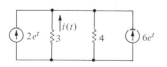

6. Find the current $i(t)$:

7. Find the electrical power flow *out* of the source:

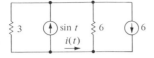

Series Networks

8. Find the voltage $v(t)$:

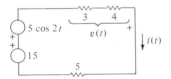

9. Find the voltage $v(t)$:

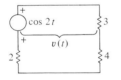

10. Find the electrical power flow into the 4-Ω resistor:

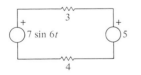

Practical Problems

Resistor Color Code

The color code of Table 1-1 is used for marking electronic component values A color code for the percentage tolerance of components, Table 1-2, is also used.

Fixed-composition type resistors are constructed of a cylinder of resistive material, each end of which makes contact with a wire terminal. The resis-

Table 1-1 Color Code

Color	Significant figure	Decimal multiplier
Black	0	$10^0 = 1$
Brown	1	$10^1 = 10$
Red	2	$10^2 = 100$
Orange	3	10^3
Yellow	4	10^4
Green	5	10^5
Blue	6	10^6
Violet	7	10^7
Gray	8	$10^{-2} = 0.01$
White	9	$10^{-1} = 0.1$

Table 1-2 Component Tolerence

Color	Percent tolerance
Gold	± 5
Silver	± 10
No color	± 20

tive material is surrounded by an insulating case. This type of resistor is coded with four color bands as indicated in the sketch of Figure 1-34(a). Bands A and B give the significant figures of the resistance, band C gives the decimal multiplier to those first two digits, and band D gives the percentage tolerance. For instance, the resistor depicted in Figure 1-34(b) has value $25 \times 10^2 = 2500\ \Omega \pm 5\%$.

1. For a resistor with color code as shown in Figure 1-34(c), what is the resistance and what is the tolerance? Specify the range of resistance within which the coded resistor must fall.

2. What is the color code for a 23 Ω $\pm 10\%$ resistor?

Figure 1-34 Component color code

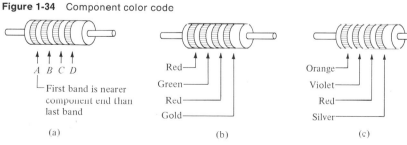

(a) (b) (c)

A group of resistors packaged in a dual in-line package (DIP), the package which is commonly used for integrated circuits. This unit replaces the 10 discrete resistors, above. (*Photo courtesy of Beckman Instruments, Inc., Helipot Division*)

Preferred Resistor Values

Table 1-3 shows the significant figures of standard composition resistor values for each of the three common tolerances. The standard values are placed approximately 40% apart for $\pm 20\%$ tolerance resistors, 20% apart for $\pm 10\%$ tolerance, and 10% apart for $\pm 5\%$ tolerance.

3.   What is the closest $\pm 10\%$ standard value to 4075 Ω?

4.   A certain application requires the use of a composition resistor with a resistance that lies between 11,450 and 15,200 Ω. Specify a standard value and tolerance for the resistor, using the largest possible resistor tolerance for greatest economy.

Resistor Power Rating

In addition to their resistance, physical resistors are also characterized by their power rating, the maximum average power they are designed to dis-

Table 1-3 Significant Figures of
Standard Values

±20%	+10%	+5%
10	10	10
		11
	12	12
		13
15	15	15
		16
	18	18
		20
22	22	22
		24
	27	27
		30
33	33	33
		36
	39	39
		43
47	47	47
		51
	56	56
		62
68	68	68
		75
	82	82
		91
100	100	100

sipate under normal conditions. Except possibly for very short intervals of time, $p_{into\ R}$ should not exceed the resistor power rating.

Standard power ratings for composition type resistors are $\frac{1}{4}$, $\frac{1}{2}$, 1, and 2 W. Larger sized resistors, frequently made of a length of resistance wire wound on a tubular bobbin (called *wirewound* resistors), are available for applications requiring greater power dissipation.

5. What is the maximum constant voltage that may be safely applied to a 100-Ω $\frac{1}{2}$-W resistor? What is the maximum safe constant current for the same resistor?

6. Two resistors, of resistance 3 and 10 Ω, are connected in series so that the same current will flow through each. If the 3-Ω resistor power rating should be 100 W, what is the smallest acceptable power rating for the 10-Ω resistor?

7. What is the smallest acceptable power rating for the 50 Ω resistor in the network of Figure 1-35?

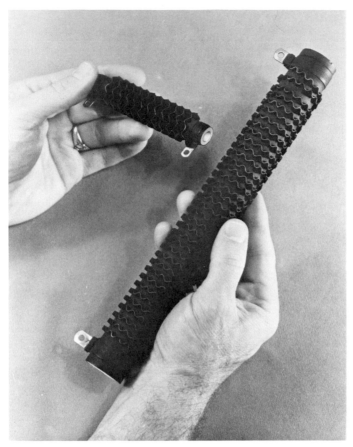

Power resistors, capable of dissipating relatively large amounts of electrical power as heat. (*Photo courtesy of Dale Electronics, Inc.*)

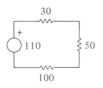
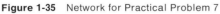

Figure 1-35 Network for Practical Problem 7

Rheostats and Adjustable Resistors

Adjustable resistors are given the descriptive symbol shown in Figure 1-36. Often these resistors are constructed so that a metallic contact (represented by the arrow on the symbol) may be moved along a length of resistive material. If the device is intended to be adjusted frequently—for instance,

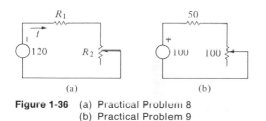

Figure 1-36   (a) Practical Problem 8
(b) Practical Problem 9

if it is connected to a knob on an instrument panel or to a control motor shaft—it is called a *rheostat*. If it is intended to be adjusted infrequently—for instance, by loosening a screw and moving a contact plate—the device is called an *adjustable resistor*.

The specified resistance R of a rheostat or adjustable resistor is its maximum resistance.

These devices are rated according to the maximum average power dissipation at the full resistance setting. At other than the full resistance setting, the device is capable of dissipating less power, the allowable power dissipation being proportional to the resistance setting. The power rating may be found by multiplying the full resistance by the square of the maximum current.

8. It is desired to choose a resistor R_1 and a rheostat R_2 in the network of Figure 1-36(a) so that the rheostat is capable of adjusting the current i from 1 A to 5 A. Find values for R_1 and R_2.

9. What is the smallest acceptable power rating for the rheostat in the network of Figure 1-36(b)?

Resistor Models of Physical Devices

The resistor is also useful as a model for electrical devices, besides physical resistors, that convert electrical energy into other forms of energy. Lamps, lengths of wire, resistance heaters, even electric motors under the proper circumstances may be modeled by a resistor.

10. A certain 110-V lamp draws 0.91 A when lit. What is the resistance of the lamp when lit? What is the electrical power flow into the lamp when it is connected to a constant 110-V source?

11. A certain electric heater consists of two identical 0.3-Ω heating elements which are simply lengths of nichrome wire strung on an insulating, heat-resistant holder. The two electrical elements are connected in series for low heat and in parallel for high heat. If the power source is a constant 120 V, compare the electrical power flows (into heat) at the low and at the high settings.

12. A constant 220-V source is connected to eight 220-V, 500-W lamps connected in parallel. What current must the source provide?

Wire Gauge

The resistance of a conductor of uniform cross-sectional area A and length L is

$$R = \frac{\rho L}{A}$$

where ρ is the resistivity of the conducting material.

The American Wire Gauge (AWG) is a common way of specifying the diameter of wires of circular cross section. The wire diameter is often given

Table 1-4 Data for Standard Annealed Solid Copper Wire

AWG gauge number	Diameter in mils	Area in circular mils	Resistance in Ω/1000 ft	Weight in lb/1000 ft	Typical max allowable current in amperes*
00	365	133,100	0.080	403	225
0	325	105,500	0.100	319	200
1	289	83,690	0.126	253	120
2	258	66,370	0.159	201	94.8
3	229	52,640	0.201	159	75.2
4	204	41,740	0.253	126	59.6
5	182	33,100	0.319	100	47.3
6	162	26,250	0.403	79.5	37.5
7	144	20,820	0.508	63.0	29.7
8	128	16,510	0.541	50.0	23.6
9	114	13,090	0.808	39.6	18.7
10	102	10,380	1.02	31.4	14.8
11	91	8,234	1.28	24.9	11.8
12	81	6,530	1.62	19.8	9.33
13	72	5,178	2.04	15.7	7.40
14	64	4,107	2.58	12.4	5.87
15	57	3,257	3.25	9.86	4.65
16	51	2,583	4.09	7.82	3.69
17	45	2,048	5.16	6.20	2.93
18	40	1,624	6.51	4.92	2.32
19	36	1,288	8.21	3.90	1.84
20	32	1,022	10.4	3.09	1.46
21	28.5	810	13.1	2.45	1.16
22	25.3	642	16.5	1.95	0.918
23	22.6	510	20.8	1.54	0.728
24	20.1	404	26.2	1.22	0.577
25	17.9	320	33.0	0.97	0.458

* Calculated on the basis of 700 circular mils per ampere.

in mils, which are 10^{-3} in. The cross-sectional area is then specified in *circular mils*, which is the square of the diameter in mils. Table 1-4 gives data for copper wire.

13. Specify the AWG gauge number of the smallest diameter of standard annealed solid copper wire for which 225 ft has a resistance less than 0.25 Ω.

Resistor Temperature Coefficient

The resistance of a material generally depends somewhat on its temperature. A linear (or straight-line) approximation to a resistance, R as a function of temperature, T, is

$$R(T_2) = R(T_1)[1 + \alpha(T_2 - T_1)]$$

The number α is called the *temperature coefficient* of the resistor. It may be positive, indicating that the resistance increases with temperature, or for some materials negative, indicating a decrease in resistance with increasing temperature.

A *thermistor* is a resistive device which is especially constructed to have a large temperature coefficient. Thermistors are useful for temperature measurement.

Testing digital computer processors on the factory floor before final assembly. Other computer equipment (center) is used to speed the extensive tests. (*Photo courtesy of Digital Equipment Corp.*)

14. A certain resistor has a temperature coefficient of $\alpha = 0.021\,°C$ at $30°C$. What is its resistance at $45°C$ if the resistance is $470\,\Omega$ at $30°C$?

Advanced Problems

Network Specification in a Computer Program

1. Carefully describe a method by which you could, without ambiguity, easily specify a network in words, not a drawing. When such a specification is given in a mutually understood format, it might be used to enter the description of a network into a digital computer.

Resistor Models

2. Resistor models that have negative values of resistance are useful as models for some electronic devices. For the network of Figure 1-37(a), which contains a negative resistor, find i.

3. Resistor models for which the resistance varies with time are useful as models for some physical devices. For the network of Figure 1-37(b), which contains a time varying resistor, find $v(t)$.

4. Resistor models for which the resistance depends on some other signal, such as a network voltage or current, are useful in modeling some electronic systems. For the network of Figure 1-37(c), which contains a current-controlled resistor, find the two possible solutions for v.

5. A short circuit is a length of (ideal) conductor between two terminals; an open circuit exists when there is no connection between the terminals. A short circuit may be thought of as a voltage source of 0 V or as a resistor of $0\,\Omega$. What are analogous models for an open circuit?

Electrical Power Flow

6. The electrical power flow out of a certain two-terminal element is

$$P_{out}(t) = i^4(t)$$

Figure 1-37 (a) Advanced Problem 2
(b) Advanced Problem 3
(c) Advanced Problem 4

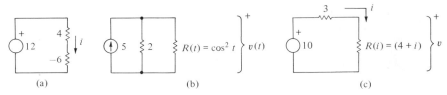

where $i(t)$ is the element current. What is the sink reference voltage-current relation for this element?

7. The power dissipated by a certain 10-Ω resistor as a function of time is

$$P_{\text{into } 10 \, \Omega}(t) = |\sin t|$$

Graph four of the infinite number of possible different resistor currents.

8. Two resistors are connected end-to-end (in series) so that the same current flows through each. The voltage across one resistor is twice that across the other. How do the power flows into the two resistors compare?

CHAPTER TWO
Source-Resistor Network Solutions Using Equivalent Circuits

2.1 Introduction

Perhaps the most powerful technique in network solution is one of transformation. A complicated network is transformed to simpler, *equivalent* networks through one or more stages until its solution can be found easily.

This transformation or equivalent circuit method is very commonly used for networks in all areas of electrical engineering because it is often the easiest solution method and because it retains an identification with the original network which is lacking if the problem is immediately converted to equations.

When you complete this chapter, you should know

1. What it means to say two elements are equivalent to one another.
2. How to construct equivalent elements for current sources in parallel, voltage sources in series, and other simple series and parallel equivalences.
3. How to find equivalent resistances for resistors in series and for resistors in parallel.
4. How to perform a Thévenin-Norton transformation.

5. How sources may always be substituted for known voltages and currents in a network.
6. How to use equivalent circuits to solve involved network problems.

2.2 Simple Equivalences

2.2.1 The Meaning of Two-Terminal Element Equivalence

A combination of elements that has two terminals is termed a *two-terminal network*. Two 2-terminal networks are said to be equivalent if both have the same voltage-current relations at their terminals.

If a combination of elements is replaced with an equivalent in a network, all voltages and currents (signals) within the two networks are the same, *external* to the portion replaced. A network may be simplified in this way, often to the point where a two-node or single-loop equation may be written.

The following is a listing of some simple equivalences. A general relation is implied even though some examples show only a specific number of elements for simplicity. For clarity, the terminals of the equivalents are indicated now with small circles.

2.2.2 Current Sources in Parallel

Figure 2-1 illustrates how current sources in parallel with one another may be combined into a single equivalent current source.

2.2.3 Voltage Sources in Series

Voltage sources in series with one another may be combined into a single equivalent voltage source, as indicated in Figure 2-2.

2.2.4 Current Source in Series with Anything

In the equivalence of Figure 2-3, the current $i(t)$ enters one terminal and leaves the other. Whether or not the series element is present makes a

Figure 2-1 Equivalent current sources in parallel

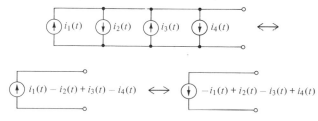

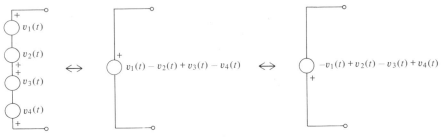

Figure 2-2 Equivalent voltage sources in series

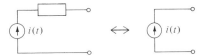

Figure 2-3 Equivalent of a current source in series with any element

difference to the current source and, of course, to the series element; but external to the current source and series element, there may as well only be the current source present.

The series element may be a two-terminal combination of more simple elements, as in the example of Figure 2-4.

2.2.5 Voltage Source in Parallel with Anything

The terminal voltage is $v(t)$ in Figure 2-5, whether or not the parallel element is present. Of course, it makes a difference to the source whether or not the parallel element is present; the source must supply any additional current that flows through the element. It obviously makes a difference to the

Figure 2-4 Current source in series with an element composed of a combination of simpler elements

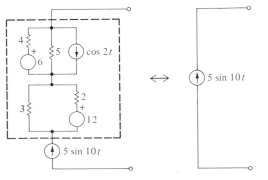

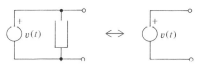

Figure 2-5 Equivalent of a voltage source in parallel with any element

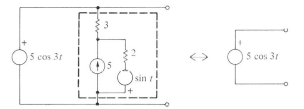

Figure 2-6 Voltage source in parallel with an element composed of a combination of simpler elements

parallel element whether or not it is present. External to the combination, however, the presence or absence of the parallel element is of no consequence.

The parallel element may be a two-terminal combination of simpler elements, as in Figure 2-6.

2.2.6 Series and Parallel Equivalences

The order of any elements in a series connection may be changed. Figure 2-7 illustrates this.

If a resistor is turned end for end, it still has the same voltage-current relation. An element such as a resistor is *bilateral*. Turning a source end for end in a network will change the network, however. Sources are *nonbilateral*. So it must be understood that in equivalents involving sources, as in Figure 2-8, the senses of these nonbilateral elements are to be preserved.

Similarly, providing the senses of nonbilateral elements are preserved, the order of any elements in a parallel connection may be changed. An illustration is given in Figure 2-9.

Figure 2-7 Equivalent interchanges of the order of elements in series

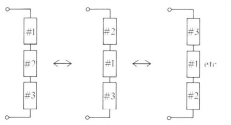

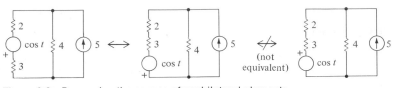

Figure 2-8 Preserving the senses of nonbilateral elements

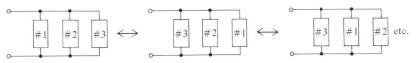

Figure 2-9 Equivalent interchanges of the order of elements in parallel

D2-1

Find the indicated signals by using equivalent circuits to reduce the problem to a single-loop or two-node problem, then solving:

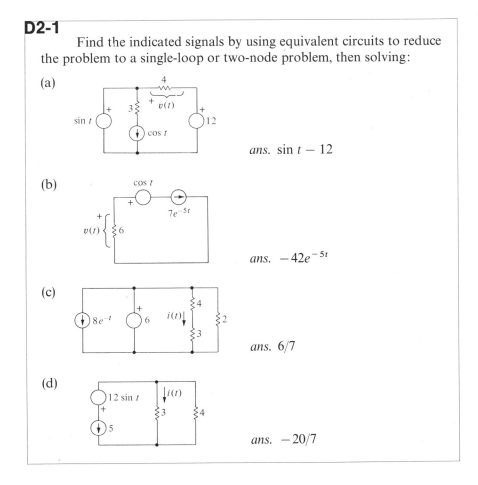

(a)

ans. $\sin t - 12$

(b)

ans. $-42e^{-5t}$

(c)

ans. $6/7$

(d)

ans. $-20/7$

2.3   Equivalent Resistors

2.3.1   Resistors in Series

Any two-terminal combination of resistors is equivalent to a single resistor. For resistors in series, the equivalent single resistor has resistance which is just the sum of the individual resistances. In Figure 2-10 it is demonstrated that the series resistors and the appropriate single resistor have the same voltage-current relation; thus they are equivalent.

It should be noted that the equivalences presented here are between one two-terminal combination of more simple elements and another two-terminal element or combination of elements; and the resulting networks are equivalent only so far as voltages and currents external to the transformed portions are concerned.

For example, resistors R_1 and R_2 in the network of Figure 2-11 cannot be replaced by an equivalent resistor of resistance $(R_1 + R_2)$ because of the connection between R_1 and R_2. The dashed box containing R_1 and R_2 has *three* terminals. R_3 and R_4, however, may be replaced by an equivalent resistor so far as any voltage or current external to that element is concerned.

The two networks in Figure 2-11 (which differ only so far as the dashed boxes to the right are concerned) are equivalent; they have the same voltages

Figure 2-10   Equivalent resistance of resistors in series

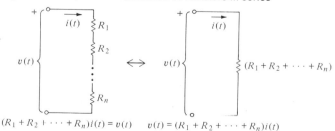

$$(R_1 + R_2 + \cdots + R_n)i(t) = v(t) \qquad v(t) = (R_1 + R_2 + \cdots + R_n)i(t)$$

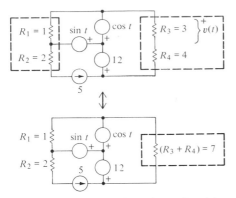

Figure 2-11 Resistors in series and resistors not in series

and currents external to the dotted boxes. The two networks are not equivalent so far as the voltage $v(t)$, within the box, is concerned, however. That
voltage does not appear in the second network.

2.3.2 Resistors in Parallel

The equivalent resistance for a set of resistors in parallel is the inverse of
the sum of the inverses of the individual resistances. In Figure 2-12, it is
shown that the parallel resistors and the appropriate single resistor have the
same voltage-current relation at their terminals.

The parallel resistor equivalence for just two resistors in parallel is

$$R_{\text{equiv}} = \cfrac{1}{\cfrac{1}{R_1} + \cfrac{1}{R_2}} = \frac{R_1 R_2}{R_1 + R_2}$$

This "product over the sum" relationship is well worth committing to
memory because of the algebraic manipulation it saves each time it is applied.
The corresponding relationships for three or more resistors in parallel are
not "product over sum." They are more complicated, so the "inverse of the
sum of the inverses" is probably as good a relationship to start with as any
other for more than two resistors in parallel.

The equivalent resistance of a parallel combination of resistors is always
less than any of the individual resistances in the combination.

Figure 2-12 Equivalent resistance of resistors in parallel

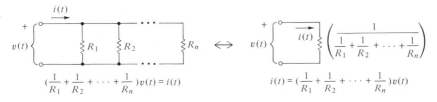

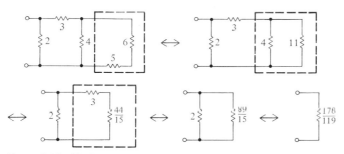

Figure 2-13 Finding the equivalent resistance of a two-terminal combination of resistors

2.3.3 Conductance

The inverse of resistance is called *conductance* and has symbol G. A resistor with resistance $R = 5\ \Omega$ has conductance $G = \frac{1}{5}\,\Omega^{-1}$. The unit Ω^{-1} is commonly termed the "mho" (ohm spelled backwards) for convenience.

The equivalent conductance of several resistors in parallel is the sum of the individual resistor conductances:

$$G_{equiv} = \frac{1}{R_{equiv}} = \frac{1}{R_1} + \frac{1}{R_2} + \cdots = G_1 + G_2 + \cdots$$

Dealing in terms of conductances instead of resistances is of advantage for networks where all of the elements are in parallel, but there is no advantage in more general (and more interesting) networks since the equivalent conductance of resistors in series is analogous to resistances in parallel:

$$G_{equiv} = \frac{1}{R_{equiv}} = \frac{1}{R_1 + R_2 + \cdots} = \frac{1}{\dfrac{1}{G_1} + \dfrac{1}{G_2} + \cdots}$$

2.3.4 Two-Terminal Combinations of Resistors

The series and parallel resistor equivalences allow easy solution for equivalent resistances for more complicated two-terminal combinations of resistors. An example is given in Figure 2-13.

There are, however, two-terminal combinations of resistors that cannot be so reduced. In the two-terminal network of Figure 2-14, no resistor is

Figure 2-14 A two-terminal combination of resistors in which no resistors are in series or in parallel with one another

in series or in parallel with any other resistor. Series and parallel combinations are not sufficient to find single equivalent resistors for *all* possible two-terminal resistor combinations, even though the equivalents exist.

D2-2

Find the equivalent resistances of the following two-terminal resistor networks:

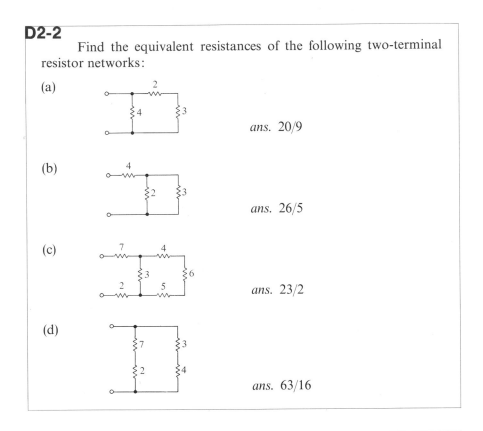

(a)

ans. 20/9

(b)

ans. 26/5

(c)

ans. 23/2

(d)

ans. 63/16

D2-3

Find the indicated signals using equivalent circuits. There are three signals to find in each network.

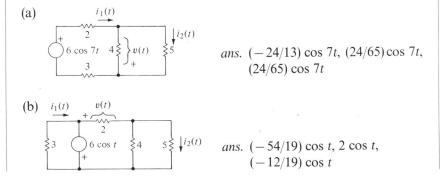

(a)

ans. $(-24/13) \cos 7t$, $(24/65) \cos 7t$, $(24/65) \cos 7t$

(b)

ans. $(-54/19) \cos t$, $2 \cos t$, $(-12/19) \cos t$

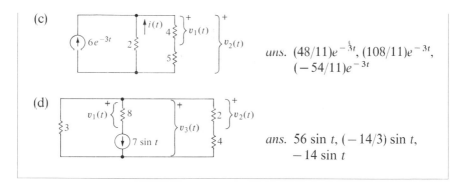

(c)

ans. $(48/11)e^{-3t}$, $(108/11)e^{-3t}$, $(-54/11)e^{-3t}$

(d)

ans. $56 \sin t$, $(-14/3) \sin t$, $-14 \sin t$

Resistors in Series and Parallel

Series: $R_{\text{equiv}} = R_1 + R_2 + \cdots$

Parallel: $R_{\text{equiv}} = \dfrac{1}{\dfrac{1}{R_1} + \dfrac{1}{R_2} + \cdots}$

For two resistors in parallel,

$$R_{\text{equiv}} = \frac{R_1 R_2}{R_1 + R_2}$$

2.4 The Thévenin-Norton Equivalence

One of the more interesting equivalences is between a voltage source in series with a resistance and a current source in parallel with a resistance. The former is called a *Thévenin element* and the latter is called a *Norton element*. Each is illustrated in Figure 2-15.

The two-terminal networks of Figure 2-15 are equivalent providing the relation between $v(t)$ and $i(t)$ is the same in both. For the Thévenin element,

$$R_T i(t) - v(t) - v_T(t)$$

$$v(t) = v_T(t) + R_T i(t)$$

For the Norton element,

$$\left(\frac{1}{R_N}\right) v(t) = i_N(t) + i(t)$$

$$v(t) = R_N i_N(t) + R_N i(t)$$

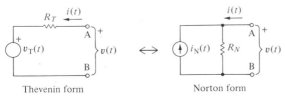

Thevenin form   Norton form

Figure 2-15 Thévenin and Norton elements

Thus the two are equivalent if

$$\begin{cases} v_T(t) = R_N i_N(t) \\ R_T = R_N \end{cases}$$

Since the Thévenin and Norton elements contain sources, the senses of which must be preserved, this equivalence holds only if terminal A (closest to the plus sign on v_T) is replaced by the other terminal A (toward which the reference for i_N points) and terminal B is replaced by the other terminal B.

Specific examples of conversions between Thévenin and Norton elements are given in Figure 2-16.

A Thévenin-Norton or a Norton-Thévenin conversion, replacing one element by its equivalent, is often very useful in finding network solutions. For the network of Figure 2-17, the indicated Thévenin-Norton transformation gives a resulting single-loop network, for which

$$(2 + 3)i(t) = \cos 4t + 18$$

$$i(t) = \frac{1}{5} \cos 4t + \frac{18}{5}$$

Figure 2-16 Examples of Thévenin-Norton and Norton-Thévenin conversions

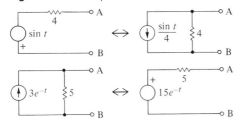

Figure 2-17 Example of Norton-Thévenin conversion in network solution

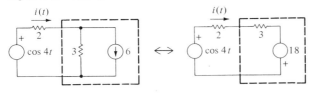

D2-4

Use the Thévenin-Norton element equivalence to find the indicated signals:

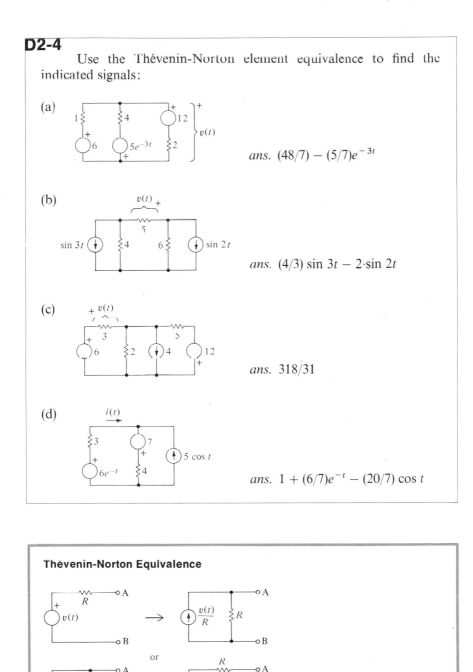

(a)

ans. $(48/7) - (5/7)e^{-3t}$

(b)

ans. $(4/3)\sin 3t - 2 \cdot \sin 2t$

(c)

ans. $318/31$

(d)

ans. $1 + (6/7)e^{-t} - (20/7)\cos t$

Thévenin-Norton Equivalence

or

2.5 Substitution of Sources for Known Voltages and Currents

Suppose that the voltage across a certain element in a network $v(t)$ is known. Then connecting a voltage source with source function $v(t)$ across the terminals of that element, as in Figure 2-18, will not affect the rest of the network. The original voltage was $v(t)$ and, with the source, it is still $v(t)$. Whatever the voltage-current relations of the element and of the original network, since the voltage is unchanged, the current $i(t)$ is unchanged. The current through the source is thus zero.

The voltage source in parallel with the element is equivalent, so far as the rest of the network is concerned, to just the voltage source. Thus any element in a network may be replaced by a voltage source with source function the element voltage, without changing any network voltage or current.

Similarly, any element in a network may be replaced by a current source with source function the element current, without changing any network voltage or current, as indicated in Figure 2-19.

These equivalences, known as the *substitution theorem*, are particularly useful in consolidating results found by using equivalent circuits and in

Figure 2-18   Substituting a voltage source for an element

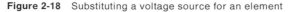

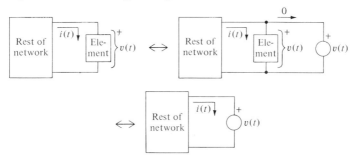

Figure 2-19   Substituting a current source for an element

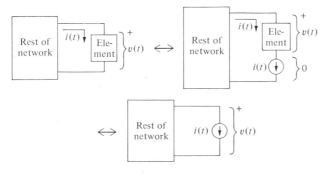

picturing how any element affects a network. Whenever an element's voltage or current is known, so far as the rest of the network is concerned, the element may be considered to be a source.

D2-5

The voltage or current of the element that is represented by a box is given. Find the other indicated signal:

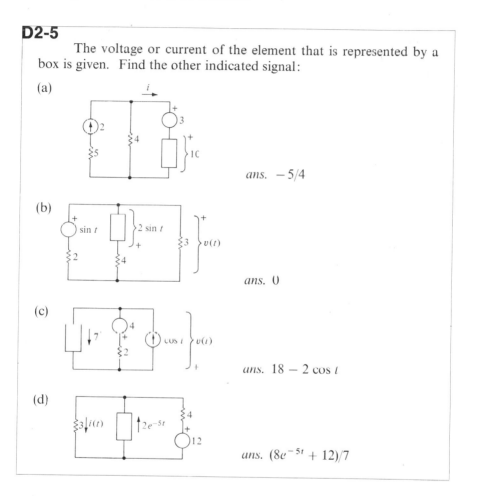

(a)

ans. $-5/4$

(b)

ans. 0

(c)

ans. $18 - 2\cos t$

(d)

ans. $(8e^{-5t} + 12)/7$

Substitution of Sources for Known Voltages and Currents

If any element in a network is replaced by a voltage source with source function equal to the element voltage, each network voltage and current is unchanged.

If any element in a network is replaced by a current source with source function equal to the element current, each network voltage and current is unchanged.

2.6   Solution of Networks Using Equivalent Circuits

2.6.1   The Approach

The equivalent circuits described so far may be used to find voltages and currents in almost any network, the exceptions being those networks that require three-terminal or, generally, multiterminal equivalences such as two-terminal combinations of resistors where no resistor is in series or parallel with any other resistor. Further equivalences (called *delta-wye transformations*) may be developed for these special cases, and this will be done in Chapter Four.

The approach in solving a network using equivalent circuits is to reduce the problem, using equivalences, to a two-node or single-loop problem, solving for the node-to-node voltage or the loop current. Then, working from this solution, often using the current-divider and voltage-divider rules, we find the voltages and currents desired in the original network.

With this method, there are numerous ways to proceed, depending on one's experience and skill in achieving a solution with minimal effort.

2.6.2   First Example

Figure 2-20 describes a network in which it is desired to find the voltage $v(t)$. With the indicated series of transformations, the problem is finally reduced to an equivalent single-loop network. For that network,

$$\left(6 + 4 + \frac{6}{5} + 5\right)i(t) = -7 \sin 8t - 50$$

$$i(t) = \frac{5}{81}(-7 \sin 8t - 50)$$

$$v(t) = 6i(t) = -\frac{30}{81}(7 \sin 8t + 50)$$

Figure 2-20   First example of network solution via equivalent circuits

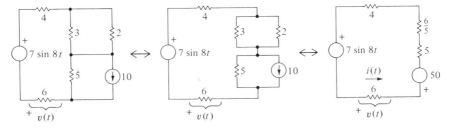

2.6.3 Second Example

In Figure 2-21 it is desired to find $i(t)$. The transformations to an equivalent parallel network are such that $i(t)$ does not appear in the final network. Instead, the voltage $v(t)$ is first found:

$$\left(\frac{1}{3} + \frac{1}{4} + \frac{1}{5}\right)v(t) = \frac{3}{2}e^{-7t} - \frac{\sin t}{3} - \frac{12}{5}$$

$$v(t) = \frac{90}{47}e^{-7t} - \frac{20}{47}\sin t - \frac{144}{47}$$

Returning to the original network, now knowing $v(t)$,

$$5i(t) = v(t) + 12$$

$$i(t) = \frac{18}{47}e^{-7t} - \frac{4}{47}\sin t + \frac{84}{47}$$

2.6.4 Third Example

To find $v(t)$ and $i_2(t)$ in the network of Figure 2-22, the indicated transformations are chosen. A two-node equation is then solved for $v_1(t)$:

$$\left(\frac{1}{5} + \frac{5}{26}\right)v_1(t) = 9 - 8\cos t$$

$$v_1(t) = \frac{130}{51}(9 - 8\cos t)$$

Then

$$i_1(t) = \frac{v_1(t)}{5} = \frac{26}{51}(8\cos t - 9)$$

Figure 2-21 Second equivalent circuit example

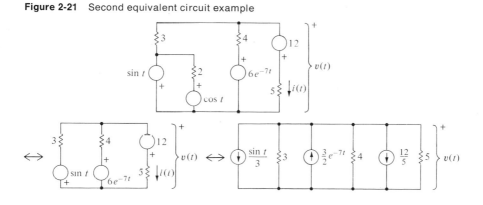

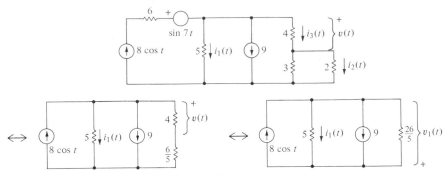

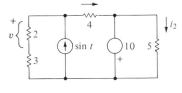

Figure 2-22 Third equivalent circuit example

Returning to the previous equivalent network, using the voltage-divider rule,

$$v(t) = \frac{4}{4 + 6/5} v_1(t) = \frac{100}{51} (8 \cos t - 9)$$

Returning to the original network,

$$i_3(t) = \frac{v(t)}{4} = \frac{25}{51} (8 \cos t - 9)$$

Using the current-divider rule,

$$i_2(t) = \frac{3}{5} i_3(t) = \frac{15}{51} (8 \cos t - 9)$$

D2-6

Use equivalent circuits to find the indicated signals. There are three signals to find in each network.

(a)

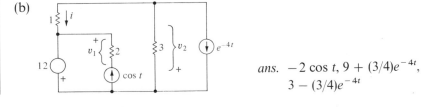

ans. $(8/9) \sin t - (20/9),$
$(5/9) \sin t + (10/9), -2$

(b)

ans. $-2 \cos t, 9 + (3/4)e^{-4t},$
$3 - (3/4)e^{-4t}$

(c)

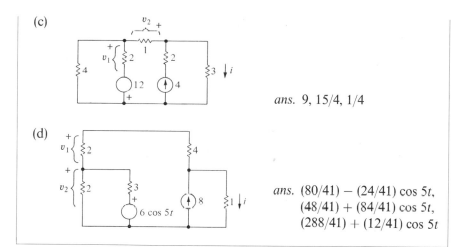

ans. 9, 15/4, 1/4

(d)

ans. (80/41) − (24/41) cos 5*t*,
(48/41) + (84/41) cos 5*t*,
(288/41) + (12/41) cos 5*t*

Network Solution Using Equivalent Circuits

Use equivalent circuits to simplify the network until a two-node or single-loop network results. Then, using signals found in the simplified network and previous equivalences as necessary, find other signals of interest in the original network.

Chapter Two Problems

Basic Problems

Equivalent Circuits

1.   Find $v(t)$ using equivalent circuits:

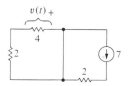

2.   Find, v, i_1, and i_2 using equivalent circuits:

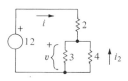

3. Find $v(t)$ using equivalent circuits:

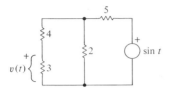

4. Find $v(t)$ and $i(t)$ using equivalent circuits:

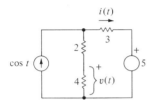

5. Find $i(t)$ using equivalent circuits:

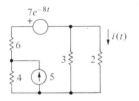

6. Find $v(t)$ and $i(t)$ using equivalent circuits:

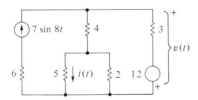

7. Find $i(t)$ using equivalent circuits:

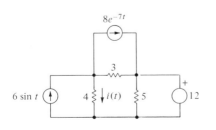

8. Find $v(t)$, $i_1(t)$, and $i_2(t)$ using equivalent circuits:

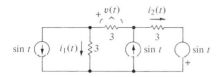

9. Find $v_1(t)$, $v_2(t)$, and $i(t)$ using equivalent circuits:

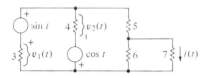

Equivalent Resistance

10. Find the equivalent resistance of the two-terminal combination of resistors:

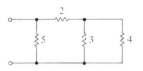

Practical Problems

Ammeters

A permanent magnet type ammeter consists basically of a permanent magnet in proximity to a coil of wire through which the current to be measured flows. When current flows through a coil of wire, the coil becomes an electromagnet which is attracted to or repelled from the permanent magnet, depending on the direction of current flow. The coil is restrained by a spring and the mechanism is usually designed so that the coil's deflection is proportional to the coil current.

Since the ammeter coil consists of a length of wire, a simple model for the ammeter is a resistance, called the *internal resistance* of the meter. In other words, to a first approximation at least, the meter "looks" like a resistor to the circuit in which it is connected.

Permanent magnet ammeters are often called *dc ammeters* because their most common use is for the measurement of constant or nearly constant currents. Such currents are called "direct current" (or dc).

1. Compare the current i in the circuit of Figure 2-23(a) with the current i' of Figure 2-23(b) which would be measured if an ammeter with 2 Ω internal

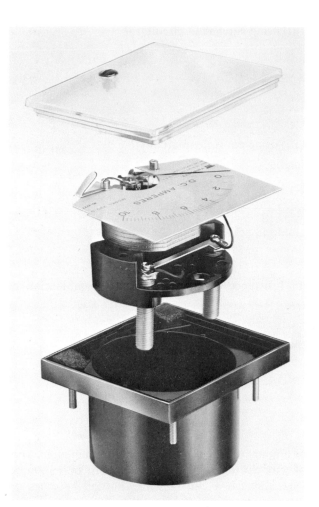

Internal view of an ammeter. The movement has full-scale sensitivity of 1 mA. A shunt (the rectangular bar of conductor mounted between the meter terminal studs) is used to reduce the sensitivity to 10 A full scale.   (*Photo courtesy of the Triplett Corp.*)

Figure 2-23   Practical Problem 1

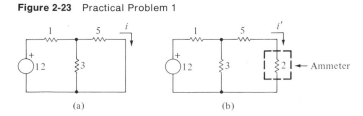

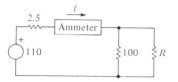

Figure 2-24 Practical Problem 2

resistance were inserted to measure the current. Find the percent of error due to the effect of the meter,

$$\left| \frac{i - i'}{i} \right| \times 100$$

2.   An ammeter with internal resistance 0.5 Ω is inserted in an electrical network as shown in Figure 2-24 and reads 6.25 A.
(a)   What is the value of the resistance R?
(b)   What would the percent error,

$$\left| \frac{R - R'}{R} \right| \times 100$$

in the calculated value of resistance R' be if the ammeter resistance were neglected?

Ammeter Shunts

A convenient method of adjusting the scale of a permanent magnet ammeter so that the meter is less sensitive than it otherwise would be is to place a resistor, called a *meter shunt* resistor, in parallel with the meter coil, as in Figure 2-25. A fraction of the net current i through the shunt and coil combination flows through the ammeter coil itself:

$$i_{coil} = \left(\frac{R_{shunt}}{R_{shunt} + R_{coil}} \right) i$$

Thus if the shunt resistance is made equal to the coil resistance, an ammeter which would otherwise read 5 A full scale would, with such a shunt, read 10 A full scale.

Figure 2-25 Analysis of an ammeter shunt

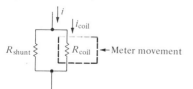

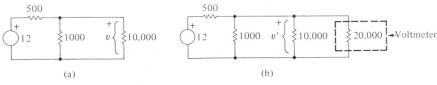

(a)   (b)

Figure 2-26   Practical Problem 4

Although an ammeter shunt reduces the sensitivity of the ammeter, it also reduces the overall internal resistance of the instrument, thus reducing its effect on the network under test.

3.   For an ammeter movement that reads 1 milliampere (1 mA) full scale and has an internal resistance of 250 Ω,

(a)   What should the shunt resistance be if the instrument, with the shunt, is to read 2.5 A full scale?

(b)   What is the internal resistance of this 2.5-A ammeter?

(c)   What electrical power will the shunt resistor be required to dissipate when the ammeter reads full scale?

Voltmeters

If the internal resistance of a permanent magnet ammeter is deliberately made very large, by winding the coil with many turns of fine wire, for instance, it may be used to measure voltage. The meter responds to its coil current, which is proportional to the voltage across the instrument's terminals.

As with the ammeter, a simple model for such a dc voltmeter is a resistance, the internal resistance of the meter.

4.   Compare the voltage v in the circuit of Figure 2-26(a) with the voltage v' in (b) which would be measured if a voltmeter with 20 kilohms (kΩ) internal resistance were attached to measure the voltage. Find the percent of error due to the effect of the meter,

$$\left| \frac{v - v'}{v} \right| \times 100$$

5.   A voltmeter with internal resistance 5000 Ω is attached in an electrical network as shown in Figure 2-27 and reads 32.5 V.

Figure 2-27   Practical Problem 5

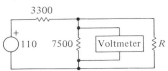

Digital multimeter with liquid crystal display. The rotary switch selects various ranges of voltage current and resistance measurement.   (*Photo courtesy of Dana Laboratories, Inc.*)

(a)   What is the value of the resistance R?

(b)   What would the percent error,

$$\left| \frac{R - R'}{R} \right| \times 100$$

in the calculated value of resistance R' be if the voltmeter resistance were neglected?

Voltmeter Scale Change

A convenient method of adjusting the scale of a permanent magnet voltmeter so that the meter is less sensitive than it otherwise would be is to place a resistor in series with the meter coil, as in Figure 2-28. For a fixed voltage

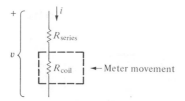

Figure 2-28 Analysis of voltmeter resistance

across the resistor and coil combination, less current will flow through the coil the higher this series resistance:

$$i = \frac{v}{R_{series} + R_{coil}}$$

Thus if the series resistance is made equal to the coil resistance, a voltmeter that would otherwise read 2 V full scale would, with such a series resistance, read 4 V full scale.

Although the series resistance reduces the sensitivity of the voltmeter, it also increases the overall resistance of the instrument, thus reducing its effect on the network under test.

It is common to construct permanent magnet voltmeters with standard internal resistances (including the series resistor) such as 20,000 ohms per volt of full scale reading. A 20,000 Ω/V voltmeter that reads 15 V full scale would have an internal resistance of $15(20,000) = 300$ kΩ.

6. For a 20,000 Ω/V voltmeter that reads 0.5 V full scale, what series resistance should be added to make the instrument read 25 V full scale?

7. Show that if a voltmeter internal resistance is R Ω/V of full-scale reading, the addition of a series resistor results in the same number of ohms per volt of the new full-scale reading.

Battery Internal Resistance

An adequate model for a battery under most circumstances consists of a constant voltage source in series with a resistor. The source voltage is called the *open-circuit voltage* or simply the battery voltage and the resistor value for a given battery is called its *internal resistance.*

8. An ammeter with 0.7 Ω internal resistance is connected across a 110-V battery with a 0.65-Ω internal resistance. What current flows?

9. A 24-V battery is formed by connecting two 12-V batteries in series. If the 12-V batteries have internal resistances of 0.15 and 0.21 Ω, respectively, what is the internal resistance of the 24-V battery?

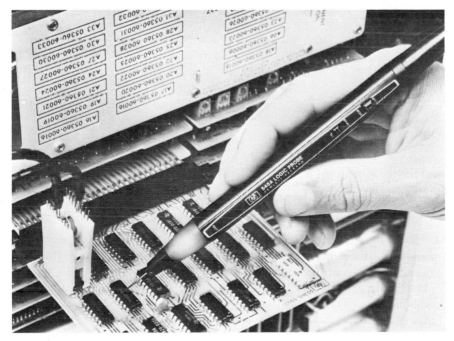

A digital logic probe being used to test integrated circuits. The device senses the voltage between its tip and a wire which connects to the other end of the pencil-like probe. If the voltage exceeds a certain amount (indicating a logic "one" in digital circuits), a lamp near the tip lights.   (*Photo courtesy of Hewlett-Packard, Inc.*)

10.   A 12.4-V battery with an internal resistance of 0.38 Ω is recharged by connecting it to a battery charger that may be modeled by a 14.2-V source in series with 10.5 Ω. Find the battery current. When recharging, the electrical power flow is into the battery, of course, so the battery current is in the opposite direction from that when the battery is supplying power.

Advanced Problems

Element Equivalence

1.   An element's voltage-current relation is unchanged if it is turned end for end, providing the voltage-current relation itself is unchanged if v is replaced by $-v$ and i is replaced by $-i$. This is to say that the element is equivalent to itself turned end for end. Such an element is said to be *bilateral*. For example, in the sink reference voltage-current relation for the resistor, $v = Ri$, replacing v by $-v$ and i by $-i$ still gives $v = Ri$.

   Apply this test to demonstrate that (nonzero) voltage sources and current sources are *not* bilateral elements. Would an element with voltage-current relation

$$v = 10i^2$$

be bilateral?

Equivalent Resistors

2. Show that the equivalent resistance of three resistors, R_1, R_2, and R_3, in parallel is

$$R_{equiv} = \frac{R_1 R_2 R_3}{R_1 R_2 + R_2 R_3 + R_1 R_3}$$

3. Why is there no particular advantage to combining parallel resistors in a two-node network before writing the two-node equation?

4. Show that the sum of the electrical power flows into a parallel connection of resistors at any instant of time is equal to the electrical power flow into the equivalent resistor.

It is true in general that the net electrical power flow into any two-terminal combination of resistors is the same as the electrical power flow into the equivalent resistor.

Thévenin-Norton Transformation

5. If a Thévenin element, a voltage source in series with a resistor, is replaced by this Norton equivalent in a network, show that
(a) The electrical power flows into the Thévenin element and into the Norton element are identical.
(b) The electrical power flows into the Thévenin resistor and into the Norton resistor are not necessarily the same.

The Thévenin-Norton equivalence is only for voltages and currents external to the elements. The net power flows into the Thévenin and Norton elements may be expressed in terms of external voltages and currents, but the Thévenin and Norton resistor power flows involve a voltage or a current *internal* to the element.

CHAPTER THREE
Systematic Simultaneous Equations

3.1   Introduction

Step-by-step methods of expressing network problems, no matter how complicated, as a set of mathematical equations are important in several ways. Certain types of problems, for instance those involving controlled sources, are very difficult to handle in any other manner. When a large network problem is subjected to computer-aided analysis, efficient programming requires a systematic procedure which is lacking in equivalent circuit methods. Also, well-known properties of the equations describing networks imply certain important properties of the networks themselves.

It is quite easy, of course, to write all of the equations for a network. Simply define every different current and voltage in the network. Then write Kirchhoff's current law at each junction, Kirchhoff's voltage law around each loop, and Ohm's law for each resistor. The resulting set of equations may be solved for any network voltage or current. After all, the only laws for source-resistor networks are those of Kirchhoff and Ohm.

The difficulties with this "brute force" method are many. There are a large number of variables (every network voltage and current) and a large number of equations. It will be very helpful to deal in terms of a smaller set of variables and with a much smaller number of simultaneous equations.

Most important, not all applications of Kirchhoff's laws yield independent equations. This is quite evident in two-node network problems in which the application of Kirchhoff's current law at either node gives the same equation. Once equations are written, determining which of them are independent of the rest can be quite involved.

In this chapter, two systematic methods are presented for writing sets of independent equations in terms of a relatively small number of variables. These methods are extensions of the two-node and single-loop approaches of Chapter One.

When you complete this chapter, you should know

1. How to solve simultaneous linear algebraic equations using Cramer's rule, evaluating determinants by Laplace expansion when necessary.
2. How to write systematic simultaneous nodal equations.
3. How to accommodate voltage sources in nodal equations.
4. How to write systematic simultaneous mesh equations.
5. How to accommodate current sources in mesh equations.
6. How to handle controlled sources in nodal equations and in mesh equations.

3.2 Cramer's Rule

3.2.1 Solution in Terms of Determinants

The solution of a set of n independent linear algebraic equations in n variables,

$$a_{11}x_1 + a_{12}x_2 + \cdots + a_{1n}x_n = y_1$$
$$a_{21}x_1 + a_{22}x_2 + \cdots + a_{2n}x_n = y_2$$
$$\vdots$$
$$a_{n1}x_1 + a_{n2}x_2 + \cdots + a_{nn}x_n = y_n$$

is given by Cramer's rule,

$$x_i = \frac{\Delta_i}{\Delta}$$

where Δ is the determinant of the equations,

$$\Delta = \begin{vmatrix} a_{11} & a_{12} & \cdots & a_{1n} \\ a_{21} & a_{22} & \cdots & a_{2n} \\ \vdots & \vdots & \vdots & \vdots \\ a_{n1} & a_{n2} & \cdots & a_{nn} \end{vmatrix}$$

and Δ_i is the same as Δ except that the ith column is replaced by the column of "knowns":

ith column

$$\Lambda_i = \begin{vmatrix} a_{11} & \cdots & y_1 & \cdots & a_{1n} \\ a_{21} & \cdots & y_2 & \cdots & a_{2n} \\ \vdots & \vdots & \vdots & \vdots & \vdots \\ a_{n1} & \cdots & y_n & \cdots & a_{nn} \end{vmatrix}$$

For example, in

$$\begin{cases} 2x_1 - x_2 - x_3 = 8\cos t \\ 3x_1 + 2x_2 \quad\quad = 0 \\ -x_1 \quad\quad + 4x_3 = 12 \end{cases}$$

$$x_1 = \frac{\begin{vmatrix} 8\cos t & -1 & -1 \\ 0 & 2 & 0 \\ 12 & 0 & 4 \end{vmatrix}}{\begin{vmatrix} 2 & -1 & -1 \\ 3 & 2 & 0 \\ 1 & 0 & 4 \end{vmatrix}}$$

$$x_2 = \frac{\begin{vmatrix} 2 & 8\cos t & -1 \\ 3 & 0 & 0 \\ -1 & 12 & 4 \end{vmatrix}}{\begin{vmatrix} 2 & -1 & -1 \\ 3 & 2 & 0 \\ -1 & 0 & 4 \end{vmatrix}}$$

$$x_3 = \frac{\begin{vmatrix} 2 & -1 & 8\cos t \\ 3 & 2 & 0 \\ -1 & 0 & 12 \end{vmatrix}}{\begin{vmatrix} 2 & -1 & -1 \\ 3 & 2 & 0 \\ -1 & 0 & 4 \end{vmatrix}}$$

3.2.2 Cofactors and the Laplace Expansion

The *cofactor* of any element of a square array is the determinant of the array formed by deleting the row and column of that element and multiplying b one or minus one, as given by the "checkerboard" of signs.

$$\begin{vmatrix} + & - & + & - & \cdots \\ - & + & - & + & \cdots \\ + & - & + & - & \cdots \\ \vdots & & & & \end{vmatrix}$$

For example, the cofactor of the element in row 2, column 3, in

$$\begin{vmatrix} 1 & -2 & 3 \\ 4 & 5 & ⑥ \\ 7 & 8 & -9 \end{vmatrix}$$

is

$$\Delta_{23} = -\begin{vmatrix} 1 & -2 \\ 7 & 8 \end{vmatrix} = -22$$

└ from the "checkerboard" pattern

The value of a 2 × 2 (two rows by two columns) determinant is

$$\begin{vmatrix} c_1 & c_2 \\ c_3 & c_4 \end{vmatrix} = c_1 c_4 - c_2 c_3$$

Values of the determinants of larger square arrays may be found by *Laplace expansion*.

A determinant may be Laplace expanded along any row or any column. The value of the determinant is the sum of the products of each element in the row or column times their cofactors. The following is an example of Laplace expansion:

$$\begin{vmatrix} -1 & 2 & 3 & -4 \\ 5 & 6 & 0 & 8 \\ 0 & 0 & 9 & 1 \\ 4 & 0 & -3 & -2 \end{vmatrix} = -2\begin{vmatrix} 5 & 0 & 8 \\ 0 & 9 & 1 \\ 4 & -3 & -2 \end{vmatrix} + 6\begin{vmatrix} -1 & 3 & -4 \\ 0 & 9 & 1 \\ 4 & -3 & -2 \end{vmatrix}$$

$$= -2\left[5\begin{vmatrix} 9 & 1 \\ -3 & -2 \end{vmatrix} + 4\begin{vmatrix} 0 & 8 \\ 9 & 1 \end{vmatrix}\right]$$

$$+ 6\left[9\begin{vmatrix} -1 & -4 \\ 4 & -2 \end{vmatrix} - \begin{vmatrix} -1 & 3 \\ 4 & -3 \end{vmatrix}\right]$$

$$= -2[5(-15) + 4(-72)] + 6[9(18) - (-9)]$$

3.2.3 Example

Consider the set of three linear algebraic equations in three variables

$$\begin{cases} x_1 - 2x_2 + x_3 = 4\cos 3t \\ -x_1 + 3x_2 \quad\quad = -12 \\ 2x_1 \quad\quad + 4x_3 = 0 \end{cases}$$

Using Cramer's rule, the solution for x_1, x_2, and x_3 are as follows:

$$x_1(t) = \frac{\Delta_1}{\Delta} = \frac{\begin{vmatrix} 4\cos 3t & -2 & 1 \\ -12 & 3 & 0 \\ 0 & 0 & 4 \end{vmatrix}}{\begin{vmatrix} 1 & -2 & 1 \\ -1 & 3 & 0 \\ 2 & 0 & 4 \end{vmatrix}} = \frac{48\cos 3t - 96}{2\begin{vmatrix} -2 & 1 \\ 3 & 0 \end{vmatrix} + 4\begin{vmatrix} 1 & -2 \\ -1 & 3 \end{vmatrix}} = -2$$

$$x_2(t) = \frac{\Delta_2}{\Delta} = \frac{\begin{vmatrix} 1 & 4\cos 3t & 1 \\ -1 & -12 & 0 \\ 2 & 0 & 4 \end{vmatrix}}{-2} = \frac{16\cos 3t - 24}{-2}$$

$$x_3(t) = \frac{\Delta_3}{\Delta} = \frac{\begin{vmatrix} 1 & -2 & 4\cos 3t \\ -1 & 3 & -12 \\ 2 & 0 & 0 \end{vmatrix}}{-2} = \frac{-24\cos 3t + 48}{-2}$$

3.2.4 Difficulties with High Order

For more than three equations in three variables, Cramer's rule with the Laplace expansion of determinants becomes very inefficient in comparison to other solution methods.

The long-range interest in Cramer's rule here is twofold. First, it is simple and straightforward to apply in the case of a small number of equations, say two or three. It would be wise to seek digital computer aid in the solution of larger problems. Second, Cramer's rule is a very good theoretical tool for showing network properties, the subject of Chapter Four.

D3-1

Solve the following simultaneous linear algebraic equations using Cramer's rule:

(a) $\begin{cases} -2x_1 + \dot{x}_2 = 0 \\ 8x_1 - 3x_2 = -4 \end{cases}$ ans. $-2, -4$

(b) $\begin{cases} x_1 - 3x_2 = -\sin t \\ -x_1 + x_2 = 12 \end{cases}$ ans. $(\frac{1}{2})\sin t - 18, (\frac{1}{2})\sin t - 6$

(c) $\begin{cases} 3x_1 - x_2 - 3x_3 = -5 \\ -x_1 + 4x_2 = 0 \\ -3x_1 + 7x_3 = 6 \end{cases}$ ans. $-68/41, -17/41, -6/41$

(d) $\begin{cases} 2x_1 + x_2 + x_3 = 0 \\ 7x_1 + 8x_2 - x_3 = 0 \\ -6x_1 + 9x_2 + x_3 = 0 \end{cases}$ *ans.* 0, 0, 0

Cramer's Rule

The solution of a set of n independent linear algebraic equations in n variables,

$$\begin{cases} a_{11}x_1 + a_{12}x_2 + \cdots + a_{1n}x_n = y_1 \\ a_{21}x_1 + a_{22}x_2 + \cdots + a_{2n}x_n = y_2 \\ \vdots \\ a_{n1}x_1 + a_{n2}x_2 + \cdots + a_{nn}x_n = y_n \end{cases}$$

is given by

$$x_i = \frac{\Delta_i}{\Delta}$$

3.3 Systematic Nodal Equations

3.3.1 Node-to-Node Voltages and Notation

For a network with n nodes, the voltages between any one of the nodes and each of the remaining $(n - 1)$ nodes form a convenient set of independent variables. For example, in the network of Figure 3-1(a), which has five nodes, a set of four such node-to-node voltages is $v_1(t)$, $v_2(t)$, $v_3(t)$, and $v_4(t)$.

The network for this example has been drawn in a manner that makes the nodes especially evident, and this is generally helpful. It is convenient, too, to define the node-to-node voltages all with one end of the reference polarity bracket at the same node and with the plus signs of the reference polarities at the other ends of the brackets, as shown in Figure 3-1(b).

Figure 3-1 A network with five nodes

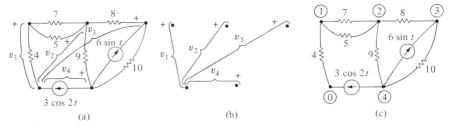

(a)   (b)   (c)

Instead of cluttering up the drawing with all those brackets, the notation of numbering the nodes 0, 1, 2, ..., shown in Figure 3-1(c), is adopted. The zero node (or *reference* node) is the node that is common to all the voltages. The voltage v_3, for example, is the voltage between node 3 and node 0, with the plus sign of the voltage reference polarity at node 3.

If these four voltages were known, any other voltage or current in the network would be quite easy to find. The indicated signals in Figure 3-2 are related to the node voltages and source functions as follows:

$$i_1 = \frac{v_1}{4}$$

$$v_5 = v_3 - v_2$$

$$i_2 = i_3 - 6 \sin t = \frac{v_3 - v_4}{10} - 6 \sin t$$

3.3.2 Equations for Each Independent Node

An independent equation may be written for each of the numbered nodes, 1, 2, ... For the first node in the example under discussion, Figure 3-3(a),

$$i_a + i_b + i_c = 0$$

Substituting for the resistor currents in terms of the node voltages,

$$\frac{v_1}{4} + \frac{v_1 - v_2}{5} + \frac{v_1 - v_2}{7} = 0$$

For the second node, Figure 3-3(b),

$$i_d + i_e + i_f + i_g = 0$$

$$\frac{v_2 - v_1}{7} + \frac{v_2 - v_1}{5} + \frac{v_2 - v_3}{8} + \frac{v_2 - v_4}{9} = 0$$

For the third node, Figure 3-3(c),

$$i_h + i_i = 6 \sin t$$

$$\frac{v_3 - v_2}{8} + \frac{v_3 - v_4}{10} = 6 \sin t$$

Figure 3-2   Finding other network signals from the node voltages

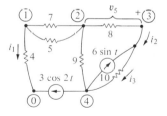

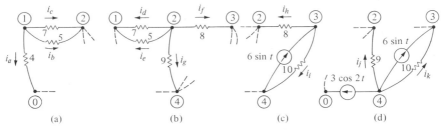

Figure 3-3 Node-by-node analysis of the example network

For the fourth node, Figure 3-3(d),

$$i_j + i_k = -6 \sin t - 3 \cos 2t$$

$$\frac{v_4 - v_2}{9} + \frac{v_4 - v_3}{10} = -6 \sin t - 3 \cos 2t$$

Rearranging these four equations gives

$$\begin{cases} (\frac{1}{4} + \frac{1}{5} + \frac{1}{7})v_1 - (\frac{1}{5} + \frac{1}{7})v_2 - (0)v_3 - (0)v_4 = 0 \\ -(\frac{1}{5} + \frac{1}{7})v_1 + (\frac{1}{5} + \frac{1}{7} + \frac{1}{8} + \frac{1}{9})v_2 - (\frac{1}{8})v_3 - (\frac{1}{9})v_4 = 0 \\ -(0)v_1 - (\frac{1}{8})v_2 + (\frac{1}{8} + \frac{1}{10})v_3 - (\frac{1}{10})v_4 = 6 \sin t \\ -(0)v_1 - (\frac{1}{9})v_2 - (\frac{1}{10})v_3 + (\frac{1}{9} + \frac{1}{10})v_4 = -6 \sin t - 3 \cos 2t \end{cases}$$

which may be solved for the node-to-node voltages v_1, v_2, v_3, and v_4.

If an equation were to be written for the zero node, it would be a linear combination of the other equations and thus redundant.

3.3.3 General Form of the Equations

In general, systematic nodal equations have the form

$$g_{11}v_1 - g_{12}v_2 - g_{13}v_3 - \cdots = i_1$$
$$-g_{21}v_1 + g_{22}v_2 - g_{23}v_3 - \cdots = i_2$$
$$-g_{31}v_1 - g_{32}v_2 + g_{33}v_3 - \cdots = i_3$$
$$\vdots$$

where

g_{11} = sum of inverses of resistances connected directly to node 1

g_{22} = sum of inverses of resistances connected directly to node 2

g_{33} = sum of inverses of resistances connected directly to node 3

$\vdots$

and the other coefficients are of the form

g_{mn} = sum of inverses of resistances connected directly between nodes m and n

= g_{nm}

For example,

g_{12} = sum of inverses of resistances connected directly between nodes 1 and 2

g_{13} = sum of inverses of resistances connected directly between nodes 1 and 3

g_{21} = sum of inverses of resistances connected directly between nodes 2 and 1

g_{23} = sum of inverses of resistances connected directly between nodes 2 and 3

The driving function terms involve the source currents as follows:

$$i_1 = \begin{pmatrix} \text{sum of currents from current} \\ \text{sources connected directly to} \\ \text{node 1, with reference} \\ \text{directions toward node 1} \end{pmatrix} - \begin{pmatrix} \text{sum of currents from current} \\ \text{sources connected directly to} \\ \text{node 1, with reference} \\ \text{directions away from node 1} \end{pmatrix}$$

$$i_2 = \begin{pmatrix} \text{sum of currents from current} \\ \text{sources connected directly to} \\ \text{node 2, with reference} \\ \text{directions toward node 2} \end{pmatrix} - \begin{pmatrix} \text{sum of currents from current} \\ \text{sources connected directly to} \\ \text{node 2 with reference} \\ \text{directions away from node 2} \end{pmatrix}$$

$$i_3 = \begin{pmatrix} \text{sum of currents from current} \\ \text{sources connected directly to} \\ \text{node 3, with reference} \\ \text{directions toward node 3} \end{pmatrix} - \begin{pmatrix} \text{sum of currents from current} \\ \text{sources connected directly to} \\ \text{node 3, with reference} \\ \text{directions away from node 3} \end{pmatrix}$$

$\vdots$

Each equation is the expression of Kirchhoff's current law at one node, with resistor currents written in terms of the node voltages.

3.3.4 Example

Another example network is given in Figure 3-4. The systematic nodal equations, in terms of the indicated node voltages, are as follows:

$$\begin{cases} (\frac{1}{5} + \frac{1}{6} + \frac{1}{4})v_1 - (\frac{1}{6})v_2 - (\frac{1}{4})v_3 - (0)v_4 = 9 \sin t - 12 + \cos 10t \\ -(\frac{1}{6})v_1 + (\frac{1}{2} + \frac{1}{3} + \frac{1}{6} + \frac{1}{7})v_2 - (\frac{1}{2} + \frac{1}{3})v_3 - (\frac{1}{7})v_4 = 12 \\ -(\frac{1}{4})v_1 - (\frac{1}{2} + \frac{1}{3})v_2 + (\frac{1}{2} + \frac{1}{3} + \frac{1}{4})v_3 - (0)v_4 = -\cos 10t \\ -(0)v_1 - (\frac{1}{7})v_2 - (0)v_3 + (\frac{1}{7} + \frac{1}{8})v_4 = 0 \end{cases}$$

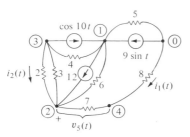

Figure 3-4   Another example network

Once these equations are solved for v_1, v_2, v_3, and v_4, then

$$i_1 = -\frac{v_4}{8}$$

$$i_2 = \frac{v_3 - v_2}{2}$$

$$v_5 = v_2 - v_4$$

Note the symmetry of the coefficients about the diagonal elements. Any resistor connected between nodes a and b is also connected between nodes b and a.

The negative *coupling* terms in each equation are also contained within the positive *self* terms. That is, the self terms in each equation are equal to or larger than the negative sum of the coupling terms.

If a current source is connected between two nodes for which equations are written, the source function appears with a plus sign in one equation and a minus sign in the other since the reference direction of the source must be away from one node and toward the other. These properties provide an easy (although incomplete) check on the equations.

D3-2

Find the indicated voltages by writing and solving systematic nodal equations:

(a)

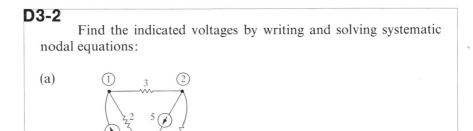

ans. $128/9$, $-4/9$

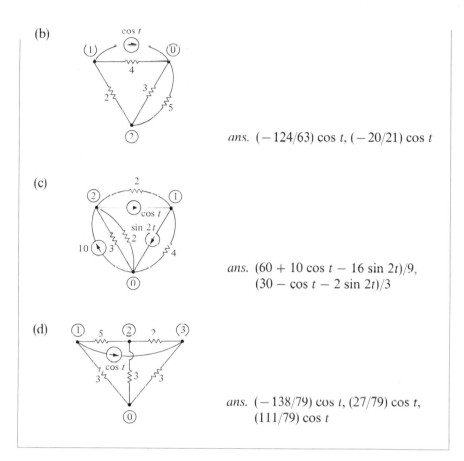

(b)

ans. $(-124/63)\cos t, (-20/21)\cos t$

(c)

ans. $(60 + 10\cos t - 16\sin 2t)/9,$
$(30 - \cos t - 2\sin 2t)/3$

(d)

ans. $(-138/79)\cos t, (27/79)\cos t,$
$(111/79)\cos t$

D3-3

Write systematic nodal equations, solve for the node voltages, then find the two indicated signals:

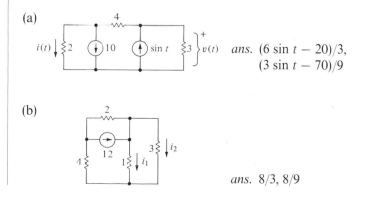

(a)

ans. $(6\sin t - 20)/3,$
$(3\sin t - 70)/9$

(b)

ans. $8/3, 8/9$

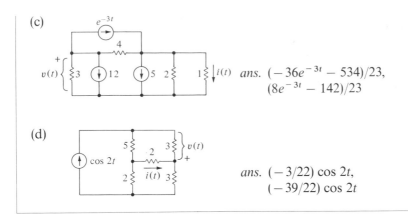

(c)

$v(t)$ ans. $(-36e^{-3t} - 534)/23,$
$(8e^{-3t} - 142)/23$

(d)

ans. $(-3/22) \cos 2t,$
$(-39/22) \cos 2t$

Systematic Simultaneous Nodal Equations

Node voltages are indicated on the network diagram by numbering the nodes 0, 1, 2, ... An equation is written for each of the nodes except the zero node, in terms of the node voltages $v_1, v_2, \ldots$.

The coefficient of v_n in the nth equation is the sum of the inverses of the resistances connected directly to node n. The coefficient of v_m, $m \neq n$, in the nth equation is the negative sum of the resistances connected directly between node n and node m.

The driving function, on the opposite side of each equation from the variables, is, for the nth equation, the sum of the current source functions for sources connected directly to node n, with reference directions toward node n minus those with reference directions away from node n.

3.4   Nodal Equations for Networks Containing Voltage Sources

If there is a voltage source between nodes, the current flowing through the source is unknown. On the other hand, the node-to-node voltage is known; it is fixed by the voltage-source function. Nodal equations for such a network may be written as in the usual case, but treating voltage sources as current sources with unknown current functions. Then the known node-to-node voltages may be added to the set of equations.

An example network is shown in Figure 3-5. For this network the currents through each voltage source are defined, and the usual nodal equations

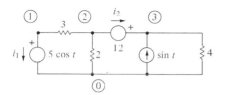

Figure 3-5   A network with voltage sources

are written in terms of these source currents. Then for each voltage source an additional equation is added to the set that expresses the known node-to-node voltage fixed by the voltage source:

$$
\begin{cases}
\tfrac{1}{3}v_1 - \tfrac{1}{3}v_2 & = -i_1 \\
-\tfrac{1}{3}v_1 + (\tfrac{1}{3} + \tfrac{1}{2})v_2 & = -i_2 \\
\quad \tfrac{1}{4}v_3 = \sin t + i_2 \\
v_1 & = 5 \cos t \\
\quad -v_2 + v_3 = 12
\end{cases}
\quad
\begin{array}{l}
\text{basic nodal equations with} \\
\text{unknown currents supplied} \\
\text{by voltage sources} \\
\text{node-to-node voltages fixed} \\
\text{by voltage sources}
\end{array}
$$

The mathematical problem is now to solve the five equations in five variables, $v_1, v_2, v_3, i_1,$ and i_2.

$$
\begin{cases}
\tfrac{1}{3}v_1 & \tfrac{1}{3}v_2 & + i_1 & = 0 \\
-\tfrac{1}{3}v_1 + (\tfrac{1}{3} + \tfrac{1}{2})v_2 & & + i_2 = 0 \\
& \tfrac{1}{4}v_3 & - i_2 = \sin t \\
v_1 & & = 5 \cos t \\
& -v_2 + v_3 & = 12
\end{cases}
$$

Although the voltage sources add to the number of these equations, the added equations are simple ones which may be easily used to eliminate variables in the other equations. For example, substituting for v_1 from the fourth equation and

$$v_3 = 12 + v_2$$

from the fifth equation gives

$$
\begin{cases}
-\tfrac{1}{3}v_2 + i_1 & = -\tfrac{5}{3} \cos t \\
\tfrac{5}{6}v_2 & + i_2 = \tfrac{5}{3} \cos t \\
\tfrac{1}{4}v_2 & - i_2 = \sin t - 3,
\end{cases}
$$

which may be further simplified by adding the bottom two equations to eliminate i_2.

D3-4

Find the indicated signals by writing and solving nodal equations:

(a)

ans. $(\sin t - 36)/7$

(b)

ans. $(-50/29) - (30/19) \cos 6t$

Voltage Sources in Nodal Equations

Define the current through each voltage source and write the systematic nodal equations as if these source currents were known.

Augment the set of equations with one equation for each voltage source, expressing a known node voltage or difference between two node voltages.

3.5 Controlled Sources and Nodal Equations

In some applications it is convenient to include voltage and current sources in which the source function is not given as a specific function of time but, rather, is expressed in terms of some other voltage or current in the network. These *controlled sources* are denoted by diamond-shaped symbols.

To write nodal equations for networks containing controlled sources, write the ordinary nodal equations then append equations, if necessary, expressing the source-controlling signals in terms of the node voltages.

An example is given in Figure 3-6. Systematic simultaneous nodal equations are written as usual, with the controlled source function simply appearing, in terms of the controlling signal i_1, as a source term. Then an

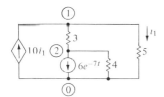

Figure 3-6 Network containing a controlled source

equation is added to the set, relating the controlling signal to the node voltages:

$$
\begin{cases}
(\tfrac{1}{3} + \tfrac{1}{5})v_1 & -\tfrac{1}{3}v_2 = 10i_1 \\
-\tfrac{1}{3}v_1 + (\tfrac{1}{3} + \tfrac{1}{4})v_2 = -6e^{-7t}
\end{cases}
\quad
\begin{array}{l}
\text{basic nodal equations with} \\
\text{controlling signal(s) as source term(s)}
\end{array}
$$

$$
i_1 = \frac{v_1}{5}
\quad
\begin{array}{l}
\text{controlling signal(s) related to} \\
\text{node voltages unless controlling} \\
\text{signal is a node voltage}
\end{array}
$$

The equations may then be rearranged before solving as follows:

$$
\begin{cases}
(\tfrac{1}{3} + \tfrac{1}{5})v_1 & -\tfrac{1}{3}v_2 - 10i_1 = 0 \\
-\tfrac{1}{3}v_1 + (\tfrac{1}{3} + \tfrac{1}{4})v_2 & = -6e^{-7t} \\
-\tfrac{1}{5}v_1 & \quad | \quad i_1 = 0
\end{cases}
$$

Substitution for i_1 from the third equation into the first equation easily simplifies the solution.

D3-5

Find the indicated signals by writing and solving nodal equations:

(a)

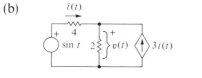

ans. $9 \cos 7t$, $(27/2) \cos 7t$

(b)

ans. $(2/3) \sin t$, $(1/12) \sin t$

Controlled Sources in Mesh Equations

Write the equations as if the controlled source functions were known.
Augment the equations with one equation for each controlling signal, relating the controlling signal to the mesh currents.

3.6 Systematic Simultaneous Mesh Equations

3.6.1 Mesh Currents for Planar Networks

A network is said to be *planar* if it can be drawn on a flat surface with no conductor crossovers, such as the network in Figure 3-7. A network with elements along the sides of a cube and the diagonals is an example of a nonplanar network.

A planar network is composed of "boxes," or *meshes*, and current flows in such a network may be indicated by placing a current symbol in each of the "boxes" such as i_1 and i_2 in Figure 3-7. By this notation it is meant, for example, that the current i_a through the 2-Ω resistor is i_1. The current i_b is $i_1 - i_2$, and the current i_c is $i_2 - i_1$.

Mesh currents are a way of indicating Kirchhoff's current law schematically. Currents through elements that are in a single "box" are mesh currents. Currents through elements that are common to two "boxes" are the difference between two of the mesh currents.

The reference directions of the mesh currents are all chosen to have the same sense, all clockwise or all counterclockwise, so that currents in elements common to two "boxes" are always the difference between two mesh currents. It is not always possible to choose the reference directions so that these currents in elements common to two "boxes" are the sum of two mesh currents, as the drawing of Figure 3-8 indicates.

Knowing the mesh currents, it is simple to find any other current in the network and thus any network voltages as well. Referring to Figure 3-9, in

Figure 3-7 Example planar network

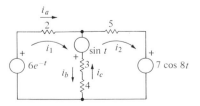

Figure 3-8   Impossibility of mesh currents always adding in elements

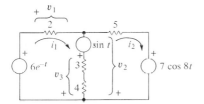

Figure 3-9   Other signals in the example network

which several signals in the example network have been indicated,

$$v_1 = 2i_1$$
$$v_2 = v_3 + \sin t = 7(i_2 - i_1) + \sin t$$

3.6.2   Equations for Each Mesh

For the example under discussion, defining each resistor voltage and apply-ing Kirchhoff's voltage law around the first mesh, shown in Figure 3-10(a),

$$-6e^{-t} + v_a - \sin t + v_b + v_c = 0$$

Substituting for the resistor voltages in terms of the mesh currents,

$$6e^{-t} + 2i_1 \quad \sin t + 3(i_1 - i_2) + 4(i_1 - i_2) = 0$$
$$(2 + 3 + 4)i_1 - (3 + 4)i_2 = 6e^{-t} + \sin t$$

Figure 3-10   Mesh-by-mesh analysis of the example network

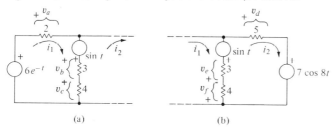

(a)   (b)

For the second mesh, Figure 3-10(b),

$$v_d + 7 \cos 8t + v_f + v_e + \sin t = 0$$
$$5i_2 + 7 \cos 8t + 4(i_2 - i_1) + 3(i_2 - i_1) + \sin t = 0$$
$$-(3 + 4)i_1 + (3 + 4 + 5)i_2 = -7 \cos 8t - \sin t$$

These two mesh equations may be solved simultaneously for i_1 and i_2.

This method is analogous to the process used to write systematic simultaneous nodal equations. For the nodal equations, the node numbering scheme expressed the Kirchhoff's voltage-law relations. Equations were then written expressing Kirchhoff's current law, using the resistor voltage-current relations to put these equations in terms of the node voltages. An equation was not written for one of the nodes (the 0 node) because one application of Kirchhoff's current law is not independent of the others.

For the mesh equations, Kirchhoff's current-law relations are expressed by the mesh current scheme. Then Kirchhoff's voltage law is applied around each "box," or mesh, using the resistor voltage-current relations to express the equations in terms of the mesh currents. Although there are other loops besides the meshes for which one can write the voltage law, the resulting equations are not independent of the equations for the meshes; they are linear combinations of the mesh equations.

3.6.3 General Form of the Equations

The systematic simultaneous mesh equations have the following general form:

$$r_{11}i_1 - r_{12}i_2 - r_{13}i_3 - \cdots = v_1$$
$$-r_{21}i_1 + r_{22}i_2 - r_{23}i_3 - \cdots = v_2$$
$$-r_{31}i_1 - r_{32}i_2 + r_{33}i_3 - \cdots = v_3$$
$$\vdots$$

where

r_{11} = sum of resistances in the mesh through which i_1 flows
r_{22} = sum of resistances in the mesh through which i_2 flows
r_{33} = sum of resistances in the mesh through which i_3 flows
$\vdots$

and the other coefficients are of the form

r_{mn} = sum of resistances common to the i_m and i_n meshes
       $= r_{nm}$

For example,

$r_{12} = $ sum of resistances common to the i_1 and i_2 meshes

$r_{13} = $ sum of resistances common to the i_1 and i_3 meshes

$r_{21} = $ sum of resistances common to the i_2 and i_1 meshes

r_{23} sum of resistances common to the i_2 and i_3 meshes

The driving function terms involve the source voltages as follows:

$$v_1 = \begin{pmatrix} \text{sum of voltage source} \\ \text{functions for sources in} \\ \text{the } i_1 \text{ mesh that have the} \\ \text{source reference with } i_1 \end{pmatrix} - \begin{pmatrix} \text{sum of voltage source} \\ \text{functions for sources in} \\ \text{the } i_1 \text{ mesh that have the} \\ \text{sink reference with } i_1 \end{pmatrix}$$

$$v_2 - \begin{pmatrix} \text{sum of voltage source} \\ \text{functions for sources in} \\ \text{the } i_2 \text{ mesh that have the} \\ \text{source reference with } i_2 \end{pmatrix} - \begin{pmatrix} \text{sum of voltage source} \\ \text{functions for sources in} \\ \text{the } i_2 \text{ mesh that have the} \\ \text{sink reference with } i_2 \end{pmatrix}$$

$$v_3 = \begin{pmatrix} \text{sum of voltage source} \\ \text{functions for sources in} \\ \text{the } i_3 \text{ mesh that have the} \\ \text{source reference with } i_3 \end{pmatrix} - \begin{pmatrix} \text{sum of voltage source} \\ \text{functions for sources in} \\ \text{the } i_3 \text{ mesh that have the} \\ \text{sink reference with } i_3 \end{pmatrix}$$

$\vdots$

Each equation is the expression of Kirchhoff's voltage law around a mesh, where resistor voltages have been expressed in terms of the mesh currents.

The systematic loop equations have symmetries similar to those of the nodal equations: The "self" terms in each equation are positive, all the "coupling" terms are negative. The array of equation coefficients is symmetric about the diagonal, and the coupling terms are contained within the self term in each equation. If a voltage source is common to two meshes, the source function appears with a plus sign in one equation and with a minus sign in the other equation.

3.6.4   Example

For the network of Figure 3-11, the systematic simultaneous mesh equations, in terms of the indicated mesh currents, are as follows:

$$\begin{cases} (2 + 5 + 6 + 3)i_1 - & (5)i_2 - & (3 + 6)i_3 - & (0)i_4 = -\sin t - \cos t \\ -(5)i_1 + (5 + 4 + 7 + 8)i_2 - & (0)i_3 - & (8)i_4 = 0 \\ -(3 + 6)i_1 - & (0)i_2 + (3 + 6 + 9)i_3 - & (9)i_4 = 12 + \cos t \\ -(0)i_1 - & (8)i_2 - & (9)i_3 + (8 + 9)i_4 = -e^{-10t} \end{cases}$$

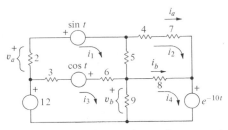

Figure 3-11   Another mesh equation example

In terms of the mesh currents,

$$i_a = i_2, \qquad v_a = -2i_1, \qquad i_b = i_4 - i_2,$$

and

$$v_b = 9(i_3 - i_4)$$

D3-6

Find the indicated mesh currents by writing and solving systematic equations:

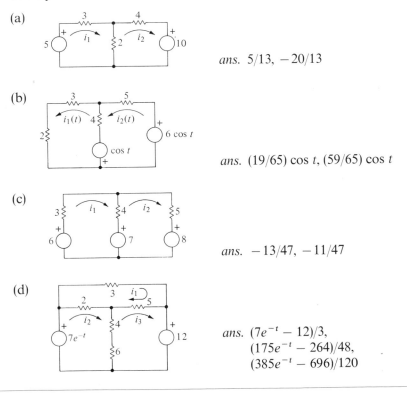

(a)

ans. $5/13, -20/13$

(b)

ans. $(19/65)\cos t, (59/65)\cos t$

(c)

ans. $-13/47, -11/47$

(d)

ans. $(7e^{-t} - 12)/3,$
$(175e^{-t} - 264)/48,$
$(385e^{-t} - 696)/120$

D3-7

Write systematic mesh equations, solve for the mesh currents, then find the two indicated signals:

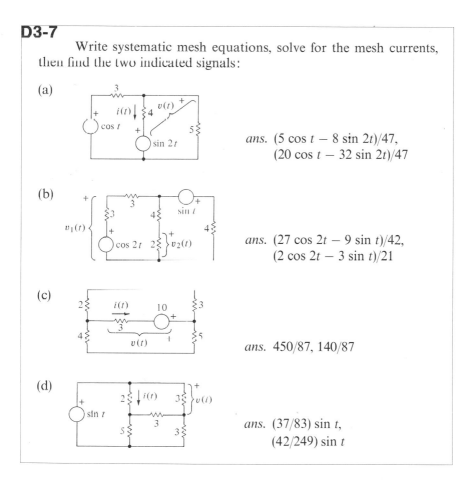

(a)

ans. $(5 \cos t - 8 \sin 2t)/47$,
$(20 \cos t - 32 \sin 2t)/47$

(b)

ans. $(27 \cos 2t - 9 \sin t)/42$,
$(2 \cos 2t - 3 \sin t)/21$

(c)

ans. $450/87$, $140/87$

(d)

ans. $(37/83) \sin t$,
$(42/249) \sin t$

Systematic Simultaneous Mesh Equations

Mesh currents are indicated, each with the same sense, inside each mesh (or "box") of the network. An equation is written for each of the meshes, in terms of the mesh currents $i_1, i_2, \ldots$.

The coefficient of i_n in the nth equation is the sum of the resistances around the nth mesh. The coefficient of i_m, $m \neq n$ in the nth equation is the negative sum of the resistances that are common to mesh n and mesh m.

The driving function, on the opposite side of each equation from the variables, is, for the nth equation, the sum of the voltage-source functions for sources in mesh n, with source reference relations with i_n, minus those with sink reference relations with i_n.

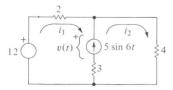

Figure 3-12 Network containing a current source

3.7 Mesh Equations for Networks Containing Current Sources

Similar to the situation with nodal equations for a network that contains voltage sources, loop equations may be written for a network such as that in Figure 3-12 which contains current sources. Before the solution, the voltage across a current source is not generally known, but a mesh current or a difference between two mesh currents is fixed by the current source.

First, indicate by symbols each of the unknown voltages across the current sources and write the systematic mesh equations, including in them those unknown current source voltages, as if they were known.

$$(2 + 3)i_1 - \quad\quad 3i_2 = 12 - v$$
$$- 3i_1 + (3 - 4)i_2 = v$$

Then, augment the set by relating each current source function to the loop currents.

$$i_2 - i_1 = 5 \sin 6t$$

Finally, rearrange the equations before solving as follows;

$$\left\{\begin{array}{l} 5i_1 - 3i_2 + v = 12 \\ - 3i_1 + 7i_2 - v = 0 \\ \quad i_1 - \quad i_2 \quad\quad = 5 \sin 6t \end{array}\right.$$

basic mesh equations with unknown voltages supplied by current sources

current(s) fixed by current source(s)

Since the current source equations are always relatively simple, they may be easily used to simplify the set. For example, the third equation may be solved for i_1 and substituted into the first two equations above.

D3-8

Find the indicated signals by writing and solving mesh equations:

(a)

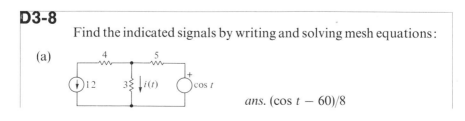

ans. $(\cos t - 60)/8$

(b)

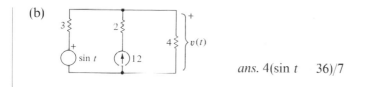

ans. 4(sin *t*   36)/7

Current Sources in Mesh Equations

Define the voltage across each current source and write the mesh equations as if these source voltages were known.

Augment the set of equations with one equation for each current source expressing a known mesh current or difference between two mesh currents.

3.8   Mesh Equations for Networks Containing Controlled Sources

If a source function depends on some other voltage or current in the network, first write the mesh equations in terms of the controlling signal. In the example of Figure 3-13(a), a current source is present, so the unknown voltage supplied by that source is defined as v_2 in Figure 3-13(b). Mesh equations are written as usual, treating the current source as a voltage source with source function v_2. Then the loop current is described in terms of the current source function:

$$\begin{cases} 3i_1 - 3i_2 = v_2 - \sin t \\ -3i_1 + 12i_2 = \sin t \\ i_1 \qquad - 10v_1 \end{cases}$$

basic mesh equations with unknown voltage supplied by current source

loop current expressed in terms of current source function

Figure 3-13   Network containing a controlled current source

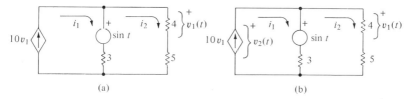

(a)   (b)

Because of the controlled source, one more equation is added to the set, expressing the controlling signal $v_1(t)$ in terms of the loop current:

$$v_1 = 4i_2$$

Rearranging the example equations, the following set results:

$$\begin{cases} 3i_1 - 3i_2 \quad\quad - v_2 = -\sin t \\ -3i_1 + 12i_2 \quad\quad\quad = \sin t \\ \quad i_1 \quad\quad - 10v_1 \quad = 0 \\ \quad\quad - 4i_2 + \quad v_1 \quad = 0 \end{cases}$$

As always, substitutions from the simpler equations into the others may be used to obtain the solution.

D3-9

Find the indicated signals by writing and solving mesh equations:

(a)

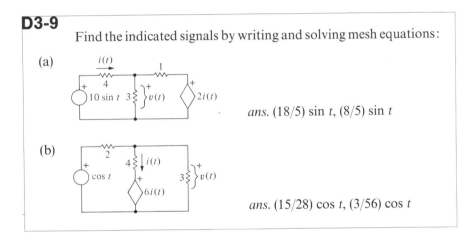

ans. $(18/5)\sin t$, $(8/5)\sin t$

(b)

ans. $(15/28)\cos t$, $(3/56)\cos t$

Controlled Sources in Nodal Equations

Write the equations as if the controlled source functions were known.
Augment the equations with one equation for each controlling signal, relating the controlling signal to the node voltages.

3.9 Loop Equations for Nonplanar Networks

3.9.1 Selection of Loop Currents

The systematic mesh equations as presented previously are not a general method for network solution as are systematic nodal equations because the network must be planar for those methods to apply. One can always find a set of independent loops about which to write equations even if the network is not planar. For a nonplanar network, however, a mesh current cannot simply be put in each of the "boxes" because, with crossovers, the "boxes" are not evident.

To determine an independent set of loops, first form the *graph* of the network under consideration. The graph is merely a picture of the network's nodes, together with lines, called *branches*, indicating where elements connect from node to node. A planar network and its graph are shown in Figure 3-14(a) and (b).

A tree is any set of branches that connect all the nodes without forming any loops. For the example network, trees are indicated with solid lines in the drawings of Figure 3-14(c). The network also has several other trees.

For a given tree, a *link* is any branch not in the tree. The links for each of the trees of Figure 3-14 are indicated with dashed lines. All of the trees for a network have the same number of links.

Given the network and a chosen tree, a set of loops about which Kirchhoff's voltage law may be applied independently consists of those loops, one for each link, that traverse a closed path through the tree and a single link each.

As with the systematic equations, a loop current may be indicated in each of these independent loops. Applying Kirchhoff's voltage law around the path of each loop current and expressing resistor voltages in terms of these currents yields a set of simultaneous algebraic equations.

3.9.2   Planar Network Example

Although the use of this general method is necessary only for nonplanar networks, it is sometimes used with planar networks instead of the systematic, one current in each "box" method. A simple example of the general method will now be applied to the planar network of Figure 3-15(a). In Figure 3-15(b)

Figure 3-14   Network, its graph and several of its trees

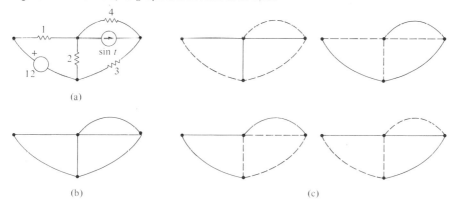

(a)

(b)   (c)

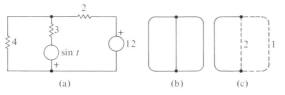

(a)   (b)   (c)

Figure 3-15   An example network, its graph and a tree of the graph

is shown the network graph; and a tree, where the links have been numbered, is selected in Figure 3-15(c).

In the network itself, redrawn in Figure 3-16, the loop current i_1 is indicated in a closed path through the tree and link 1, while i_2 is in a closed path through the tree and link 2.

Summing voltages around the outer loop and expressing them in terms of the loop currents i_1 and i_2 gives

$$v_1 + v_2 + 12 = 0$$
$$4(i_1 + i_2) + 2i_1 + 12 = 0$$

Summing voltages around the smaller loop,

$$v_1 + v_3 - \sin t = 0$$
$$4(i_1 + i_2) + 3i_2 - \sin t = 0$$

Rearranging,

$$\begin{cases} 6i_1 + 4i_2 = -12 \\ 4i_1 + 7i_2 = \sin t \end{cases}$$

3.9.3   Nonplanar Network Example

A more involved example, in which the general method is applied to a non-planar network, is now given. For the network of Figure 3-17(a), a tree has been selected and the links numbered as in Figure 3-17(b).

Figure 3-16   Example network with chosen loop currents

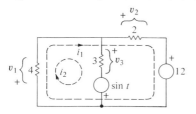

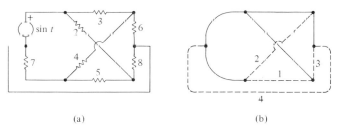

(a)   (b)

Figure 3-17 Nonplanar network and tree with numbered links

The loop currents corresponding to each link are placed, in Figure 3-18, in such a manner that only the ith current flows through the ith link.

Kirchhoff's voltage law is now written around each of the loops defined by the loop currents, expressing resistor voltages in terms of the loop currents:

$$\begin{cases}
7(i_1 + i_2) - \sin t + 2(i_1 - i_3) + 5i_1 = 0 \\
7(i_1 + i_2) - \sin t + 3(i_2 + i_3 + i_4) + 4i_2 = 0 \\
8i_3 + 2(i_3 - i_1) + 3(i_2 + i_3 + i_4) + 6(i_3 + i_4) = 0 \\
-\sin t + 3(i_2 + i_3 + i_4) + 6(i_3 + i_4) = 0
\end{cases}$$

Or,

$$\begin{cases}
14i_1 + 7i_2 - 2i_3 &- \sin t \\
7i_1 + 14i_2 + 3i_3 + 3i_4 &- \sin t \\
-2i_1 + 11i_2 + 19i_3 + 9i_4 &= 0 \\
3i_2 + 9i_3 + 9i_4 &= \sin t
\end{cases}$$

Figure 3-18 Example network and selected loop currents

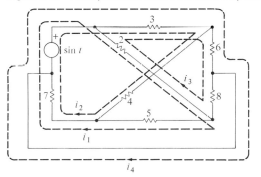

D3-10

Write loop equations in terms of the indicated loop currents, then solve for the indicated voltage:

(a)

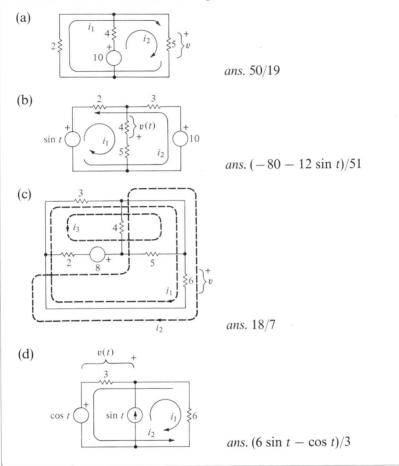

ans. 50/19

(b)

ans. $(-80 - 12 \sin t)/51$

(c)

ans. 18/7

(d)

ans. $(6 \sin t - \cos t)/3$

D3-11

Write (but do not solve) loop equations for the following network. Show your choice of loop-current variables clearly:

Loop Equations for Nonplanar Networks

For nonplanar networks, draw the network graph, form a tree, and define loop currents so that a different single loop current flows through each link. Then write Kirchhoff's voltage law around each of the loops defined by the loop currents, expressing resistor voltages in terms of the loop currents.

Chapter Three Problems

Basic Problems

Simultaneous Linear Algebraic Equations

1. Solve, using Cramer's rule:

(a) $\begin{cases} 12x_1 - 3x_2 - x_3 = 8 \cos t \\ -3x_1 + 5x_2 = 6 - 8 \cos t \\ -x_1 + 9x_3 = -6 \end{cases}$

(b) $\begin{cases} 6x_1 - 3x_2 - x_3 = \cos 2t \\ -3x_1 + 5x_2 - 2x_3 - 8 - \cos 2t \\ -x_1 - 2x_2 + 8x_3 = 0 \end{cases}$

Systematic Nodal Equations

2. Write and solve systematic nodal equations for v_1, v_2, and v_3:

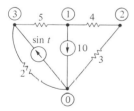

3. Find $v(t)$ and $i(t)$ using systematic nodal equations:

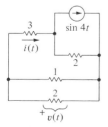

4. Find $i(t)$ using systematic nodal eqautions:

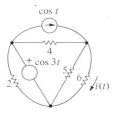

5. Find $v(t)$ using systematic nodal equations:

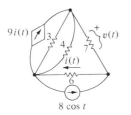

6. Find $i_1(t)$ and $i_2(t)$ using systematic nodal equations:

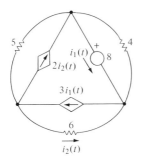

Systematic Mesh Equations

7. Write and solve systematic mesh equations for i_1, i_2, and i_3:

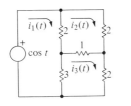

8. Find v and i using systematic mesh equations:

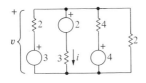

9. Find $i(t)$ using systematic mesh equations:

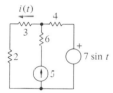

10. Find $v(t)$ using systematic mesh equations:

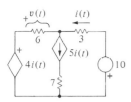

11. Find $v_1(t)$ and $v_2(t)$ using systematic mesh equations:

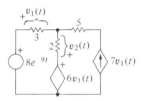

Practical Problems

Potentiometers

A potentiometer is an adjustable resistor similar to a rheostat, but where the movable contact and both ends of the resistive material are available for electrical connection. The symbol for a potentiometer is shown in the network diagrams of Figure 3-19.

The indicated resistance R is the end-to-end resistance of the resistive material. Commonly, the device is designed so that the resistance between one end and the movable contact (or *arm*) is proportional to the angular

Figure 3-19 (a) Practical Problem 1
(b) Practical Problem 2

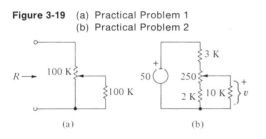

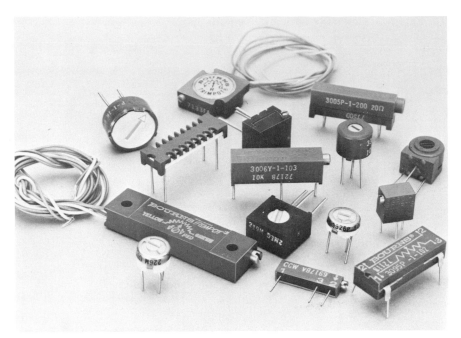

An assortment of screwdriver adjustable potentiometers. In most of these the contact arm is moved along a helical coil of resistance wire. (*Photo courtesy of Trimpot Products Division of Bourns, Inc.*)

position of a shaft on which the arm is mounted. The potentiometer is then said to have a *linear taper*. In other potentiometers, the end-to-arm resistance is made to vary in some other way with the arm position. For example, in volume controls the resistance taper is approximately logarithmic because perceived audio loudness varies logarithmically.

1. A linear taper potentiometer is connected to a resistor as shown in Figure 3-19(a). Find the resistance between the indicated terminals as a function of the potentiometer arm location and sketch the result.

2. Potentiometers are often used for adjustment of electronic equipment. In the network of Figure 3-19(b) a potentiometer controls the voltage v. What is the range over which v may be adjusted by the potentiometer? If the source voltage may vary as much as $\pm 10\%$, will it always be possible to adjust v to 18 V?

Potentiometric Voltage Measurement

Errors caused by the effects of voltmeter internal resistance on a network being measured can be eliminated by using a *potentiometric* measurement.

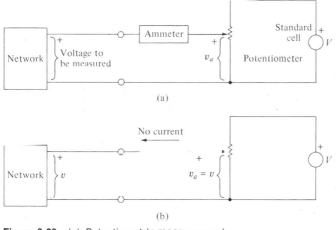

(a)

(b)

Figure 3-20   (a) Potentiometric measurement
(b) Potentiometer at balance

In this method the unknown voltage is compared with a known voltage, often the voltage of a battery. Batteries especially constructed to supply stable, predictable voltages are called *standard cells*.

A potentiometer is used to develop an adjustable voltage v_a and is connected through an ammeter to the network to be measured as indicated in the drawing of Figure 3-20(a).

Zero current flows through the ammeter when the potentiometer voltage r_a equals the voltage to be measured, as indicated in Figure 3-20(b). At that setting of the potentiometer slider, the potentiometer and the network may as well be disconnected from one another. The potentiometer voltage is then equal to the open circuit network voltage and is proportional to the slider position.

3. For the potentiometric voltage apparatus of Figure 3-21, which has a linear taper potentiometer, what is the voltage v if the potentiometer slider arm is one-third of the way from bottom to top of the resistance element? What fraction of the distance from bottom to top is the potentiometer slider when $v = 6.3$?

Figure 3-21   Practical Problem 3.

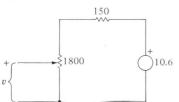

A radio station studio. The large knobs on the control board adjust the volume of the various audio sources via potentiometers with logarithmic tapers. (*Photo courtesy of the Radio Corporation of America*)

Resistance Measurement

One very fundamental way to measure the resistance of a resistor is to connect it in a circuit so that a current flows through it, then measure both the resistor voltage and the resistor current. The ratio of the sink reference voltage to current is the resistance, as diagrammed in Figure 3-22(a).

Figure 3-22 (a) Determining a resistance from voltage and current measurements
(b) Basic ohmmeter circuit

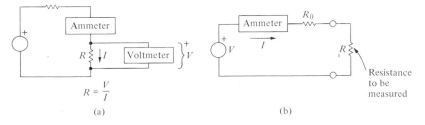

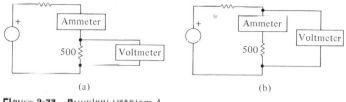

(a)   (b)

Figure 3-23   Practical Problem 4

Because the measurement of resistance is so fundamental, special instruments called *ohmmeters* have been designed to give a direct reading of resistance on a single meter. Ohmmeters consist basically of a battery or other
fixed, constant voltage source in series with a resistor and an ammeter, as
shown in Figure 3-22(b).
The ammeter current is

$$I = \frac{V}{R_0 + R}$$

and the scale of the ammeter may be calibrated in terms of R.
Different ranges of resistance R may be measured by changing the fixed
resistance R_0.

4. The ammeter has an internal resistance of $1 \, \Omega$ and the voltmeter's
internal resistance is $1000 \, \Omega$. In each of the two situations of Figure 3-23,
what is the ratio of the voltmeter reading to the ammeter reading? If the
internal resistances of the two meters are each known, their effects may be
taken into account.

5. In each of the two situations of Figure 3-24, the ammeter internal resistance is $2.5 \, \Omega$, the voltmeter internal resistance is $5000 \, \Omega$, and the measured
current and voltage are indicated. The voltage source V and the resistance
R_0 are unknown. Find R_1 and R_2, taking the meter internal resistances
into account.

6. An ohmmeter is constructed from a 1.5-V battery with an internal
resistance of $1.2 \, \Omega$ in series with an ammeter with an internal resistance of

Figure 2-24   Practical Problem 5

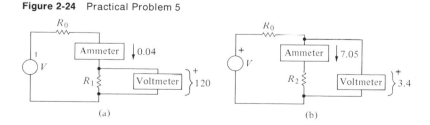

(a)   (b)

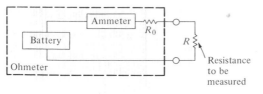

Figure 3-25 Practical Problem 6

125 Ω in series with a resistor R_0 as indicated in Figure 3-25. What value should the resistor R_0 have if the ammeter is to register a full scale reading of 10 mA when the measured resistance R is 0 Ω (a short circuit)? This is the maximum current through a measured resistance R, and care must be taken in practice to avoid damaging delicate devices such as transistors with excessive ohmmeter currents.

With this value for R_0, what measured resistances R will make the ammeter read three-quarters, one-half, and one-quarter of its full scale reading?

Resistance Bridge

A resistance bridge is a method of measuring an unknown resistance by comparing it to a known resistance.

The circuit drawn in Figure 3-26(a) is known as a *Wheatstone bridge*. Zero current, I, flows through the ammeter when the voltage across the ammeter is zero, that is, when $v_1 = v_2$. Under that condition, the ammeter may as well be removed from the circuit since no current flows through it. This situation is shown in Figure 3-26(b).

The voltages $v_1 = v_2$ are then given by the voltage divider rule,

$$v_1 = \frac{R_2 V}{R_1 + R_2} = v_2 = \frac{R_4 V}{R_3 + R_4}.$$

Figure 3-26 (a) The Wheatstone bridge
(b) Wheatstone bridge at balance
(c) Adjustable Wheatstone bridge for resistance measurement

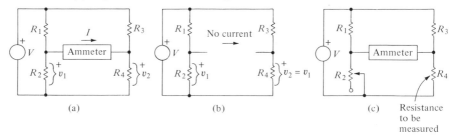

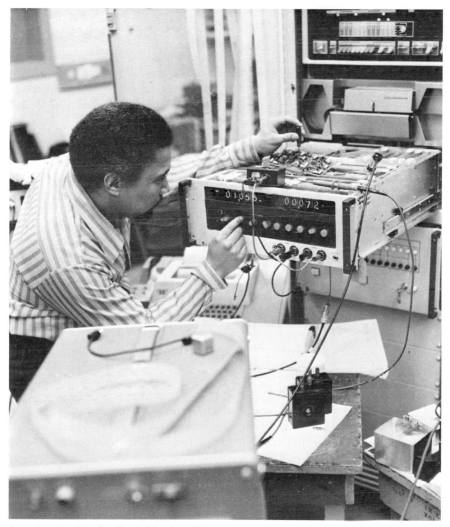

An automatically adjusting bridge is calibrated at the factory.   (*Photo courtesy of GenRad, Inc.*)

hence there is zero ammeter current when

$$\frac{R_2}{R_1 + R_2} = \frac{R_4}{R_3 + R_4}$$

$$R_2R_3 + R_2R_4 - R_1R_4 \mid R_2R_4$$

$$R_2R_3 = R_1R_4$$

or

$$\frac{R_1}{R_2} = \frac{R_3}{R_4}$$

Under this condition, the bridge is said to be *balanced*.

If $R_1 = R_3$, R_2 is adjustable and R_4 is the resistance to be measured, as as in the arrangement of Figure 3-26(c), the ammeter reads zero (the bridge is balanced) when R_2 is adjusted to be equal to R_4. In general, at balance,

$$R_2 = \frac{R_1}{R_3} R_4$$

so that various ratios (R_1/R_3) will give various proportional relationships between the measured resistance and the adjustable balancing resistance.

7. The adjustable resistor in the Wheatstone bridge of Figure 3-27(a) has a linear taper. What fraction of the full slider travel is the slider setting when the bridge is balanced?

8. Choose resistances R_1 and R_3 so that the bridge of Figure 3-27(b) may measure resistances R in the range 0–100 Ω. Repeat for the range 0–1000 Ω.

Advanced Problems

Linear Algebraic Equations

1. Show that if all driving functions (the "knowns") in a linearly independent set of n simultaneous linear algebraic equations in n variables are proportional to some function $f(t)$, then the solution of those equations (the "unknowns") is proportional to $f(t)$.

This is to say that if all the sources in a network are constant, all voltages and currents in the network are constant. If all sources vary as e^{3t}, all voltages and currents in the network vary as e^{3t}. If a network has a single source, all network voltages and currents are proportional to that source function.

Figure 3-27 (a) Practical Problem 7
(b) Practical Problem 8

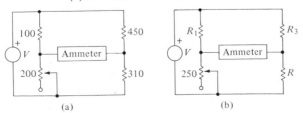

(a)   (b)

2. Show that if all driving functions in a linearly independent set of n simultaneous linear algebraic equations in n variables are multiplied by a constant k, then the solution of those equations is multiplied by k.

This is to say that if all the fixed sources in a network are doubled, all voltages and currents in the network are doubled.

Network Equations

3. A set of nonsystematic equations for the network of Figure 3-28(a) is as follows:

$$\begin{cases}
v_1 + v_2 = 0 \\
v_1 - v_3 = 0 \\
v_1 + 12 - v_4 = 0 \\
v_2 + v_3 = 0 \\
v_2 - 12 + v_4 = 0 \\
v_3 + 12 - v_4 = 0 \\
\cos t + i_2 + i_3 - i_4 = 0 \\
i_1 - i_4 = 0 \\
-i_1 + i_3 + i_2 + \cos t = 0 \\
v_2 = 4i_2 \\
v_3 = -5i_3 \\
v_4 = 3i_4
\end{cases}$$

Solve these equations for i_3.

4. The voltage and the current in every element in the network of Figure 3-28(b) have been defined. Write the equations for Kirchhoff's voltage law around *every* loop in the network, write the equations for Kirchhoff's current law at *every* node in the network, and write the equations for Ohm's law for *every* resistor in the network.

Having converted the network problem to a mathematical problem, solve this set of simultaneous linear algebraic equations for v_3.

Figure 3-28   (a) Advanced Problem 3
(b) Advanced Problem 4

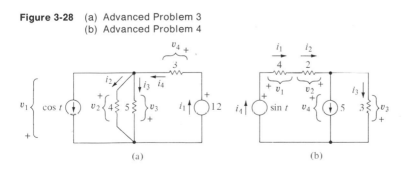

(a)   (b)

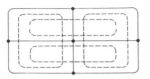

Figure 3-29 Advanced Problem 6

Controlled Sources

5. Show that if a network contains only resistors and controlled sources that have source functions proportional to other voltages and currents in the network, all network voltages and currents must be zero.

Network Topology

6. Construct a network with the graph shown in Figure 3-29 and show that loop equations for the four loops indicated are not linearly independent of one another. Find another set of four loops, the equations for which are linearly independent.

7. Show that for the graph of any network, the number of branches in a tree is one less than the number of nodes.

Ladder Networks

8. A network of the type sketched in Figure 3-30(a) is called a *ladder network*. To solve such a network using systematic simultaneous equations generally involves quite a few mesh or nodal equations.

An easier method of solution is as follows. Start at the end of the network farthest from the source and *assume* some nonzero resistor current, $i_{assumed}$, say 1 A, as in Figure 3-30(b). Using this current, the voltage across the next top-to-bottom resistor may be found. Using that voltage, the next current may be found, and so on.

Figure 3-30 (a) A ladder network
(b) Calculating voltages and currents using an assumed current
(c) Comparing calculated source voltage with the actual source voltage

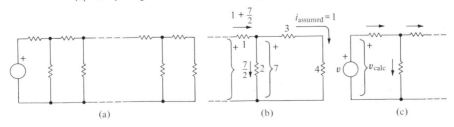

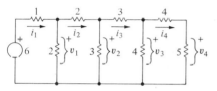

Figure 3-31 Advanced Problem 8

Eventually, the currents through the resistors nearest the source are found and a source voltage, v_{calc}, is found that would produce the originally assumed current and all the other voltages and currents found from the assumed current. This situation is illustrated in Figure 3-30(c).

Unless a very lucky choice was made for the assumed current, v_{calc} will not be the same as the actual source voltage v. If v_{calc} would produce $i_{assumed}$, what current i, in the farthest resistor, is produced by the actual source voltage v? All of the actual voltages and currents in the network are

$$\frac{v}{v_{calc}}$$

times as large as the values calculated based upon $i_{assumed}$.

Use this method to find each of the indicated voltages and currents in the ladder of Figure 3-31.

Repeat for a source voltage function, instead, of $6 \sin t$.

CHAPTER FOUR
Source-Resistor Network Properties

4.1 Introduction

Having characterized source-resistor networks by simultaneous linear algebraic equations in the last chapter, some important properties of the networks that follow from corresponding characteristics of their equations are now investigated.

These properties are fundamental to a deeper understanding of networks and are most important in network design and in the subsequent analysis of more involved networks here.

When you complete this chapter, you should know

1. How to use nodal and mesh equations to find the equivalent resistance of any two-terminal combination of resistors, including those in which some of the resistors are neither in series nor in parallel with other resistors.
2. How to find the equivalent resistance of any two-terminal combination of resistors and controlled sources.
3. The superposition property and how to use it for sources, for source components, and in the presence of controlled sources.

4. How to find Thévenin and Norton equivalents, and how to use them in the solution of networks.
5. The maximum power transfer theorem and how to use it.
6. What network transfer ratios are and how to find and use them.
7. Delta-wye transformation methods.

4.2 Equivalent Resistances

4.2.1 Equivalent Resistance Using Nodal Equations

Nodal equations offer a method of calculating equivalent resistances in those cases in which resistors are neither in series nor in parallel with one another, as in the two-terminal network of Figure 4-1.

Consider applying some current $i(t)$ through the element terminals, as in Figure 4-2, then finding the sink reference (for the element) voltage $v_1(t)$. The voltage $v_1(t)$ will be proportional to $i(t)$, and their ratio is the equivalent resistance of the element. A voltage source can be applied, measuring the current instead; but if nodal equations are to be written, they will be simpler with an applied source current.

For this example network, the systematic nodal equations are

$$\begin{cases} (\frac{1}{1} + \frac{1}{4})v_1 - & (\frac{1}{1})v_2 - & (\frac{1}{4})v_3 = i(t) \\ (\frac{1}{1})v_1 + (\frac{1}{1} + \frac{1}{2} + \frac{1}{3})v_2 & (\frac{1}{3})v_3 = 0 \\ -(\frac{1}{4})v_1 - & (\frac{1}{3})v_2 + (\frac{1}{3} + \frac{1}{4} + \frac{1}{5})v_3 = 0 \end{cases}$$

where the source current $i(t)$, although not specified, is treated as a known quantity. Every network voltage and current will be proportional to $i(t)$, whatever it is.

Figure 4-1 Two-terminal combination of resistors

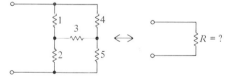

Figure 4-2 Current applied to the resistor network

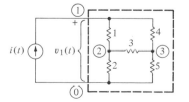

Simplifying the equations by eliminating the fractions,

$$\begin{cases} 5v_1 - 4v_2 - v_3 = 4i(t) \\ -6v_1 + 11v_2 - 2v_3 = 0 \\ -15v_1 - 20v_2 + 47v_3 = 0 \end{cases}$$

$$v_1(t) = \frac{\begin{vmatrix} 4i(t) & -4 & -1 \\ 0 & 11 & -2 \\ 0 & -20 & 47 \end{vmatrix}}{\begin{vmatrix} 5 & -4 & -1 \\ -6 & 11 & -2 \\ -15 & -20 & 47 \end{vmatrix}} = \frac{4i(t)\begin{vmatrix} 11 & -2 \\ -20 & 47 \end{vmatrix}}{5\begin{vmatrix} 11 & -2 \\ -20 & 47 \end{vmatrix} + 4\begin{vmatrix} -6 & -2 \\ -15 & 47 \end{vmatrix} - \begin{vmatrix} -6 & 11 \\ -15 & -20 \end{vmatrix}}$$

$$= \frac{1908 i(t)}{5(477) + 4(-312) - (285) = 852}$$

The equivalent resistance of the two-terminal network is then the ratio

$$R_{\text{equiv}} = \frac{v_1(t)}{i(t)} = \frac{1908}{852}$$

If desired, a specific applied current $i(t)$ such as 1 or 10 or $\sin t$ could be used in the calculation.

4.2.2 Equivalent Resistance Using Mesh Equations

Mesh equations may also be used to compute equivalent resistances of two-terminal combinations of resistors where the individual resistors are neither in series nor in parallel. The two-terminal network of Figure 4-3(a) is an example.

A voltage is applied to the network in Figure 4-3(b). The equivalent resistance is the ratio

$$R_{\text{equiv}} = \frac{v(t)}{i_1(t)}$$

Figure 4-3 Two-terminal combination of resistors with applied voltage

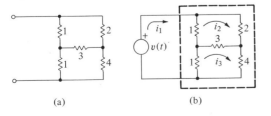

(a)   (b)

Systematic mesh equations are solved for i_1 as follows:

$$\begin{cases} 2i_1 - i_2 - i_3 = v(t) \\ -i_1 + 6i_2 - 3i_3 = 0 \\ -i_1 - 3i_2 + 8i_3 = 0 \end{cases}$$

$$i_1 = \frac{\begin{vmatrix} v & -1 & -1 \\ 0 & 6 & -3 \\ 0 & -3 & 8 \end{vmatrix}}{\begin{vmatrix} 2 & 1 & 1 \\ -1 & 6 & -3 \\ -1 & -3 & 8 \end{vmatrix}} = \frac{v\begin{vmatrix} 6 & -3 \\ -3 & 8 \end{vmatrix}}{2\begin{vmatrix} 6 & -3 \\ -3 & 8 \end{vmatrix} + \begin{vmatrix} -1 & -3 \\ -1 & 8 \end{vmatrix} - \begin{vmatrix} -1 & 6 \\ -1 & -3 \end{vmatrix}}$$

$$= \frac{39v}{78 - 11 + 9}$$

Then

$$R_{\text{equiv}} = \frac{v(t)}{i_1(t)} = \frac{76}{39}$$

4.2.3 Equivalent Resistances for Networks Involving Controlled Sources

A two-terminal network composed of only resistors and controlled sources is equivalent to a single resistor, providing that the controlling signals for the sources are voltages or currents within the two terminal network.

The equivalent resistance of such a network may be found by methods similar to those of the previous sections in which a voltage or current source is connected to the element terminals and the ratio of the sink reference terminal voltage to terminal current is found.

An example network with resistors and a controlled source is given in Figure 4-4(a). In Figure 4-4(b) a current is applied to the network terminals. Systematic nodal equations are as follows:

$$\begin{cases} (\tfrac{1}{2})v_1 - (\tfrac{1}{2})v_2 = i(t) + 3v_2 \\ -(\tfrac{1}{2})v_1 + (\tfrac{1}{2} + \tfrac{1}{4})v_2 = -3v_2 \end{cases}$$

or

$$\begin{cases} v_1 - 7v_2 = 2i(t) \\ -2v_1 + 15v_2 = 0 \end{cases}$$

Then

$$v_1(t) = \frac{\begin{vmatrix} 2i(t) & -7 \\ 0 & 15 \end{vmatrix}}{\begin{vmatrix} 1 & -7 \\ -2 & 15 \end{vmatrix}} = 30i(t)$$

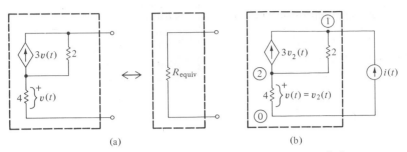

(a)

(b)

Figure 4-4 Two-terminal network involving resistors and a controlled source

and

$$R_{equiv} = \frac{v_1(t)}{i(t)} = 30 \,\Omega$$

Since controlled sources may supply energy to a network, it is possible that within the element more energy is supplied than is dissipated, in which event the equivalent resistance will be negative. This situation is illustrated by the example of Figure 4-5 which is solved by applying a voltage to the terminals, in Figure 4-5(b), and finding the sink reference terminal current.

Systematic mesh equations for the network are

$$\begin{cases} 3i_1 - 3i_2 = v(t) \\ -3i_1 + 7i_2 = 5i_2 \end{cases}$$

giving

$$i_1 = \frac{\begin{vmatrix} v(t) & -3 \\ 0 & 2 \end{vmatrix}}{\begin{vmatrix} 3 & -3 \\ -3 & 2 \end{vmatrix}} = \frac{2v(t)}{-3}$$

Figure 4-5 Two-terminal network composed of resistors and a controlled source, with an applied voltage

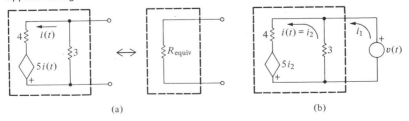

(a)

(b)

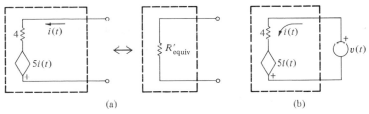

Figure 4-6 Simplified resistance calculation

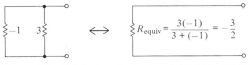

Figure 4-7 Original equivalent resistance found by parallel combination

Then

$$R_{equiv} = \frac{v(t)}{i_1(t)} = -\frac{3}{2}$$

Because the 3-Ω resistor may be taken out of the "box" containing the rest of the elements and the controlling signal $i(t)$, the problem of Figure 4-5 could have been simplified by considering it to consist of the 3-Ω resistor in parallel with the rest of the element, the simpler element being that in Figure 4-6, for which

$$4i(t) = v(t) + 5i(t)$$
$$v(t) = -i(t)$$
$$R'_{equiv} = -1$$

To obtain the net resistance of the entire network of the original problem, the parallel combination of 3 and -1 Ω is computed as shown in Figure 4-7.

D4-1

   Find the equivalent resistance of the following two-terminal resistor network using nodal equations:

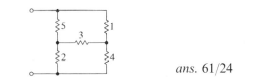

ans. 61/24

D4-2

Find the equivalent resistance of the following two-terminal resistor network using mesh equations:

ans. 179/66

D4-3

Find the equivalent resistance of each of the following networks:

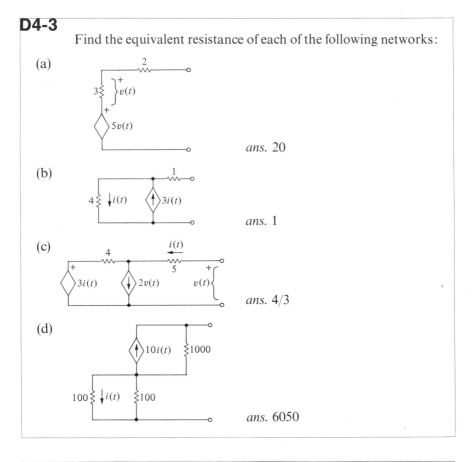

(a)

ans. 20

(b)

ans. 1

(c)

ans. 4/3

(d)

ans. 6050

Equivalent Resistances

Any two-terminal network consisting of only resistors and controlled sources, which contains the controlling voltages and currents for the controlled sources, is equivalent to a single resistor.

> The equivalent resistance may be found by solving the network problem in which a voltage or current source is placed across the element terminals and the sink reference ratio of terminal voltage to terminal current is found.
>
> If controlled sources are present in the network, the equivalent resistance may be a negative number.

4.3   Superposition of Sources

4.3.1   Signal Components Due to Individual Sources

Linear equations have the property that if their driving functions (the "knowns," usually placed to the right of the equal signs) are decomposed into the sum of two or more component parts, the solution consists of a corresponding sum of component parts, each due to one driving function component acting separately.

For example, the solution to the equations

$$\begin{cases} x_1 - 2x_2 + x_3 = 4 \cos 3t \\ -x_1 + 3x_2 \quad\quad = \quad 12 \\ 2x_1 \quad\quad + 4x_3 = 0 \end{cases}$$

may be found by solving

$$\begin{cases} x_1' - 2x_2' + x_3' = 4 \cos 3t \\ -x_1' + 3x_2' \quad\quad = 0 \\ 2x_1' \quad\quad + 4x_3' = 0 \end{cases}$$

then solving

$$\begin{cases} x_1'' - 2x_2'' + x_3'' = 0 \\ -x_1'' + 3x_2'' \quad\quad = -12 \\ 2x_1'' \quad\quad + 4x_3'' = 0 \end{cases}$$

The solutions for x_1, x_2, and x_3 in the original equations are then

$$x_1 = x_1' + x_1''$$
$$x_2 = x_2' + x_2''$$
$$x_3 = x_3' + x_3''$$

In the linear equations describing an electrical network if the driving function terms due to each individual, fixed network source are superimposed, the results are sets of equations, each set describing a single-source network, with all but one source replaced by zero. In other words a network may be solved by solving a succession of single-source problems in which all but one fixed source at a time is set to zero. The response due to all of the sources acting together is then the sum of the single-source network solutions.

The solution for any network voltage or current then consists of a sum of terms, one term due to each fixed source acting separately. Each voltage and each current is said to consist of a sum of component parts, one component due to each separate source.

4.3.2 Source Superposition

In an electrical network a voltage source is set to zero by replacing the source by a *short circuit* (a conductor, zero voltage). To set a current source to zero, replace it with an *open* circuit (remove it, leaving no connection, zero current). A network containing several sources may be solved by solving it once for each source alone, setting the other sources to zero, then summing each of the single-source solutions to obtain the solution when all sources are present.

An example is shown in Figure 4-8. The component of $i(t)$ due to the voltage source is

$$i'(t) = \frac{3e^{-2t}}{9}$$

The component of $i(t)$ due to the current source is

$$i''(t) = -\tfrac{5}{9} \cdot 6 \sin 7t$$

so

$$i(t) = i'(t) + i''(t) = \frac{1}{3} e^{-2t} - \frac{10}{3} \sin 7t$$

Figure 4-8 Superposition of sources

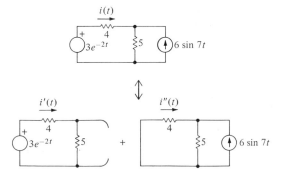

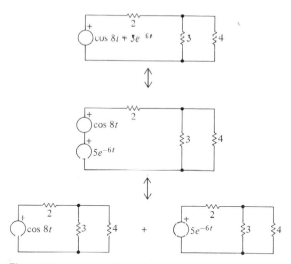

Figure 4-9 Superposition of voltage-source components

4.3.3 Superposition of Source Components

A complicated source function may sometimes be simplified by expressing that source as the sum of simpler sources, as with the voltage source in Figure 4-9. A current-source example is shown in Figure 4-10.

4.3.4 Source Superposition with Controlled Sources

Controlled sources cannot be superimposed because they appear in network equations in the same way as do resistors, not in the manner of fixed sources. All controlled sources should be left in all component solutions, as in the example of Figure 4-11.

Figure 4-10 Superposition of current-source components

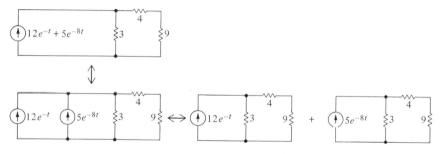

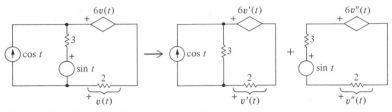

Figure 4-11   Superposition with a controlled source

With source-resistor networks, superposition of sources and of source components has little to offer as a solution method. Solving several different networks for signal components is usually more difficult than solving the original network just once.

The purpose here is to gain an understanding that will be fundamental in later work with more involved networks. Note that theoretically one could concentrate on networks with only a single fixed source, since all problems with multiple fixed sources may be reduced to a set of single-source problems.

D4-4

Find the indicated signals by superimposing the sources:

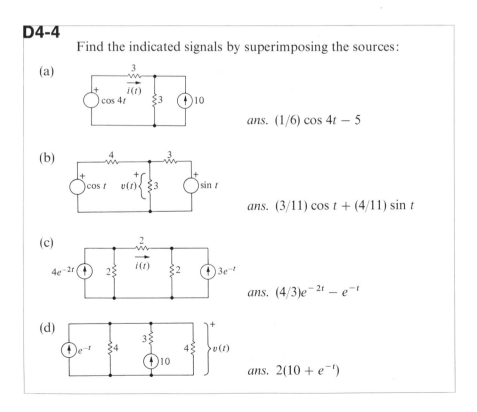

(a)

ans. $(1/6) \cos 4t - 5$

(b)

ans. $(3/11) \cos t + (4/11) \sin t$

(c)

ans. $(4/3)e^{-2t} - e^{-t}$

(d)

ans. $2(10 + e^{-t})$

D4-5

Superimpose the fixed sources to find the indicated signals:

(a)

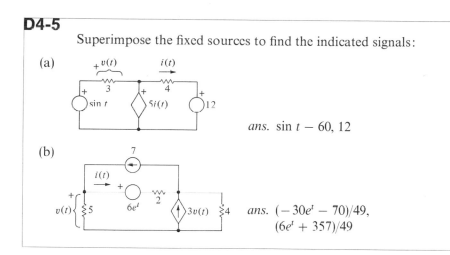

ans. sin *t* − 60, 12

(b)

ans. (−30*e^t* − 70)/49,
(6*e^t* + 357)/49

Superposition of Sources

A network containing several fixed sources may be solved by solving it once for each source alone, setting all other fixed sources to zero, then summing the solutions to obtain the entire solution. Controlled sources must be left intact in each of the component problems.

To set a voltage source to zero, replace it with a short circuit; a current source is set to zero by replacing it with an open circuit.

4.4   The Thévenin Equivalent

4.4.1   Derivation of the Thévenin Equivalent

Any two-terminal network that contains only sources and resistors is equivalent (that is, has the same voltage-current relation at its terminals) to its *Thévenin equivalent*, consisting of a voltage source in series with a resistor, Figure 4-12. [Leon Thévenin (1857–1926) was a French telegraph engineer.] The voltage-source function in the Thévenin equivalent is called the *Thévenin*

Figure 4-12   Network and its Thévenin equivalent

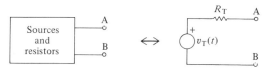

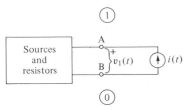

Figure 4-13   Two-terminal network with current applied at the terminals

voltage and the resistance in the Thévenin equivalent is called the *Thévenin resistance*.

Suppose a current $i(t)$ is applied to the original two-terminal network. Imagine writing nodal equations for the network and the external current source, choosing the 0 node and the 1 node (in addition to others in the "box") for convenience as shown in Figure 4-13. The relation between the element voltage and current is then the relation between $v_1(t)$ and $i(t)$.

Nodal equations are chosen here because they are easily written for any network, not just the planar ones. Using Cramer's rule, the solution for v_1 has the form

$$v_1(t) = \frac{\Delta_1}{\Delta}$$

where Δ is just a number, the determinant of the equations; and the elements of the determinant Δ_1 are numbers except for the first-column elements, which are source functions. The determinant Δ_1 may be expanded as

$$\Delta_1 = \begin{vmatrix} i(t) + f_1(t) & \# & \cdots & \# \\ f_2(t) & \# & \cdots & \# \\ f_n(t) & \# & \cdots & \# \end{vmatrix} = i(t)\Delta_{11} + g(t)$$

where $g(t)$ is some function of time that depends on the sources within the "box." Then

$$v_1(t) = \frac{\Delta_{11}}{\Delta} i(t) + \frac{g(t)}{\Delta}$$

The voltage-current relation of the Thévenin element, Figure 4-14, is

$$v_1(t) = R_T i(t) + v_T(t)$$

which is identical to that of the "box" for the correct choice of R_T and $v_T(t)$.

Figure 4-14   Thévenin network with current applied at the terminals

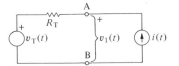

Figure 4-15 Open-circuit voltage of the Thévenin equivalent and of the network

4.4.2 Calculations for Networks without Controlled Sources

Fortunately, in many cases it is not necessary to write and solve systematic equations to find the Thévenin equivalent. Consider measuring the voltage at the terminals of the Thévenin equivalent, with nothing connected externally to the terminals, as in Figure 4-15. There is no current through the Thévenin resistance, so there is no voltage across it, and the terminal voltage is the Thévenin voltage. The same measurement on the original network must give the same result. Thus the voltage across the original network terminals, with nothing connected externally to the terminals (called the *open-circuit* terminal voltage), is its Thévenin voltage, $v_T(t)$.

If all the sources (except controlled sources) in the original network, Figure 4-16(a), are set to zero as in Figure 4-16(b), the solution for the terminal voltage $v_1(t)$, which in general is

$$v_1(t) = R_T i(t) + v_T(t)$$

becomes

$$v_1(t) = R_T i(t)$$

or

$$\frac{v_1(t)}{i(t)} = R_T$$

This is to say that the resistance looking into the terminals with all fixed sources set to zero is the Thévenin resistance, as in Figure 4-16(c).

An example of Thévenin equivalent calculation is shown in Figure 4-17. The Thevenin voltage is the network's open-circuit voltage, and the Thévenin resistance is the resistance at the terminals when the fixed source is set to zero.

Note that if the plus sign on the voltage reference polarity is closest to the A terminal when the calculation is made from the original network, the

Figure 4-16 Setting fixed sources to zero to find Thévenin resistance

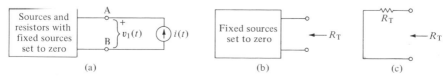

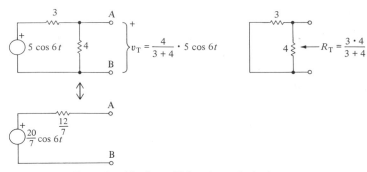

Figure 4-17   Example of finding a Thévenin equivalent

plus sign on the Thévenin voltage source in the equivalent is closest to the A terminal.

Another example Thévenin equivalent is found in Figure 4-18, where it is indicated that reversal of the Thévenin source polarity reverses the algebraic sign of the source function.

The only situation in which a Thévenin equivalent cannot be found is when the two-terminal source-resistor network is only a current source or is equivalent to only a current source. Then the open-circuit voltage cannot be found (one may think of it as being infinite) and the Thévenin resistance is infinite, as indicated in Figure 4-19.

Figure 4-18   Another Thévenin equivalent example

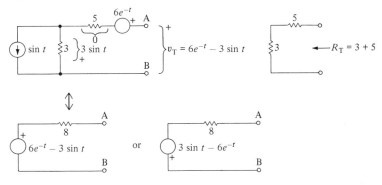

Figure 4-19   Attempting to determine the open-circuit voltage of a current source

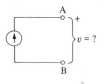

4.4.3 Networks with Controlled Sources

When it is desired to find the Thévenin equivalent of a two-terminal element containing controlled sources, the Thévenin resistance may be found by setting all fixed sources to zero, but leaving the controlled sources active. The resistance looking into the terminals then involves the equivalent resistance of a network with controlled sources in it, as in Section 4.2.3.

An example two-terminal network containing a controlled source is shown in Figure 4-20(a). Its open-circuit Thévenin voltage is given by

$$v(t) = 8v(t) - \sin t$$

$$7v(t) - \sin t$$

$$v_T(t) = v(t) = \frac{\sin t}{7}$$

In Figure 4-20(b), the fixed source in the network is set to zero, and a current $i(t)$ is applied to the network terminals in order to calculate the Thévenin resistance. The terminal voltage in Figure 4-20(b) is given by

$$\tfrac{1}{2}v = 4v + i$$

$$-\tfrac{7}{2}v = i$$

$$R_T = \frac{v}{i} = -\frac{2}{7}$$

a negative value, which is possible because of the presence of the controlled source.

The complete Thévenin equivalent is shown in Figure 4-20(c).

Figure 4-20 Thévenin equivalent of a network with a controlled source

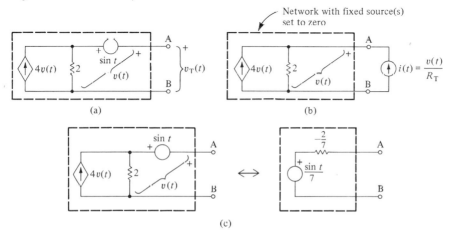

(a)

(b)

(c)

D4-6

Find the Thévenin equivalents of the following two-terminal networks. Indicate which is the A and which is the B terminal of your equivalent.

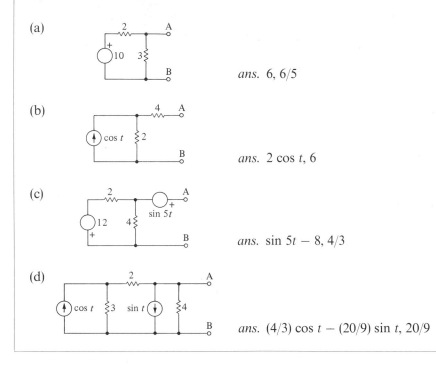

(a)

ans. 6, 6/5

(b)

ans. 2 cos t, 6

(c)

ans. sin 5t − 8, 4/3

(d)

ans. (4/3) cos t − (20/9) sin t, 20/9

D4-7

Find the Thévenin equivalents of the following networks that contain controlled sources. Indicate which is the A and which is the B terminal of your equivalent.

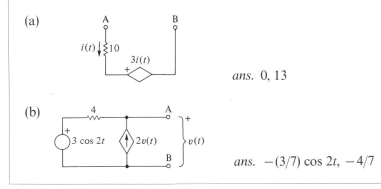

(a)

ans. 0, 13

(b)

ans. −(3/7) cos 2t, −4/7

The Thévenin Equivalent

Any two-terminal combination of sources and resistors is equivalent to a single voltage source (the Thévenin voltage) in series with a single resistor (the Thévenin resistance).

The Thévenin voltage is the two-terminal network's open-circuit voltage and the Thévenin resistance is the resistance looking into the terminals when all fixed sources are set to zero.

4.5 The Norton Equivalent

The equivalence of Thévenin and Norton networks (Section 2.4) shows that any two-terminal combination of sources and resistors is also equivalent to a current source in parallel with a resistor, Figure 4-21. An easy way to calculate the Norton equivalent is first to find the Thévenin equivalent and then convert to the Norton form.

The Norton resistance is the same as the Thévenin resistance and may be calculated in the same way: Set all sources in the network, except controlled sources, to zero, then find the resistance looking into the terminals. The Norton source current is the Thévenin source voltage divided by the Thévenin resistance.

Alternatively, the Norton source current is the *short-circuit* network current, the current through the terminals when the network terminals are connected together, as shown in Figure 4-22. Note that the sense in which the reference direction for $i_N(t)$ is chosen for the measurement on the original network determines the sense of the Norton source in the Norton equivalent.

Figure 4-21 Norton equivalent of a two-terminal source-resistor network

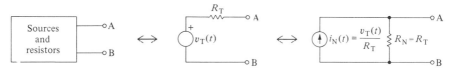

Figure 4-22 Norton current as the short-circuit current of the network

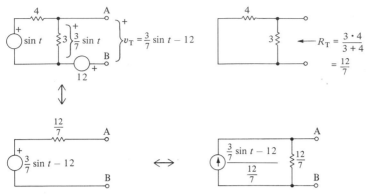

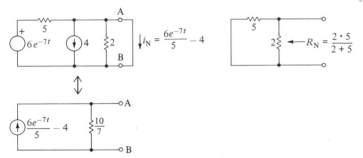

Figure 4-23 Finding the Norton equivalent by conversion of the Thévenin equivalent

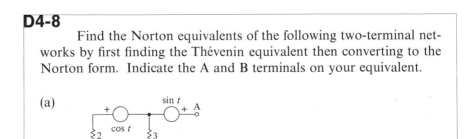

Figure 4-24 Finding the Norton equivalent directly, using the network short-circuit current

Figure 4-23 shows an example of finding the Norton equivalent by first finding the Thévenin equivalent.

Figure 4-24 is an example of finding the Norton equivalent directly.

The only situation for which a Norton equivalent does not exist is when the network is a voltage source or equivalent to just a voltage source.

D4-8

Find the Norton equivalents of the following two-terminal networks by first finding the Thévenin equivalent then converting to the Norton form. Indicate the A and B terminals on your equivalent.

(a)

ans. $(5 \sin t - 3 \cos t)/6$

(b)

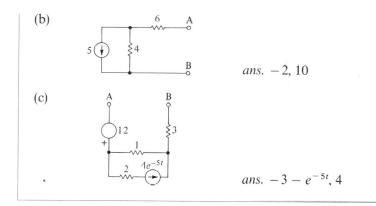

ans. $-2, 10$

(c)

ans. $-3 - e^{-5t}, 4$

D4-9

Find the Norton equivalents of the following two-terminal net-works directly, finding the short-circuit network current. Indicate the A and B terminals on your equivalent.

(a)

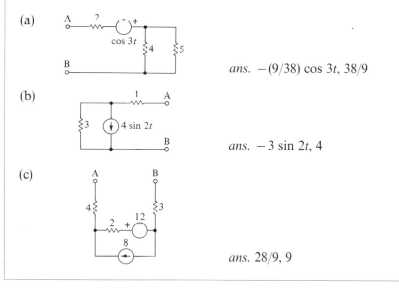

ans. $-(9/38) \cos 3t, 38/9$

(b)

ans. $-3 \sin 2t, 4$

(c)

ans. $28/9, 9$

The Norton Equivalent

Any two-terminal combination of sources and resistors is equivalent to a single current source (the Norton current) in parallel with a single resistor (the Norton resistance).

> The Norton current is the network's short-circuit current and the Norton resistance is identical to the Thévenin resistance, which is the resistance looking into the terminals when all fixed sources are set to zero.
>
> The Norton equivalent may be found by first finding the Thévenin equivalent, then transforming the Thévenin element to an equivalent Norton element.

4.6 Using Thévenin and Norton Equivalents in Network Solutions

The Thévenin and Norton equivalents offer one of the most powerful tools in the catalog of equivalent circuits for network solution. The voltage across, or current through, any element may be found by replacing all the rest of the network connected to that element by its Thévenin or its Norton equivalent. For the example network in Figure 4-25,

$$i(t) = \frac{\frac{3}{5}\cos t - 12}{5 + \frac{6}{5}} = \frac{3\cos t - 60}{31}$$

Or several intermediate Thévenin or Norton equivalents, or both, may be used on the way to a solution, as in the network of Figure 4-26, for which

$$\left(\frac{1}{9} + \frac{1}{18} + \frac{3}{4}\right) v(t) = \sin 8t + 12 - \frac{\cos 8t}{3}$$

Figure 4-25 Using a Thévenin equivalent for network solution

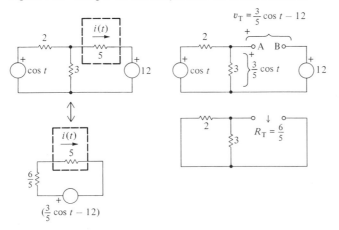

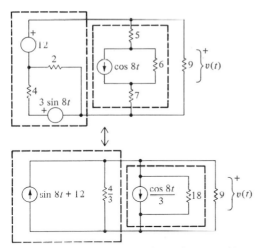

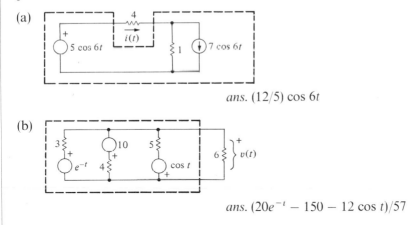

Figure 4-26   Using intermediate Thévenin or Norton equivalents in network solution

D4-10

Find the indicated signal using a Thévenin equivalent of the portion of the network in the dashed "box."

(a)

ans. $(12/5) \cos 6t$

(b)

ans. $(20e^{-t} - 150 - 12 \cos t)/57$

D4-11

Find the indicated signal using a Norton equivalent of the portion of the network in the dashed "box."

(a)

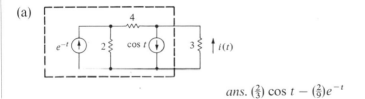

ans. $(\frac{2}{3}) \cos t - (\frac{2}{9})e^{-t}$

(b)

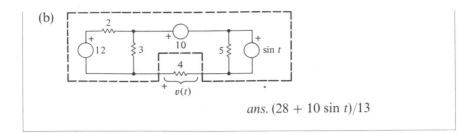

ans. $(28 + 10 \sin t)/13$

4.7 The Maximum Power Transfer Theorem

The Thévenin and Norton equivalents are important, too, from another standpoint. They indicate that any complicated model of a two-terminal device which involves just sources and resistors can be reduced to a single-source, single-resistor model. In other words, the most complicated two-terminal source-resistor model may be replaced by its Thévenin or Norton equivalent.

A common problem is the following: Given a two-terminal source-resistor network, find the resistance which, when connected across the network terminals, results in the maximum electrical power transfer into that resistance. The two-terminal network might model an audio amplifier, for instance, and the resistance across its terminals might model a loudspeaker. One would be interested in getting the maximum amount of electrical power out of the amplifier into the loudspeaker.

So far as voltages and currents external to the two-terminal network are concerned, it may be replaced by its Thévenin equivalent, as in Figure 4-27.

If the resistance connected across the network terminals, R_L (often called the *load resistance* in such problems), is very large, the current through R_L is small and the electrical power flow into R_L,

$$p_{\text{into } R_L}(t) = i^2(t)R_L$$

will be small in comparison to what it could be were R_L smaller.

For very small R_L, $p_{\text{into } R_L}(t)$ will likewise be relatively small, because of the smallness of R_L itself. For some value of R_L between $R_L = 0$ and $R_L = \infty$

Figure 4-27 Finding power transfer into the load resistor

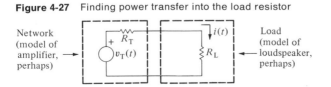

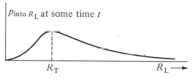

Figure 4-28 Power into the load resistance, as a function of load resistance

there will be a value of R_L for which $p_{into\ R_L}(t)$ will be maximized. Figure 4-28 is a sketch of power into the load resistance, as a function of the load resistance.

In terms of the Thévenin voltage and resistance,

$$p_{into\ R_L} = i^2 R_L = \left[\frac{v_T(t)}{(R_T + R_L)}\right]^2 R_L$$

The maximum of $p_{into\ R_L}$ is given by the value of R_L for which

$$\frac{\partial p_{into\ R_L}}{\partial R_L} = \frac{(R_T + R_L)^2 v_T^2 - v_T^2 R_L \cdot 2(R_T + R_L)}{(R_T + R_L)^4}$$

$$= \frac{(R_T + R_L - 2R_T)v_T^2}{(R_T + R_L)^3} = 0$$

$$R_L = R_T$$

To obtain maximum electrical power transfer into R_L, choose R_L equal to the Thévenin (or Norton) resistance of the rest of the network connected to the load resistance, as in the example of Figure 4-29, for which

$$R_L = \frac{13}{3}$$

for maximum power into R_L.

One common misconception is that the maximum power transfer theorem applies in reverse: That is, one should choose the Thévenin resistance equal

Figure 4-29 Example of Thévenin resistance calculation, for maximizing power transfer into load resistance

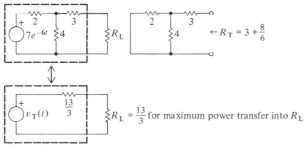

to the load resistance in order to maximize power transfer to the load. Not so. Differentiation of the power-flow equation with respect to R_T gives the quite obvious result that, if R_T is adjustable and R_L is not, R_T should be as small as possible for maximum electrical power transfer into R_L.

Another common misconception is that for maximum power transfer, the efficiency of the power transfer is 50% since equal powers are dissipated in R_T and in R_L. The Thévenin equivalent is equivalent so far as external voltages and currents are concerned, but it is *not* generally equivalent so far as *internal* electrical power is concerned. The efficiency may be less than 50%, as simple examples will demonstrate.

D4-12

Find R so that $p_{\text{into } R}(t)$ is maximum:

(a)

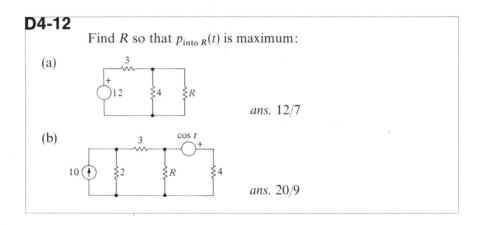

ans. 12/7

(b)

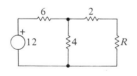

ans. 20/9

D4-13

What is the power flow out of the source when R is chosen to maximize $p_{\text{into } R}(t)$?

ans. 936/55

The Maximum Transfer Theorem

Maximum electrical power is transferred into an adjustable resistor R when R is chosen to equal the Thévenin (or Norton) resistance of the two-terminal combination of sources and resistors that is connected to R.

4.8   Transfer Ratios for Resistive Networks

For a single-source source-resistor network, every voltage and every current in the network is proportional to the source function and thus to each other. The constants of proportionality are known as the *transfer ratios* of the network. In general, a network transfer ratio is

$$T = \frac{\text{voltage or current of interest}}{\text{source function or another voltage or current of interest}}$$

For example, the transfer ratio that relates $v_1(t)$ to $v(t)$ in the network of Figure 4-30(a) is

$$T_1 = \frac{v_1(t)}{v(t)} = \frac{R_1}{R_1 + R_2}$$

For the network of Figure 4-30(b), the transfer ratio that relates $i_1(t)$ to $i(t)$ is

$$T_2 = \frac{i_1(t)}{i(t)} = \frac{-R_2}{R_1 + R_2}$$

The network of Figure 4-30(c) provides a more involved example of transfer ratio calculation. Systematic nodal equations for the network are

$$\begin{cases} (\tfrac{1}{4} + \tfrac{1}{2} + \tfrac{1}{3})v_1 - (\tfrac{1}{2} + \tfrac{1}{3})v_2 = 0 \\ -(\tfrac{1}{2} + \tfrac{1}{3})v_1 + (\tfrac{1}{2} + \tfrac{1}{3} + \tfrac{1}{5})v_2 = i(t) \end{cases}$$

or

$$\begin{cases} 13v_1 - 10v_2 = 0 \\ -25v_1 + 31v_2 = 30i(t) \end{cases}$$

The voltage v_1 is, in terms of the source, $i(t)$,

$$v_1(t) = \frac{\begin{vmatrix} 0 & -10 \\ 30i(t) & 31 \end{vmatrix}}{\begin{vmatrix} 13 & -10 \\ -25 & 31 \end{vmatrix}} = \frac{300}{153}i(t)$$

Figure 4-30   Transfer ratio examples

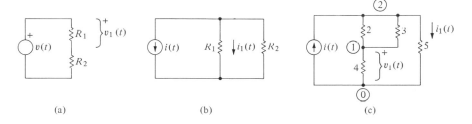

(a)　　　　　　　　(b)　　　　　　　　(c)

so the transfer ratio of the two is

$$T_1 = \frac{v_1(t)}{i(t)} = \frac{300}{153} = \frac{100}{51}$$

The voltage v_2 is

$$v_2(t) = \frac{\begin{vmatrix} 13 & 0 \\ -25 & 30i(t) \end{vmatrix}}{153} = \frac{390}{153} i(t)$$

and the current $i_1(t)$ is, in terms of $i(t)$,

$$i_1(t) = \frac{v_2(t)}{5} = \frac{78}{153} i(t)$$

The transfer ratio of the two is

$$T_2 = \frac{i_1(t)}{i(t)} = \frac{78}{153}$$

Since all network signals are proportional to $i(t)$, they are proportional to one another, so the ratio of any two signals is a constant. The ratio of i_1 to v, for instance, is

$$T_3 = \frac{i_1(t)}{v(t)} = \frac{i_1(t)/i(t)}{v(t)/i(t)}$$

$$= \frac{T_2}{T_1} = \frac{78}{300} = \frac{13}{50}$$

D4-14

Find the transfer ratios:

(a) $T_1 = \dfrac{v_1}{v}$

(b) $T_2 = \dfrac{i_3}{v}$

(c) $T_3 = \dfrac{v_2}{v}$

(d) $T_4 = \dfrac{v_2}{i_2}$

(e) $T_5 = \dfrac{i_3}{i_1}$

ans. $15/47$, $3/47$, $6/47$, $-6/5$, $3/8$

Transfer Ratios

In a single-source source-resistor network, every voltage and every current is proportional to the source function. Thus the ratio of any such network voltage or current to any other network voltage or current is a constant. Such ratios are called *transfer ratios*.

4.9 The Delta-Wye Transformation

4.9.1 Delta-Wye Equivalence

One of the most basic *three*-terminal network equivalents is that of three resistors connected in delta and in wye, as in Figure 4-31.

Although there are three voltages and three currents involved in a three-terminal network, only two of the voltages and two of the currents are independent of one another; the third voltage and the third current may each be expressed in terms of the other two, using Kirchhoff's laws, as in Figure 4-32. To show that two three-terminal elements are equivalent to one another, it is sufficient to show that for any v_1 and v_2, i_1 and i_2 will be the same in the two elements.

Figure 4-31 Equivalent delta and wye resistor networks
   (a) Delta connection of resistors
   (b) Wye connection of resistors

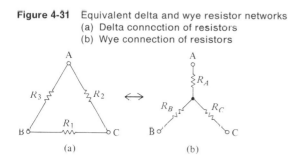

(a)   (b)

Figure 4-32 Two independent voltages and two independent currents in a three-terminal network

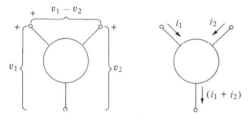

4.9.2 Delta-to-Wye Conversion

To show the delta-wye equivalence and the relationships between the resistances in the two elements, consider first the voltage-current relations in the wye element, as in Figure 4-33(a).

The equations expressing i_1 and i_2 in terms of v_1 and v_2 are easily obtained by writing systematic loop equations, as follows;

$$\begin{cases} (R_A + R_C)i_1 - R_C i_2 = v_1 \\ -R_C i_1 + (R_B + R_C)i_2 = -v_2 \end{cases}$$

The three-terminal delta network, with the same voltages and currents indicated is shown in Figure 4-33(b). The relations between i_1 and i_2, and v_1 and v_2, are found by first expressing the resistor currents in terms of the voltages, as indicated on Figure 4-33:

$$\begin{cases} i_1 = \dfrac{v_1 - v_2}{R_3} + \dfrac{v_1}{R_2} = \left(\dfrac{1}{R_2} + \dfrac{1}{R_3}\right)v_1 - \dfrac{1}{R_3}v_2 \\ i_2 = \dfrac{v_1 - v_2}{R_3} - \dfrac{v_2}{R_1} = \dfrac{1}{R_3}v_1 - \left(\dfrac{1}{R_1} + \dfrac{1}{R_3}\right)v_2 \end{cases}$$

Simplifying,

$$\begin{cases} R_2 R_3 i_1 = (R_2 + R_3)v_1 - R_2 v_2 \\ R_1 R_3 i_2 = R_1 v_1 - (R_1 + R_3)v_2 \end{cases}$$

Solving for v_1 and v_2 in terms of i_1 and i_2, the following equations result:

$$\begin{cases} v_1 = \dfrac{\begin{vmatrix} R_2 R_3 i_1 & -R_2 \\ R_1 R_3 i_2 & -(R_1 + R_3) \end{vmatrix}}{\begin{vmatrix} (R_2 + R_3) & -R_2 \\ R_1 & -(R_1 + R_3) \end{vmatrix}} \\[4mm] \quad = \dfrac{(R_1 R_2 R_3 + R_2 R_3{}^2)i_1 - R_1 R_2 R_3 i_2}{R_1 R_3 + R_2 R_3 + R_3{}^2} \\[6mm] v_2 = \dfrac{\begin{vmatrix} (R_2 + R_3) & R_2 R_3 i_1 \\ R_1 & R_1 R_3 i_2 \end{vmatrix}}{\begin{vmatrix} (R_2 + R_3) & -R_2 \\ R_1 & -(R_1 + R_3) \end{vmatrix}} \\[4mm] \quad = \dfrac{-R_1 R_2 R_3 i_1 + (R_1 R_2 R_3 + R_1 R_3{}^2)i_2}{R_1 R_3 + R_2 R_3 + R_3{}^2} \end{cases}$$

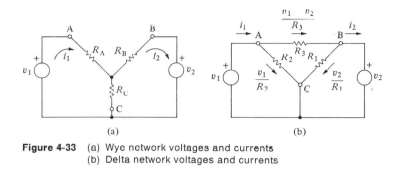

Figure 4-33   (a) Wye network voltages and currents
(b) Delta network voltages and currents

Comparing the two sets of equations—one set for the wye network, the other for the delta—the voltage-current relationships are seen to be identical if and only if

$$\begin{cases} R_A = \dfrac{R_2 R_3}{R_1 + R_2 + R_3} \\[2ex] R_B = \dfrac{R_1 R_3}{R_1 + R_2 + R_3} \\[2ex] R_C = \dfrac{R_1 R_2}{R_1 + R_2 + R_3} \end{cases}$$

Each of the above equations is of the form

$$R_i = \frac{\text{product of the two resistances connected to terminal } i}{\text{sum of the three resistances.}}$$

Each wye resistance is positive if the delta resistances are positive.

Figure 4-34 is an example of a specific delta-to-wye conversion.

If a delta network is part of a more complicated network such as the one in Figure 4-35, a delta-wye transformation may simplify the network considerably, as it does in this example.

Figure 4-34   Example of delta-to-wye conversion

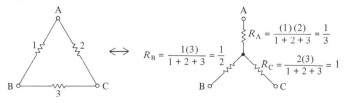

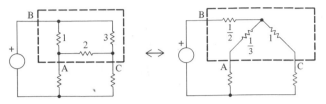

Figure 4-35   Delta-wye conversion used to simplify a network

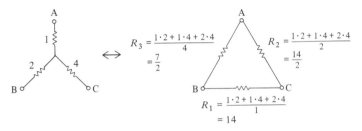

Figure 4-36   Example of wye-to-delta conversion

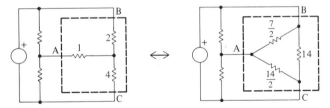

Figure 4-37   Wye-delta conversion used to simplify a network

4.9.3   Wye-to-Delta Conversion

To solve for the delta resistances in terms of the wye ones, the relations

$$\begin{cases} R_A = \dfrac{R_2 R_3}{R_1 + R_2 + R_3} \\[2ex] R_B = \dfrac{R_1 R_3}{R_1 + R_2 + R_3} \\[2ex] R_C = \dfrac{R_1 R_2}{R_1 + R_2 + R_3} \end{cases}$$

arc used, and ratios of them are formed, two at a time:

$$\begin{cases} \dfrac{R_A}{R_B} = \dfrac{\dfrac{R_2 R_3}{R_1 + R_2 + R_3}}{\dfrac{R_1 R_3}{R_1 + R_2 + R_3}} = \dfrac{R_2}{R_1} \\[4ex] \dfrac{R_A}{R_C} = \dfrac{R_3}{R_1} \\[3ex] \dfrac{R_B}{R_C} = \dfrac{R_3}{R_2} \end{cases}$$

Substituting into

$$R_C = \frac{R_1 R_2}{R_1 + R_2 + R_3} = \frac{R_1}{\dfrac{R_1}{R_2} + 1 + \dfrac{R_3}{R_2}}$$

gives

$$R_C = \frac{R_1}{\dfrac{R_B}{R_A} + 1 + \dfrac{R_B}{R_C}}$$

$$R_1 = \frac{R_A R_B + R_A R_C + R_B R_C}{R_A}$$

Similarly,

$$R_2 = \frac{R_A R_B + R_A R_C + R_B R_C}{R_B}$$

$$R_3 = \frac{R_A R_B + R_A R_C + R_B R_C}{R_C}$$

Each of these equations is of the form

$$R_i = \frac{\text{sum of products of resistances taken two at a time}}{\text{resistance connected to terminal opposite to } R_i}$$

The equivalent delta resistances are positive if the wye resistances are all positive.

Figure 4-36 is an example of wye-to-delta conversion.

As with the delta-to-wye conversion, a wye-to-delta transformation may be used to simplify a network such as one in which resistors are neither in series nor in parallel with one another. An example is shown in Figure 4-37.

D4-15

Convert the delta networks to equivalent wyes and convert the wye networks to equivalent deltas. Label the terminals A, B, and C in each equivalent network:

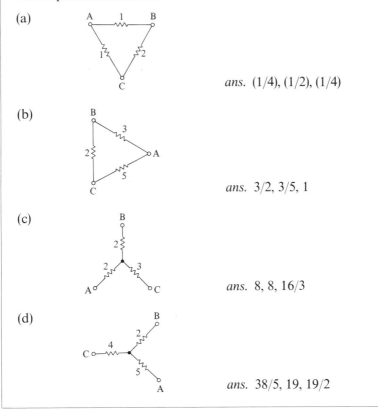

(a)

ans. (1/4), (1/2), (1/4)

(b)

ans. 3/2, 3/5, 1

(c)

ans. 8, 8, 16/3

(d)

ans. 38/5, 19, 19/2

D4-16

Use delta-wye transformation and series and parallel resistance combinations to find the equivalent resistance of the following network.

Repeat, using a wye-delta transformation.

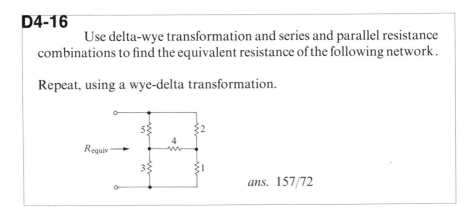

ans. 157/72

Delta-Wye and Wye-Delta Transformations

The three-terminal delta and wye resistor networks are equivalent to one another, with the following relations between the resistor values:

$$R_1 = \frac{R_A R_B + R_A R_C + R_B R_C}{R_A} \qquad R_A = \frac{R_2 R_3}{R_1 + R_2 + R_3}$$

$$R_2 = \frac{R_A R_B + R_A R_C + R_B R_C}{R_B} \qquad R_B = \frac{R_1 R_3}{R_1 + R_2 + R_3}$$

$$R_3 = \frac{R_A R_B + R_A R_C + R_B R_C}{R_C} \qquad R_C = \frac{R_1 R_2}{R_1 + R_2 + R_3}$$

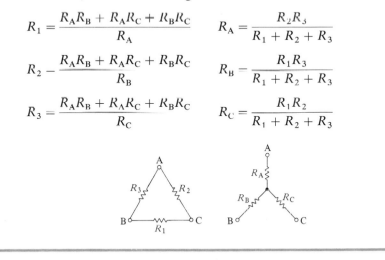

Chapter Four Problems

Basic Problems

Equivalent Resistance

1. Find the equivalent resistance of the two-terminal resistor network by using nodal equations:

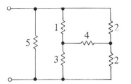

2. Find the equivalent resistance of the two-terminal resistor network by using mesh equations:

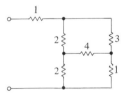

3. Find the equivalent resistance of the element below:

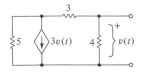

Superposition of Sources

4. Find $v(t)$ by superimposing the voltage sources:

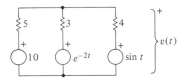

5. Find $i(t)$ by superimposing the current sources:

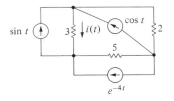

6. Superimpose the sources to find $v(t)$:

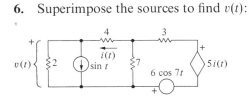

Thévenin and Norton Equivalents

7. Find the Thévenin equivalent and the Norton equivalent:

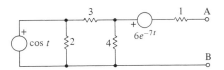

8. Find the Norton equivalent:

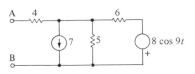

9. Find the Thévenin and the Norton equivalent:

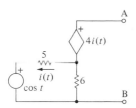

10. Find $i(t)$ using equivalent circuits, including Thévenin or Norton equivalents of the dashed elements:

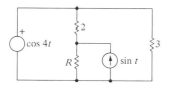

Maximum Power Transfer

11. Find the resistance R so that the electrical power flow into R is maximum. For that value of R, find $p_{\text{into } R}(t)$:

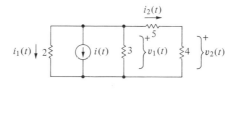

Transfer Ratios

12. Find the transfer ratios:

(a) $T_1 = \dfrac{i_1}{i}$

(b) $T_2 = \dfrac{v_1}{i}$

(c) $T_3 = \dfrac{i}{v_2}$

(d) $T_4 = \dfrac{v_1}{v_2}$

(e) $T_5 = \dfrac{i_2}{v_1}$

Delta-Wye Transformation

13. Convert the following network to an equivalent wye network. Label the A, B, and C terminals in the equivalent.

14. Convert the following network to an equivalent delta network. Label the A, B, and C terminals in the equivalent:

15. Convert the following network to an equivalent delta network. Label the A, B, and C terminals in the equivalent:

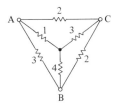

Practical Problems

Thévenin Equivalent Measurement

A convenient method of finding the Thévenin voltage and the Thévenin resistance of a two-terminal network is the following. Find the Thévenin voltage by measuring the open-circuit terminal voltage with a voltmeter with very high internal resistance. Then connect an adjustable resistance across the terminals and adjust the resistance until the terminal voltage is exactly half of the open-circuit voltage. The adjustable resistance now equals the Thévenin resistance. Figure 4-38 illustrates this.

If the voltmeter internal resistance is not large compared to the Thévenin resistance, it must be taken into account in the calculations.

1. A voltmeter with internal resistance 5000 Ω is used in the following Thévenin equivalent measurements on a network with a constant Thévenin

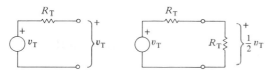

Figure 4-38   Half-voltage measurement of Thévenin resistance

A laboratory power supply. This instrument provides an adjustable constant voltage from 0 to 2000 V, provided that the power supply current is no more than 10 mA. A second adjustment causes the supply to be turned off if the current exceeds a preset value. (*Photo courtesy of Kepco, Inc.*)

voltage. With the voltmeter across the otherwise open-circuited network terminals, the meter reads 80.6 V. When a 2000-Ω resistor is placed in parallel with the meter, across the network terminals, the meter reads 48.3 V. Find the Thévenin voltage and the Thévenin resistance of the network.

2. A voltmeter with negligibly large internal resistance measures an open-circuit, constant Thévenin voltage of 62.3 V. When a 1000-Ω resistor is placed in parallel with the meter, across the network terminals, the meter reads 58.1 V.

Suppose the actual voltage at the meter terminals is 1% higher than that indicated by the meter. Find the percentage error in the Thévenin voltage,

$$\frac{(v_T)_{actual} - (v_T)_{measured}}{(v_T)_{actual}} \times 100\%$$

and the similarly defined error in the Thévenin resistance.

Repeat if, instead, the 1000-Ω resistor is actually 1% higher in resistance value.

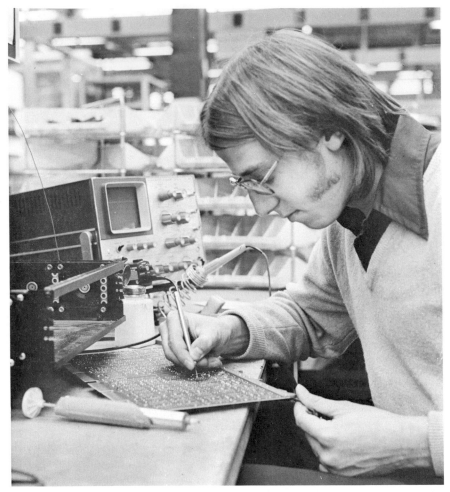

A printed circuit board is examined. The board consists of a sheet of insulating material upon which are bonded layers of copper conductors that have been shaped by a photographic etching process. (*Photo courtesy of Digital Equipment Corp.*)

Thévenin Equivalent Model

3. The voltage-current characteristic for a certain silicon junction semi-conductor diode is shown in Figure 4-39. The device has a nearly straight line relationship between voltage v and current i over a limited range of voltage and current.

Find a Thévenin equivalent model that gives an accurate approximation to the diode characteristic for voltages and currents within the ranges boardered by the dashed lines, where the curve is nearly a straight line.

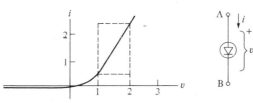

Figure 4-39 Practical Problem 3

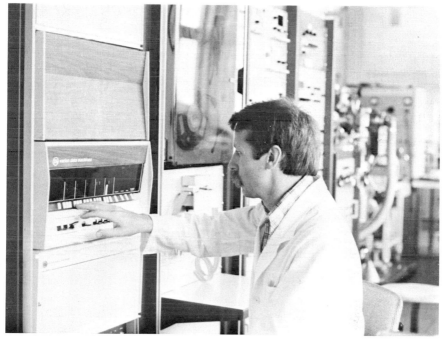

A small digital computer (minicomputer) is used for industrial process control. (*Photo courtesy of Varian Associates*)

Controlled Source Operational Amplifier Model

Controlled sources are very useful in modeling electronic devices. A highly idealized model of an *operational amplifier* is shown in Figure 4-40(a).

The device generates a voltage that is proportional to the difference between two applied voltages, v_+ and v_-. The constant of proportionality K is called the *gain* of the device and typically has a value between 10^4 and 10^6. The resistance R_i is called the *input resistance* and R_0 is called the *output resistance*.

4. For the operational amplifier with external connections as shown in Figure 4-40(b), find the voltage $v(t)$.

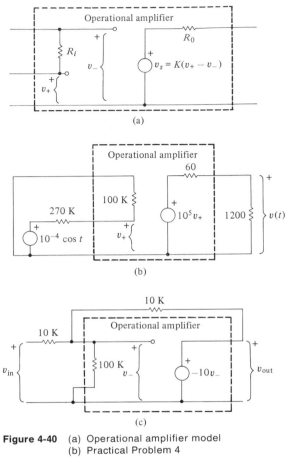

Figure 4-40   (a) Operational amplifier model
(b) Practical Problem 4
(c) Practical Problem 5

5.   For the operational amplifier connected as shown in Figure 4-40(c), find the transfer ratio

$$T = \frac{v_{\text{out}}}{v_{\text{in}}}$$

Advanced Problems

Equivalent Resistance

1.   Design a two-terminal network that has an equivalent resistance of $-100 \ \Omega$ that contains (in addition to other elements)

(a)   A voltage source controlled by a current.
(b)   A voltage controlled voltage source.

(c)   A voltage controlled current source.
(d)   A current controlled current source.

Maximum Power Flow

2.   Find the value of R that results in maximum electrical power flow into the 3-Ω resistor in Figure 4-41.

Transfer Ratios

3.   Suppose that a network contains two sources, with source functions $f_1(t)$ and $f_2(t)$. Let the transfer ratio T_1 be the ratio of some specific network voltage or current to $f_1(t)$ when $f_2 = 0$. Similarly, let T_2 be the ratio of the same network voltage or current to $f_2(t)$ when $f_1 = 0$.
    Show that the network voltage or current $g(t)$ is

$$g(t) = T_1 f_1(t) + T_2 f_2(t)$$

when both sources are nonzero.

Delta-Wye Equivalence

4.   Convert the network of Figure 4-42 to an equivalent element consisting of just three resistors in the wye configuration.

5.   Derive the relations for converting from a delta to a wye connection of resistors from the three equations equating resistances looking into each pair of terminals when the remaining terminal is not connected to anything.

Figure 4-41   Advanced Problem 2

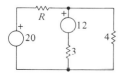

Figure 4-42   Advanced Problem 4

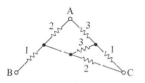

In the minds of many, the electro-magnetic telegraph is associated with the many chimerical projects constantly brought before the public and particularly with schemes so popular a year or two ago for the application of electricity as a moving power in the arts. All schemes for this purpose, I have from the first asserted, are premature and formed without proper scientific knowledge. The case, however, is entirely different in regard to the electro-magnetic telegraph. The science is now fully ripe for such an application of its principles, and I have not the least doubt, if proper means be afforded, of the perfect success of this invention.

Joseph Henry
From a letter to Samuel F. B. Morse,
Washington, D. C., 1842

The particles of an insulating dielectric whilst under induction may be compared to a series of small magnetic needles, or more correctly still to a series of small insulated conductors. If the space round a charged globe were filled with a mixture of an insulating dielectric, as oil of turpentine or air, and small globular conductors, as shot, the latter being at a little distance from each other so as to be insulated, then these would in their condition and action exactly resemble what I consider to be the condition and action of the particles of the insulating dielectric itself.

Michael Faraday
From *Experimental Researches in Electricity*
The Royal Institution, 1847

In these next chapters, energy-storing elements, inductors and capacitors, are considered in addition to sources and resistors. The equations describing networks that contain these elements are differential equations.

All of the time and effort spent on source-resistor networks will now pay added dividends. Writing systematic network equations simply involves substituting appropriate derivative or integral relations for the new elements. The Thévenin equivalent has a key role in the solution of networks that contain a single energy-storing element.

The eventual goal is to transform these more involved network problems into equivalent source-resistor network problems. Then all of the methods of source-resistor network solution may be brought to bear on general network solution.

PART TWO
Inductive and Capacitive Networks

To the Instructor:

In Chapters Five through Seven, capacitors, inductors, and coupled inductors are introduced. The systematic equations become integrodifferential ones, and the student begins learning their meaning and the principles of their solution.

Network integrodifferential equations. The same care in writing systematic equations for source-resistor networks continues for the more general networks. Inclusion of coupled inductors in both mesh and nodal equations is simplified by the introduction of simple, controlled source equivalent circuits.

Unfortunately, one cannot always count on students having a solid mathematical background in differential equations prior to the networks course. Even if they have had a course by that title, the emphasis can differ much from what they need at this point. The essentials of linear, constant-coefficient differential equations with constant driving functions are therefore developed in the concluding sections of Chapter Five. Exponential and sinusoidal driving functions are considered in following chapters, as the development of networks topics proceeds.

With this presentation of the pertinent mathematics, it is quite feasible for the student to study switched network solutions prior to or concurrently with a differential equations course. For the student with a background in differential equations, this material serves as a good review and brings our mathematical concerns into focus.

Switched networks. Switched first-order networks are then covered in detail, with many examples, using both switches and the unit step function.

In introducing switched second order series and parallel RLC networks, some compromise is expedient. On the one hand, it would be useful to have the tools of complex exponential algebra at hand. On the other, it is not very desirable to divide the subject into two separate chapters. The approach here is to provide some perspective by dealing with the trigonometric functions directly at this time, underscoring the practical importance of the exponential approach later.

Using impedance to find exponential response. The key property of exponential signals is now developed: If all network voltages and currents are exponential, varying as e^{st}, the voltage-current ratio for every network element is a constant, and each element is equivalent to a resistor.

Impedance is first used to find forced (or "steady state") network response due to exponential sources. Superposition is applied to find the response when the sources have different exponential constants, s. Then the short- and open-circuited impedance methods of finding network characteristic roots are given.

The definition of *forced* response is in accord with that used by most authors: that part of the response containing no terms of the same form as those in the homogeneous (or *natural*) solution. A few other authors use the terms "forced" and "natural" to denote zero-state and zero-input response components, which is not the case here.

CHAPTER FIVE
Network Differential Equations

5.1 Introduction

In this chapter the remaining basic network elements, the capacitor, the inductor, and coupled inductors, are introduced. The voltage-current relationships for these elements involve time derivatives and integrals, and so networks containing the elements are described by integrodifferential equations.

After discussing these new elements, methods for writing systematic simultaneous nodal and mesh equations are developed.

In preparation for solving networks described by differential equations, pertinent aspects of the subject, including solution methods, are discussed in the concluding sections of this chapter. Properties of the exponential functions that are involved in differential equation solutions are reviewed.

When you complete this chapter, you should know

1. The voltage, current, power, and energy relations for capacitors and how to use them.
2. The voltage, current, power, and energy relations for inductors and how to use them.

3. Voltage, current, winding sense, power, and energy relations for mutually coupled inductors and controlled source equivalents.
4. How to write systematic simultaneous nodal equations for networks containing resistors, capacitors, inductors, mutually coupled inductors, and fixed and controlled sources.
5. How to write systematic simultaneous mesh equations for networks containing resistors, capacitors, inductors, mutually coupled inductors, and fixed and controlled sources.
6. How to find the general solution of linear, constant-coefficient homogeneous differential equations of any order, including those with repeated and complex characteristic roots.
7. How to find the general solution of linear, constant-coefficient differential equations with constant driving functions.
8. How to apply boundary conditions to the general solution of a differential equation to obtain a specific solution.
9. The shape of the exponential function, how to graph it, and how to find and use time constants.

5.2 Voltage-Current, Power, and Energy Relations for the Capacitor

5.2.1 Defining Relations

The physical capacitor consists of two conductors in close proximity, separated by an insulator. The two conductors are called *plates*, and under the network conditions of confined current flow and negligible propagation times for disturbances, the current flowing onto one plate is identical to the current leaving the other plate.

It makes sense, then, to speak of the current flow *through* the capacitor; even though charge does not really flow between the plates, it appears at the capacitor terminals as though it does.

The symbol for a capacitor model is shown in Figure 5-1. The *capacitance* C, of the element is indicated beside the symbol for the capacitor. The unit of capacitance is the *farad* (abbreviated F). The farad is named in honor

Figure 5-1 Capacitor symbol

of Michael Faraday (1791–1867), a brilliant experimenter who made great contributions to electromagnetic theory.

The voltage across the capacitor, $v(t)$, is the charge on the plate nearest the plus sign on the voltage polarity reference, $q(t)$ [$-q(t)$ is on the other plate] divided by the capacitance:

$$v(t) = \frac{1}{C} q(t)$$

In terms of the sink reference current $i(t)$,

$$q(t) = \int_{-\infty}^{t} i(t) \, dt$$

In other words, $q(t)$ reflects the whole past history—from "way back," which is $t \to -\infty$ mathematically, to now, time t—of the rate of charge flow $i(t)$. This is known as a "running integral"; it is a function of the present time t.

Some people prefer to write integrals such as the above with a different variable of integration, as

$$q(t) = \int_{-\infty}^{t} i(x) \, dx$$

for example. This will not be done here, since there is little chance of confusing the variable of integration with the integral limits.

Sometimes the whole past history of a current is not known. What is known is the net charge on the capacitor (or the capacitor voltage, which is the charge divided by the capacitance) at some specific time t_0, and the history of the current from time t_0 on. Then the running integral may be written as

$$q(t) = \int_{-\infty}^{t_0} i(t) \, dt + \int_{t_0}^{t} i(t) \, dt = q(t_0) + \int_{t_0}^{t} i(t) \, dt$$

which holds, of course, only for times after t_0.

It is most convenient to choose the origin of the time scale to make $t_0 = 0$, giving

$$q(t) = q(0) + \int_{0}^{t} i(t) \, dt$$

after $t = 0$. There is then no information about the detailed behavior of $q(t)$ before $t = 0$.

Sink and source reference relation capacitor voltage-current relations are summarized in Figure 5-2.

Given a capacitor voltage, the capacitor current may be found by differentiation. As with the resistor, there is a minus sign in the voltage-current relationship if the voltage and current have the source reference relationship. Two examples of finding a capacitor current from its voltage are shown in

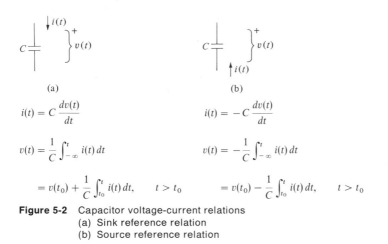

(a)

(b)

$$i(t) = C \frac{dv(t)}{dt}$$

$$i(t) = -C \frac{dv(t)}{dt}$$

$$v(t) = \frac{1}{C} \int_{-\infty}^{t} i(t)\, dt$$

$$v(t) = -\frac{1}{C} \int_{-\infty}^{t} i(t)\, dt$$

$$= v(t_0) + \frac{1}{C} \int_{t_0}^{t} i(t)\, dt, \qquad t > t_0$$

$$= v(t_0) - \frac{1}{C} \int_{t_0}^{t} i(t)\, dt, \qquad t > t_0$$

Figure 5-2   Capacitor voltage-current relations
(a) Sink reference relation
(b) Source reference relation

Figure 5-3. In Figure 5-3(b), the capacitor voltage is specified by a source, and the voltage and current have the source reference relation.

Given the entire past history of a capacitor current, the capacitor charge, and thus the capacitor voltage, may be found by integration, as in the example of Figure 5-4(a), for which

$$v(t) = -\frac{1}{2} \int_{-\infty}^{t} e^{3t}\, dt = -\frac{1}{2} \frac{e^{3t}}{3} \bigg|_{-\infty}^{t}$$

$$= -\left(\tfrac{1}{6}\right)e^{3t}$$

Figure 5-3   Determining capacitor current from the capacitor voltage

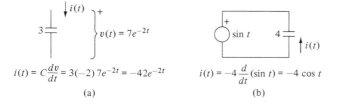

$$i(t) = C\frac{dv}{dt} = 3(-2)\,7e^{-2t} = -42e^{-2t}$$

$$i(t) = -4\frac{d}{dt}(\sin t) = -4\cos t$$

(a)

(b)

Figure 5-4   Determining capacitor voltage from the capacitor current

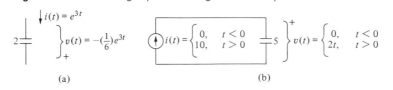

(a)

(b)

In the network of Figure 5-4(b), the given current is described piecewise, as is the solution, $v(t)$. In general,

$$v(t) = \frac{1}{C} \int_{-\infty}^{t} i(t)\, dt$$

Before $t = 0$, $i(t) = 0$, so

$$v(t) = \frac{1}{5} \int_{-\infty}^{t} 0\, dt - 0$$

After time $t = 0$, $i(t) - 10$, and

$$v(t) = \frac{1}{5} \int_{0}^{t} 10\, dt$$

$$= 2t$$

If the capacitor current is known after some time, for convenience time $t = 0$, and the capacitor voltage is known at $t = 0$, the capacitor voltage may be found after $t - 0$, as in the example of Figure 5-5. The net effect of the whole past history of the current before $t = 0$ is the capacitor voltage at $t = 0$.

$$v(0) = \frac{1}{C} \int_{-\infty}^{0} i(t)\, dt$$

After $t = 0$,

$$v(t) = \frac{1}{C} \int_{-\infty}^{0} i(t)\, dt + \frac{1}{C} \int_{0}^{t} i(t)\, dt$$

$$= v(0) + \frac{1}{C} \int_{0}^{t} i(t)\, dt = -8 + \frac{1}{7} \int_{0}^{t} \cos t\, dt$$

$$= -8 + \frac{1}{7} \sin t \Big|_{0}^{t} = -8 + \frac{1}{7} \sin t$$

Figure 5-5   Determining capacitor voltage using the initial voltage and the capacitor current thereafter

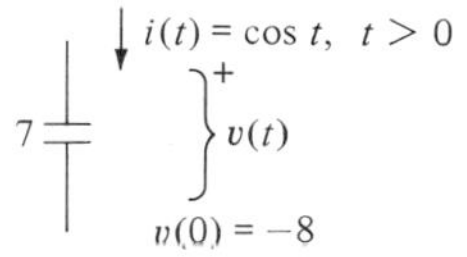

5.2.2   Capacitor Power and Energy

Capacitors are energy-storing elements. With the sink reference, the electrical power flow into a capacitor is

$$p_{into}(t) = v(t)i(t) = Cv(t)\frac{dv(t)}{dt}$$

The electrical energy stored in the capacitor is proportional to the square of the capacitor voltage, since

$$W_C(t) = \int_{-\infty}^{t} p_{into}(t)\,dt = \frac{1}{2}Cv^2(t)$$

Figure 5-6   Equivalent capacitance of capacitors in parallel

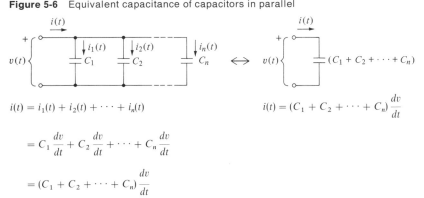

$$i(t) = i_1(t) + i_2(t) + \cdots + i_n(t)$$

$$= C_1\frac{dv}{dt} + C_2\frac{dv}{dt} + \cdots + C_n\frac{dv}{dt}$$

$$= (C_1 + C_2 + \cdots + C_n)\frac{dv}{dt}$$

$$i(t) = (C_1 + C_2 + \cdots + C_n)\frac{dv}{dt}$$

Figure 5-7   Equivalent capacitance of capacitors in series

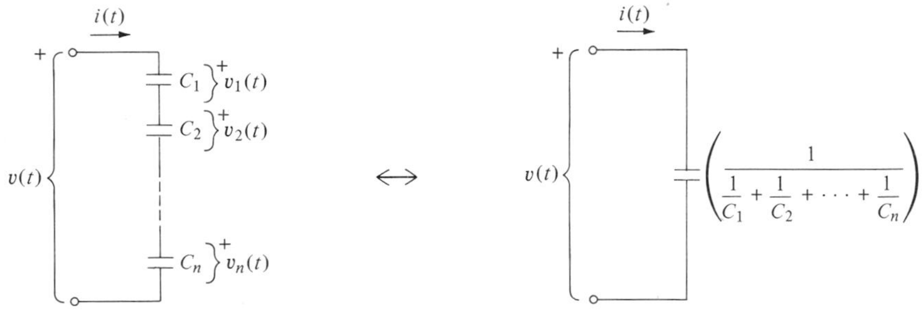

$$v(t) = \frac{1}{C_1}\int_{-\infty}^{t} i(t)\,dt + \frac{1}{C_2}\int_{-\infty}^{t} i(t)\,dt + \cdots$$

$$+ \frac{1}{C_n}\int_{-\infty}^{t} i(t)\,dt$$

$$= \left(\frac{1}{C_1} + \frac{1}{C_2} + \cdots + \frac{1}{C_n}\right)\int_{-\infty}^{t} i(t)\,dt$$

$$v(t) = \left(\frac{1}{C_1} + \frac{1}{C_2} + \cdots + \frac{1}{C_n}\right)\int_{-\infty}^{t} i(t)\,dt$$

5.2.3 Two-Terminal Combinations of Capacitors

Any two-terminal combination of capacitors is equivalent to a single capacitor. No matter how the terminal voltage is distributed from capacitor to capacitor, the terminal current is proportional to the time rate of change of that voltage.

The equivalent capacitance of several capacitors in *parallel* is just the sum of the individual capacitances, as shown in Figure 5-6. This equivalence is similar to the relation for resistors in *series*.

For capacitors in series, the equivalent capacitance is the inverse of the sum of the inverses of the individual capacitances. This equivalence is developed in Figure 5-7.

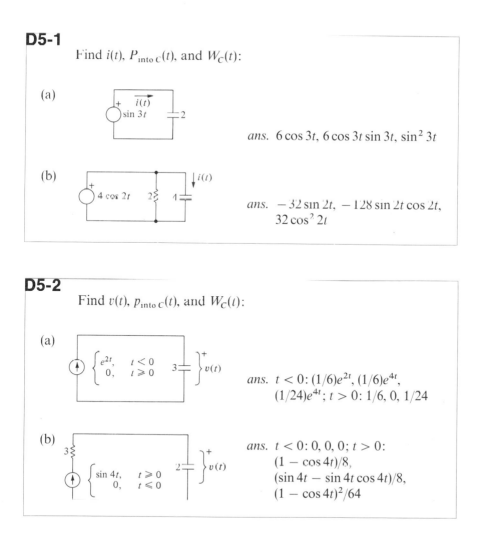

D5-1

Find $i(t)$, $P_{\text{into } C}(t)$, and $W_C(t)$:

(a)

$i(t)$

$\sin 3t$ 2

ans. $6 \cos 3t$, $6 \cos 3t \sin 3t$, $\sin^2 3t$

(b)

$4 \cos 2t$ 2 4 $i(t)$

ans. $-32 \sin 2t$, $-128 \sin 2t \cos 2t$, $32 \cos^2 2t$

D5-2

Find $v(t)$, $P_{\text{into } C}(t)$, and $W_C(t)$:

(a)

$\begin{cases} e^{2t}, & t < 0 \\ 0, & t \geqslant 0 \end{cases}$ 3 $v(t)$

ans. $t < 0$: $(1/6)e^{2t}$, $(1/6)e^{4t}$, $(1/24)e^{4t}$; $t > 0$: $1/6$, 0, $1/24$

(b)

3

$\begin{cases} \sin 4t, & t \geqslant 0 \\ 0, & t \leqslant 0 \end{cases}$ 2 $v(t)$

ans. $t < 0$: $0, 0, 0$; $t > 0$:
$(1 - \cos 4t)/8$,
$(\sin 4t - \sin 4t \cos 4t)/8$,
$(1 - \cos 4t)^2/64$

D5-3

Find $v_c(t)$ for $t \geq 0$, given that $v_c(0) = 5$:

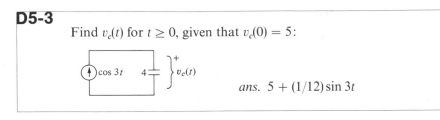

ans. $5 + (1/12)\sin 3t$

D5-4

Find the capacitors that are equivalent to the following two-terminal networks:

(a)

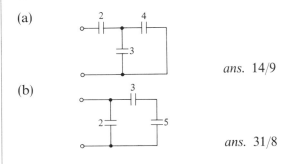

ans. $14/9$

(b)

ans. $31/8$

Capacitor Relations

A capacitor is an energy-storing element that has the sink reference voltage-current relationships

$$i(t) = C \frac{dv(t)}{dt}$$

$$v(t) = \frac{1}{C} \int_{-\infty}^{t} i(t)\, dt$$

There are minus signs in these equations if v and i have the source reference relation.

The stored energy in a capacitor is

$$W_C(t) = \tfrac{1}{2} C v^2(t)$$

where v is the capacitor voltage.

Any two-terminal combination of capacitors is equivalent to a single capacitor. The single-capacitor equivalent of capacitors in parallel is the sum of the individual capacitances. The single-capacitor equivalent of capacitors in series is the inverse of the sum of the inverses of the individual capacitances.

5.3 Voltage-Current, Power, and Energy Relations for the Inductor

5.3.1 Defining Relations

The physical inductor consists of a conductor, usually coiled, in which the magnetic field of the conductor current exerts an appreciable influence on the conductor current itself. Coiling the conductor magnifies the effect, although it is present even in a straight section of wire.

The symbol for an inductor model is shown in Figure 5-8. The *inductance*, L, of the element is indicated beside the symbol. The unit of inductance is the *henry* (abbreviated H), which is named in honor of Joseph Henry (1797–1878), an early electrical experimenter.

According to Faraday's law, the voltage induced in the inductor is

$$v(t) = \frac{d\Phi}{dt}$$

where Φ is the magnetic flux cutting the inductor. If there are no magnetic materials (such as iron) present in the vicinity of the conductor, Φ is in turn proportional to the conductor current. With the sink reference,

$$\Phi(t) = Li(t)$$

giving

$$v(t) = L\frac{di}{dt}$$

If, instead, $v(t)$ and $i(t)$ have the source reference,

$$v(t) = -L\frac{dt}{dt}$$

The reverse voltage-current relation for the inductor is analogous to the capacitor relations. With the sink reference,

$$i(t) = \frac{1}{L}\int_{-\infty}^{t} v(t)\,dt = i(t_0) + \frac{1}{L}\int_{t_0}^{t} v(t)\,dt$$

A summary of inductor voltage-current relations is given in Figure 5-9.

Figure 5-8 Inductor symbol

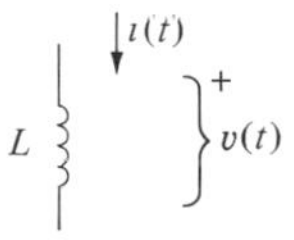

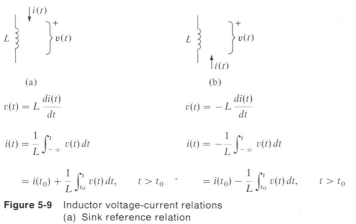

(a)   (b)

$$v(t) = L\frac{di(t)}{dt} \qquad\qquad v(t) = -L\frac{di(t)}{dt}$$

$$i(t) = \frac{1}{L}\int_{-\infty}^{t} v(t)\,dt \qquad\qquad i(t) = -\frac{1}{L}\int_{-\infty}^{t} v(t)\,dt$$

$$= i(t_0) + \frac{1}{L}\int_{t_0}^{t} v(t)\,dt, \quad t > t_0 \qquad = i(t_0) - \frac{1}{L}\int_{t_0}^{t} v(t)\,dt, \quad t > t_0$$

Figure 5-9   Inductor voltage-current relations
                    (a) Sink reference relation
                    (b) Source reference relation

    Given an inductor current, the inductor voltage may be found by differentiation. Two examples of finding an inductor voltage from the inductor current are given in Figure 5-10.

    Given the entire past history of an inductor voltage, the inductor current may be found by integration. An example is given in Figure 5-11(a). Before time $t = 0$,

$$i(t) = \frac{1}{3}\int_{-\infty}^{t} 0\,dt = 0$$

After $t = 0$,

$$i(t) = \frac{1}{3}\int_{0}^{t} 6\sin 4t\,dt = \frac{1}{3}\frac{-6\cos 4t}{4}\Big|_{0}^{t}$$

$$= \frac{-\cos 4t + 1}{2}$$

Figure 5-10   Determining inductor voltage from the inductor current

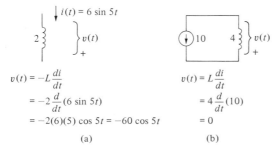

$$v(t) = -L\frac{di}{dt} \qquad\qquad\qquad v(t) = L\frac{di}{dt}$$

$$= -2\frac{d}{dt}(6\sin 5t) \qquad\qquad\quad = 4\frac{d}{dt}(10)$$

$$= -2(6)(5)\cos 5t = -60\cos 5t \qquad = 0$$

                 (a)   (b)

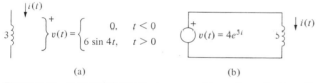

(a)   (b)

Figure 5-11 Determining inductor current from the inductor voltage

Another example of finding an inductor current from its voltage is shown in Figure 5-11(b), where the inductor voltage is specified by a source. For this network,

$$i(t) = -\frac{1}{L} \int_{-\infty}^{t} v(t)\, dt$$

$$= -\frac{1}{5} \int_{-\infty}^{t} 4e^{5t}\, dt$$

$$= -\frac{1}{5}\frac{4e^{5t}}{5}\Big|_{-\infty}^{t}$$

$$= \left(-\frac{4}{25}\right) e^{5t}$$

If the inductor voltage is known after some time, say $t = 0$, and the inductor current is known at $t = 0$, the inductor current can be found after $t = 0$. For the example of Figure 5-12, after $t = 0$,

$$i(t) = \frac{1}{L} \int_{-\infty}^{0} v(t)\, dt + \frac{1}{L} \int_{0}^{t} v(t)\, dt$$

$$= i(0) + \frac{1}{L} \int_{0}^{t} v(t)\, dt$$

$$= 10 + \frac{1}{6} \int_{0}^{t} 7t\, dt$$

$$= 10 + \left(\frac{7}{12}\right) t^{2}$$

Figure 5-12 Determining inductor current using initial inductor current and the inductor voltage thereafter

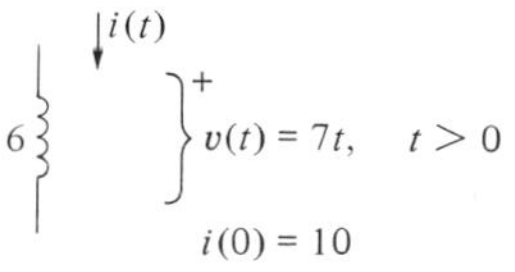

$v(t) = 7t, \quad t > 0$

$i(0) = 10$

5.3.2   Inductor Power and Energy

In the inductor, energy is stored in the magnetic field in an amount proportional to the square of the current. With the sink reference, the electrical power flow into the inductor is

$$p_{into}(t) = v(t)i(t) = Li(t)\frac{di(t)}{dt}$$

The electrical energy stored in the inductor is

$$W_L(t) = \int_{-\infty}^{t} p_{into}(t)\, dt = \frac{1}{2}Li^2(t)$$

5.3.3   Two-Terminal Combinations of Inductors

Any two-terminal combination of inductors is equivalent to a single inductor. In such a combination, however the terminal current branches through each of the individual inductors, the terminal voltage is proportional to the time rate of change of that current.

The equivalent inductance of several inductors in series is the sum of the individual inductances, as shown in Figure 5-13.

For inductors in parallel, the equivalent inductance is the inverse of the sum of the inverses of the individual inductances. This equivalence is developed in Figure 5-14.

Figure 5-13   Equivalent inductance of inductors in series

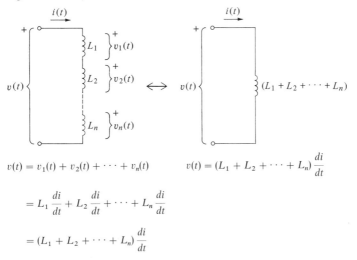

$$v(t) = v_1(t) + v_2(t) + \cdots + v_n(t)$$

$$= L_1\frac{di}{dt} + L_2\frac{di}{dt} + \cdots + L_n\frac{di}{dt}$$

$$= (L_1 + L_2 + \cdots + L_n)\frac{di}{dt}$$

$$v(t) = (L_1 + L_2 + \cdots + L_n)\frac{di}{dt}$$

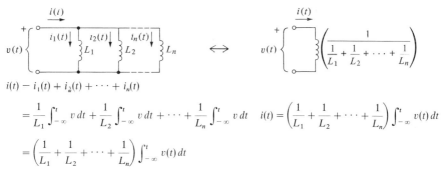

$$i(t) - i_1(t) + i_2(t) + \cdots + i_n(t)$$

$$= \frac{1}{L_1}\int_{-\infty}^{t} v\,dt + \frac{1}{L_2}\int_{-\infty}^{t} v\,dt + \cdots + \frac{1}{L_n}\int_{-\infty}^{t} v\,dt \qquad i(t) = \left(\frac{1}{L_1} + \frac{1}{L_2} + \cdots + \frac{1}{L_n}\right)\int_{-\infty}^{t} v(t)\,dt$$

$$= \left(\frac{1}{L_1} + \frac{1}{L_2} + \cdots + \frac{1}{L_n}\right)\int_{-\infty}^{t} v(t)\,dt$$

Figure 5-14   Equivalent inductance of inductors in parallel

The relations are analogous to those for resistors in series and in parallel. For two inductors in parallel,

$$L_{\text{equiv}} = \frac{L_1 L_2}{L_1 + L_2}$$

D5-5

Find $v(t)$, $p_{\text{into } L}(t)$, and $W_L(t)$:

(a)

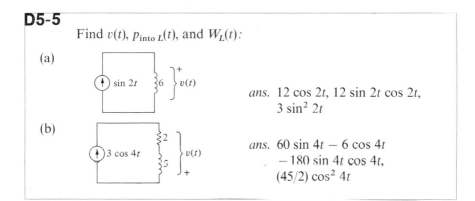

ans. 12 cos 2t, 12 sin 2t cos 2t,
   3 sin² 2t

(b)

ans. 60 sin 4t − 6 cos 4t
   − 180 sin 4t cos 4t,
   (45/2) cos² 4t

D5-6

Find $i(t)$, $p_{\text{into } L}(t)$, and $W_L(t)$:

(a)

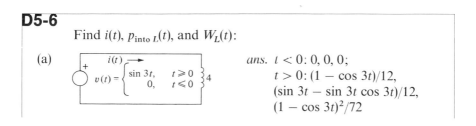

ans. $t < 0$: 0, 0, 0;
   $t > 0$: (1 − cos 3t)/12,
   (sin 3t − sin 3t cos 3t)/12,
   (1 − cos 3t)²/72

(b)

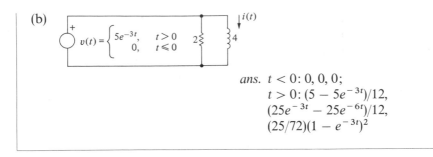

$$ans. \ t < 0: 0, 0, 0;$$
$$t > 0: (5 - 5e^{-3t})/12,$$
$$(25e^{-3t} - 25e^{-6t})/12,$$
$$(25/72)(1 - e^{-3t})^2$$

D5-7

Find $i_L(t)$ for $t \geq 0$, given that $i_L(0) = 10$:

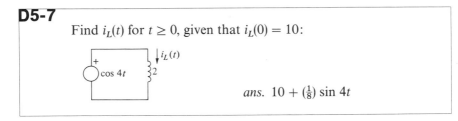

$$ans. \ 10 + (\tfrac{1}{8}) \sin 4t$$

D5-8

Find the inductors that are equivalent to the following two-terminal networks:

(a)

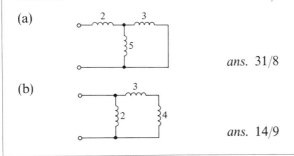

$$ans. \ 31/8$$

(b)

$$ans. \ 14/9$$

Inductor Relations

An inductor is an energy-storing element that has the sink reference voltage-current relationships

$$v(t) = L\frac{di(t)}{dt}$$

$$i(t) = \frac{1}{L}\int_{-\infty}^{t} v(t)\,dt$$

There are minus signs in these equations if v and i have the source reference relation.

The stored energy in an inductor is

$$W_L(t) = \tfrac{1}{2}Li^2(t)$$

where i is the inductor current.

Any two-terminal combination of inductors is equivalent to a single inductor. The single-inductor equivalent of inductors in series is the sum of the individual inductances. The single-inductor equivalent of inductors in parallel is the inverse of the sum of the inverses of the individual inductances.

5.4   Inductive Coupling

5.4.1   Voltage-Current Relations and Winding Senses

When a time-varying magnetic field from one inductor passes through another inductor, a voltage is induced, just as when the time-varying magnetic field is self-produced.

For two magnetically coupled coils, in the absence of magnetic materials such as iron which make the relations nonlinear, the magnetic flux cutting coil number one consists of a term proportional to coil one's current plus a term proportional to coil two's current:

$$\Phi_1(t) = L_1 i_1(t) \pm M i_2(t)$$

Similarly, the flux cutting coil number two involves a term proportional to coil two's current plus a term proportional to coil one's current:

$$\Phi_2(t) = \pm M i_1(t) + L_2 i_2(t)$$

L_1 and L_2 are called the *self-inductances* of coils one and two, respectively. The *mutual inductance*, M, is the same in the two relations. It is taken to be positive, with the algebraic sign used in the relations being dependent on the relative senses of the windings of the two coils.

The sink reference voltages in the coils are

$$v_1 = \frac{d\Phi_1(t)}{dt} = L_1\frac{di_1}{dt} \pm M\frac{di_2}{dt}$$

$$v_2 = \frac{d\Phi_2(t)}{dt} = \pm M\frac{di_1}{dt} + L_2\frac{di_2}{dt}$$

as indicated in Figure 5-15.

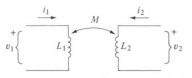

Figure 5-15 Mutually coupled inductors

The algebraic sign to be used with the mutual coupling term depends on the relative winding senses. If a positive i_2 produces flux in coil one in the same sense as does a positive i_1, the plus sign applies; if a positive i_2 produces coil-one flux that opposes the flux produced by a positive i_1, the negative sign applies. The relative senses of the coil fluxes are indicated by large dots, as in the drawings of Figure 5-16.

It is desirable to be able to write down easily the voltage-current relations for coupled inductors for any of the many possible combinations of voltage and current references (not just sink relations) and flux senses. The correct algebraic signs for the self-inductance terms are found by imagining temporarily that there is no mutual coupling.

The algebraic signs of the mutual coupling terms may be determined as follows. If the current in the opposite coil has reference direction in the sense of entering the terminal closest to the dot for that coil, the induced voltage in the first coil has sense with the plus sign closest to the dot on that coil, as in Figure 5-17(a). If both references are reversed, as in Figure 5-17(b), the first coil's mutual coupling term also appears with a plus sign.

Figure 5-16 Indicating coil flux senses with dots

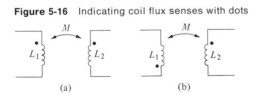

Figure 5-17 Senses of the induced voltage terms

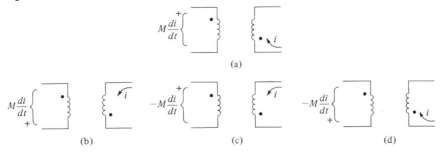

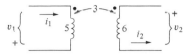

Figure 5-18 Example of determining voltage-current relations

If just one of the two references is reversed, the first coil's mutual coupling term appears with a minus sign, as shown in Figure 5-17(c) and (d).

As an example of determining voltage-current relations for two coupled inductors, consider the situation of Figure 5-18. First, the two inductor voltage-current relationships are written as if there were no mutual coupling. There is a minus sign in the first relation because v_1 and i_1 have the source reference:

$$v_1 = -5\frac{di_1}{dt}$$

$$v_2 = 6\frac{di_2}{dt}$$

Then the foregoing sign rule is used for the mutual coupling terms: i_2 has reference direction entering the dot, but v_1 has the reference plus away from the other dot, so the mutual coupling in v_1 is with a minus sign:

$$v_1 = -5\frac{di_1}{dt} - 3\frac{di_2}{dt}$$

Current i_1 has reference direction toward the dot and v_2 has plus sign nearest the opposite dot, so the mutual coupling in v_2 is with a plus sign:

$$v_2 = 3\frac{di_1}{dt} + 6\frac{di_2}{dt}$$

Another example of determining voltage-current relations is given in Figure 5-19, for which

$$v_1 = 10\frac{di_1}{dt} - 4\frac{di_2}{dt}$$

$$v_2 = 4\frac{di_1}{dt} - 8\frac{di_2}{dt}$$

Figure 5-19 Another example of determining voltage-current relations

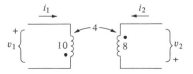

5.4.2 Power and Energy

The electrical power flow into a set of two coupled inductors is the sum of
the power flows into each of the coils. With sink reference relations for
the voltage and current at each coil,

$$p_{\text{into}} = v_1(t)i_1(t) + v_2(t)i_2(t)$$

$$= L_1 i_1 \frac{di_1}{dt} \pm M i_1 \frac{di_2}{dt} + L_2 i_2 \frac{di_2}{dt} \pm M i_2 \frac{di_1}{dt}$$

$$= \frac{d}{dt} \left(\frac{1}{2} L_1 i_1^{\,2} + \frac{1}{2} L_2 i_2^{\,2} \pm M i_1 i_2 \right)$$

where the algebraic sign associated with the mutual coupling term depends
on the relative orientation of the two coils.

The energy stored in the two coupled inductors is then

$$W_M = \int_{-\infty}^{t} p_{\text{into}}(t)\, dt$$

$$= \tfrac{1}{2} L_1 i_1^{\,2} + \tfrac{1}{2} L_2 i_2^{\,2} \pm M i_1 i_2$$

By considering various possible currents, i_1 and i_2, and using the fact that
the stored energy can never be negative, it can be shown that

$$M \le \sqrt{L_1 L_2}$$

The *coupling coefficient* of two inductors is defined by

$$k = \frac{M}{\sqrt{L_1 L_2}}$$

k can range from zero, no coupling, to unity, the maximum possible coupling.

5.4.3 Several Coupled Inductors

When more than two coils are mutually coupled, each to one another, a
great number of different sets of coupling senses are possible because the
arrangement of coils can be three dimensional. The flux senses for the
couplings between each pair of coils are indicated by separate symbols such

Figure 5-20 Voltage-current relations for several coupled inductors

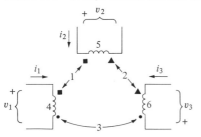

as dots, squares, and triangles. The algebraic signs of the mutual coupling terms may be determined by considering the coils two at a time.

As an example, the voltage-current relations for the three mutually coupled inductors of Figure 5-20 are as follows:

$$v_1 = 4\frac{di_1}{dt} + \frac{di_2}{dt} + 3\frac{di_3}{dt}$$

$$v_2 = \frac{di_1}{dt} + 5\frac{di_2}{dt} - 2\frac{di_3}{dt}$$

$$v_3 = -3\frac{di_1}{dt} + 2\frac{di_2}{dt} - 6\frac{di_3}{dt}$$

D5-9

Find the voltage-current relations for each set of coupled inductors:

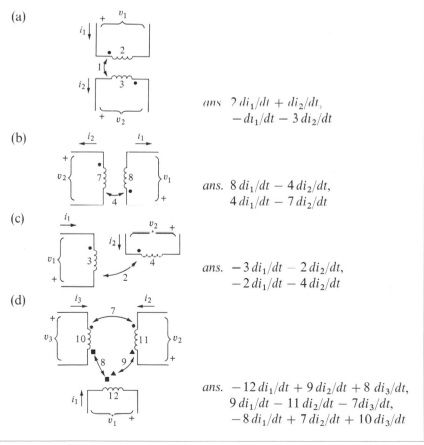

(a)

ans. $2\,di_1/dt + di_2/dt$, $-di_1/dt - 3\,di_2/dt$

(b)

ans. $8\,di_1/dt - 4\,di_2/dt$, $4\,di_1/dt - 7\,di_2/dt$

(c)

ans. $-3\,di_1/dt - 2\,di_2/dt$, $-2\,di_1/dt - 4\,di_2/dt$

(d)

ans. $-12\,di_1/dt + 9\,di_2/dt + 8\,di_3/dt$, $9\,di_1/dt - 11\,di_2/dt - 7di_3/dt$, $-8\,di_1/dt + 7\,di_2/dt + 10\,di_3/dt$

Coupled Inductors

When two inductors are magnetically coupled, the inductor voltages and currents, with the sink reference relation for each inductor, are of the form

$$\begin{cases} v_1 = L_1 \dfrac{di_1}{dt} \pm M \dfrac{di_2}{dt} \\[2mm] v_2 = \pm M \dfrac{di_1}{dt} + L_2 \dfrac{di_2}{dt} \end{cases}$$

The correct algebraic sign for the mutual coupling term, $\pm M$, depends on the relative orientations of the two coils, which is indicated by dots at one side of the symbol for each inductor.

The energy stored in a pair of mutually coupled inductors is given by

$$W_M = \tfrac{1}{2}L_1 i_1{}^2 + \tfrac{1}{2}L_2 i_2{}^2 \pm M i_1 i_2$$

The value of the mutual inductance, M, cannot exceed the geometric mean of the two self-inductances, L_1 and L_2:

$$M \le \sqrt{L_1 L_2}$$

When several inductors are mutually coupled, the mutual coupling terms may be determined by taking the inductors two at a time.

5.5 Systematic Simultaneous Mesh Equations

5.5.1 Equations for Networks without Inductive Coupling

To write systematic simultaneous equations for a network containing inductors and capacitors as well as sources and resistors, one need only write the inductor and capacitor voltages in terms of their sink reference currents, in place of the resistor voltage-current relation that would be written if the element were a resistor. The equations are most easily written in *operator notation* in which, for example,

$$\left(3 + 4\frac{d}{dt} + 5 \int_{-\infty}^{t} dt\right) i$$

stands for

$$3i + 4\frac{di}{dt} + 5 \int_{-\infty}^{4} i\, dt$$

In summing the voltages around each mesh, simply write R for each resistor, $L\,d/dt$ for each inductor, and $(1/C)\int_{-\infty}^{t} dt$ for each capacitor.

Consider the network of Figure 5-21. Systematic simultaneous mesh equations for this network are as follows:

$$
\left\{
\begin{aligned}
&\left(1 + 2\frac{d}{dt} + \frac{1}{3}\int_{-\infty}^{t} dt + 6 + 5\frac{d}{dt} + 4\right) i_1 \\
&\qquad\qquad - \left(4 + 5\frac{d}{dt}\right) i_2 - (6)i_3 = 12 \sin t + 18 \\[2mm]
&-\left(4 + 5\frac{d}{dt}\right) i_1 + \left(4 + 5\frac{d}{dt} + \frac{1}{8}\int_{-\infty}^{t} dt + 7 + \frac{1}{9}\int_{-\infty}^{t} dt\right) i_2 \\
&\qquad\qquad - \left(7 + \frac{1}{9}\int_{-\infty}^{t} dt\right) i_3 = 0 \\[2mm]
&-(6)i_1 - \left(7 + \frac{1}{9}\int_{-\infty}^{t} dt\right) i_2 + \left(7 + \frac{1}{9}\int_{-\infty}^{t} dt + 6 + 10\frac{d}{dt}\right) i_3 = -18
\end{aligned}
\right.
$$

5.5.2 Controlled Voltage-Source Equivalent for Coupled Inductors

To write systematic mesh equations for networks containing coupled inductors, it is helpful to make use of the equivalent circuit of Figure 5-22, in which the mutual coupling between the inductors is replaced by controlled sources. For each,

$$
\begin{cases}
v_1 = L_1 \dfrac{di_1}{dt} + M \dfrac{di_2}{dt} \\[3mm]
v_2 = M \dfrac{di_1}{dt} + L_2 \dfrac{di_2}{dt}
\end{cases}
$$

For arbitrary current reference directions and winding senses, the sense of the controlled source reference polarities may be found as follows. If the controlling current has reference direction into the dot, the controlled source

Figure 5-21 Planar *RLC* network without mutually coupled inductors

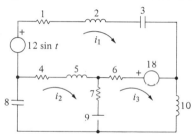

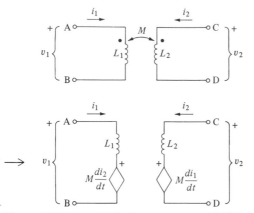

Figure 5-22   Controlled voltage-source equivalent of mutually coupled inductors

has reference polarity in the same sense as the second dot. If the controlling current has reference direction out of the dot, the controlled source has reference polarity in the opposite sense as the second dot. An example of determining the controlled source equivalent circuit for two coupled inductors is given in Figure 5-23.

5.5.3   Equations for Networks with Inductive Coupling

Mutually coupled inductors may be accommodated in mesh equations by first replacing them with the controlled voltage-source equivalent circuit of the previous section. Incorporating controlled sources into the mesh equations is done in the same manner as for source-resistor networks.

Figure 5-23   Determining senses of the controlled sources

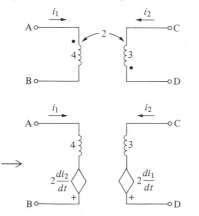

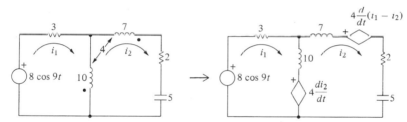

Figure 5-24 Network where mutually coupled inductors are replaced by their controlled voltage-source equivalent

Consider the network containing coupled inductors shown in Figure 5-24. The coupling has been replaced by its controlled voltage-source equivalent. Systematic simultaneous mesh equations are then

$$\begin{cases} \left(3 + 10\dfrac{d}{dt}\right)i_1 - \left(10\dfrac{d}{dt}\right)i_2 = 8\cos 9t - 4\dfrac{di_2}{dt} \\[2mm] -\left(10\dfrac{d}{dt}\right)i_1 + \left(17\dfrac{d}{dt} + 2 + \dfrac{1}{5}\int_{-\infty}^{t} dt\right)i_2 = 4\dfrac{di_2}{dt} - 4\dfrac{d}{dt}(i_1 - i_2) \end{cases}$$

or

$$\begin{cases} \left(3 + 10\dfrac{d}{dt}\right)i_1 - \left(6\dfrac{d}{dt}\right)i_2 = 8\cos 9t \\[2mm] -\left(6\dfrac{d}{dt}\right)i_1 + \left(9\dfrac{d}{dt} + 2 + \dfrac{1}{5}\int_{-\infty}^{t} dt\right)i_2 = 0 \end{cases}$$

D5-10

Write systematic integrodifferential mesh equations for the following networks in terms of the indicated currents:

(a)

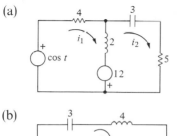

(b)

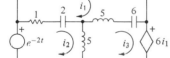

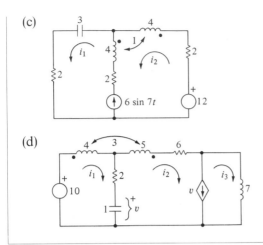

(c)

(d)

Systematic Simultaneous Mesh Equations

Mesh equations may be written in operator notation for RLC networks in the same manner as for source-resistor networks. For each inductor, instead of R, write $L\,d/dt$. For each capacitor, write $(1/C)\int_{-\infty}^{t} dt$.

When coupled inductors are involved, it is helpful to replace the mutual coupling by a controlled voltage-source equivalent circuit.

5.6 Systematic Simultaneous Nodal Equations

5.6.1 Equations for Networks without Inductive Coupling

For nodal equations, write the inductor and capacitor currents in terms of the element voltage. For a resistor, write $1/R$, for an inductor write, instead, $(1/L)\int_{-\infty}^{t} dt$. For a capacitor, write $C\,d/dt$. An example network is given in

Figure 5-25 *RLC* network without mutually coupled inductors

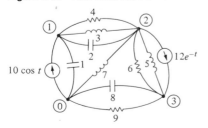

Figure 5-25, for which the systematic nodal equations are as follows:

$$\begin{cases} \left(\dfrac{d}{dt} + 2\dfrac{d}{dt} + \dfrac{1}{3}\int_{-\infty}^{t} dt + \dfrac{1}{4} \right) v_1 - \left(2\dfrac{d}{dt} + \dfrac{1}{3}\int_{-\infty}^{t} dt + \dfrac{1}{4} \right) v_2 - (0)v_3 = 10\cos t \\[2mm] -\left(2\dfrac{d}{dt} + \dfrac{1}{3}\int_{-\infty}^{t} dt + \dfrac{1}{4} \right) v_1 + \left(2\dfrac{d}{dt} + \dfrac{1}{3}\int_{-\infty}^{t} dt + \dfrac{1}{4} + \dfrac{1}{7}\int_{-\infty}^{t} dt \right. \\[2mm] \qquad \left. + \dfrac{1}{6} + \dfrac{1}{5}\int_{-\infty}^{t} dt \right) v_2 - \left(\dfrac{1}{6} + \dfrac{1}{5}\int_{-\infty}^{t} dt \right) v_3 = -12e^{t} \\[2mm] -(0)v_1 + \left(\dfrac{1}{6} + \dfrac{1}{5}\int_{-\infty}^{t} dt \right) v_2 + \left(\dfrac{1}{6} + \dfrac{1}{5}\int_{-\infty}^{t} dt + \dfrac{1}{9} + 8\dfrac{d}{dt} \right) v_3 = 12e^{-t} \end{cases}$$

The same sorts of symmetries present in the systematic algebraic source-resistor network nodal equations are present in the *operations* for each term in these integrodifferential equations.

5.6.2 Controlled Current-Source Equivalent for Coupled Inductors

The controlled source inductive-coupling equivalent of Section 5.5.2 is well-suited to mesh equations, but very poorly suited for nodal equations because it involves voltage sources controlled by network currents. A better equivalent circuit for use with nodal equations is that given in Figure 5-26. The current i_a in the equivalent is

$$i_a = i_1 + \frac{M}{L_1} i_2$$

and

$$v_1 = L_1 \frac{di_a}{dt} = L_1 \frac{di_1}{dt} + M \frac{di_2}{dt}$$

Figure 5-26 Controlled current-source equivalent of mutually coupled inductors

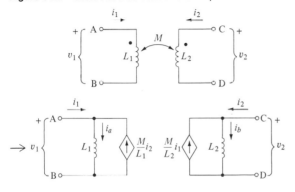

Similarly,

$$v_2 = L_2 \frac{di_b}{dt} = M \frac{di_1}{dt} + L_2 \frac{di_2}{dt}$$

as required.

5.6.3   Equations for Networks with Inductive Coupling

When it is desired to write nodal equations for a network involving mutually coupled inductors, the controlled current-source equivalent may be used to advantage.

Consider the network of Figure 5-27 which contains coupled inductors. In the equivalent network, the magnetically coupled circuits have been replaced by a controlled current-source equivalent.

The systematic nodal equations, including those relating the controlled source controlling signals to the node voltages, are as follows:

$$\left\{\begin{aligned}
&\left(4\frac{d}{dt} + \frac{1}{3}\int_{-\infty}^{t} dt + \frac{1}{8}\int_{-\infty}^{t} dt\right)v_1 - \left(\frac{1}{8}\int_{-\infty}^{t} dt\right)v_2 = \frac{i_2}{4} - \frac{2i_1}{3} \\
&-\left(\frac{1}{8}\int_{-\infty}^{t} dt\right)v_1 + \left(5 + \frac{1}{3}\int_{-\infty}^{t} dt + \frac{1}{8}\int_{-\infty}^{t} dt\right)v_2 = 6\sin 7t - \frac{i_2}{4} \\
&i_1 = \frac{1}{8}\int_{-\infty}^{t}(v_1 - v_2)\,dt \\
&i_2 = -\frac{1}{3}\int_{-\infty}^{t} v_1\,dt
\end{aligned}\right.$$

Figure 5-27   Network where mutually coupled inductors are replaced by their controlled current-source equivalent

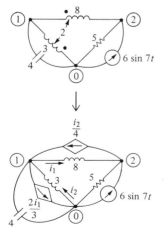

D5-11

Write systematic integrodifferential nodal equations for the following networks in terms of the indicated node voltages:

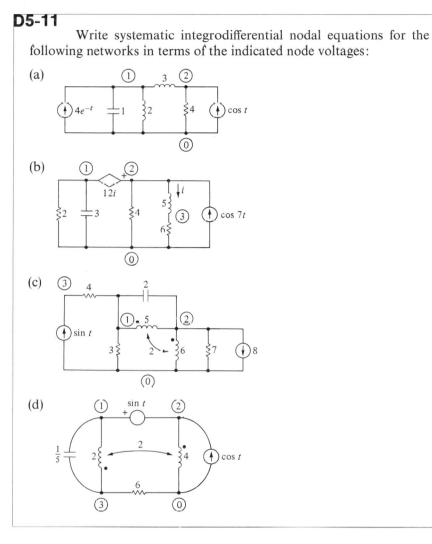

(a)

(b)

(c)

(d)

Systematic Simultaneous Nodal Equations

Nodal equations may be written in operator notation for RLC networks in the same manner as for source-resistor networks. For each capacitor, instead of $1/R$, write $C\,d/dt$. For each inductor, write $(1/L)\int_{-\infty}^{t} dt$.

When coupled inductors are involved, it is helpful to replace the coupling by a controlled current-source equivalent circuit.

5.7 Linear, Time-Invariant Differential Equations

5.7.1 Equation Form and Classification

An equation of the form

$$a_n \frac{d^n y}{dt^n} + a_{n-1} \frac{d^{n-1} y}{dt^{n-1}} + \cdots + a_1 \frac{dy}{dt} + a_0 y = f(t)$$

where $a_n, a_{n-1}, \ldots, a_1, a_0$ are constants, is called a linear, constant-coefficient differential equation. This type of differential equation is termed "linear" because the function $y(t)$ and derivatives of $y(t)$ appear linearly in the equation. It is constant coefficient (or *time-invariant* or *fixed*) because $a_n, a_{n-1}, \cdots,$ a_1, a_0 are constants, not functions of time. The function $f(t)$ is called the *driving function* of the equation.

If the highest derivative to appear in the equation is the (n)th derivative, the equation is said to be of (n)th order. For example,

$$3 \frac{dy}{dt} + 2y = \sin t$$

is a first-order linear, time-invariant differential equation and

$$-\frac{d^3 y}{dt^3} + 2 \frac{dy}{dt} = 7e^{-4t} + 10$$

is third order.

A function $y(t)$ that satisfies a differential equation is termed a solution to that equation. For example, in the equation

$$\frac{dy}{dt} - y = 0$$

$y(t) = e^t$ is a solution.

A linear differential equation has many solutions. Not only is $y(t) = e^t$ a solution to the above equation, but $y = 0$, $y = 2e^t$, and $y = 10e^t$ are also solutions. In fact, $y = Ke^t$, where K is any constant, is a solution of the equation.

The *general solution* of a differential equation is the collection of all possible solutions. It is characteristic of linear differential equations that the general solutions have *arbitrary constants* such as K, above.

Arbitrary constants in the general solution are certainly to be expected. After all, the problem of integration is the solution of the differential equation

$$\frac{dy}{dt} = f(t)$$

which is

$$y(t) = \int f(t)\, dt + K$$

The arbitrary constant K appears in the solution because a whole family of functions $y(t)$ have derivative $f(t)$. So it is with all first-order linear differential equations.

The general solution to an (n)th-order linear differential equation contains exactly n independent arbitrary constants. This is to say that for an (n)th-order equation, once a set of solutions with n independent arbitrary constants in it is found, one can stop looking for further solutions. There are no other solutions.

5.7.2   Solution of First-Order Homogeneous Equations

A *homogeneous* differential equation is an equation in which the driving function is zero. The first-order homogeneous equation is of the form

$$a_1 \frac{dy}{dt} + a_0 y = 0$$

Rearranging,

$$\frac{dy}{dt} = -\frac{a_0}{a_1} y$$

which is to ask, within the constant $(-a_0/a_1)$, what function is its own derivative? The exponential function is the one that is proportional to its derivative.

Trying the function

$$y = Ke^{st}$$

where K and s are constants, as a possible solution to the first-order homogeneous equation gives

$$a_1 \frac{dy}{dt} + a_0 y = 0$$

$$a_1 s K e^{st} + a_0 K e^{st} = 0$$

$$(a_1 s + a_0) K e^{st} = 0$$

Dividing both sides of the equation by Ke^{st},

$$a_1 s + a_0 = 0$$

$$s = -\frac{a_0}{a_1}$$

That is,

$$y(t) = Ke^{(-a_0/a_1)t}$$

satisfies the equation, as is easily checked by substitution. This solution contains one arbitrary constant and so is the general solution to the equation.

As a numerical example, consider

$$\frac{dy}{dt} + 3y = 0$$

Substituting $y = Ke^{st}$ into the equation,

$$sKe^{st} + 3Ke^{st} = 0$$
$$s + 3 = 0$$
$$s = -3$$

Thus

$$y(t) = Ke^{-3t}$$

is the general solution of this equation.

5.7.3 Solution of Higher-Order Homogeneous Equations

Exponential functions also are solutions to the higher-order homogeneous equations. For the general homogeneous equation,

$$a_n \frac{d^n y}{dt^n} + a_{n-1} \frac{d^{n-1} y}{dt^{n-1}} + \cdots + a_1 \frac{dy}{dt} + a_0 y = 0$$

substituting $y(t)$ of the form Ke^{st} gives

$$a_n s^n Ke^{st} + a_{n-1} s^{n-1} Ke^{st} + \cdots + a_1 sKe^{st} + a_0 Ke^{st} = 0$$

Dividing both sides by Ke^{st}, there results the *characteristic equation* (or *auxiliary equation*),

$$a_n s^n + a_{n-1} s^{n-1} + \cdots + a_1 s + a_0 = 0$$

It is easy to write the characteristic equation directly from the original differential equation.

The characteristic equation is an (n)th-order polynomial in s, which has n solutions or *roots*, $s_1, s_2, \ldots, s_n$. In other words, there are n values of s for which Ke^{st} is a solution to an (n)th-order homogeneous equation.

Since the equation is linear, not only are $Ke^{s_1 t}$ and $Ke^{s_2 t}$, and so on, solutions to the equation, but sums of solutions are also solutions. In fact, the arbitrary constants involved with each exponential term can be different numbers, so that any function of the form

$$y(t) = K_1 e^{s_1 t} + K_2 e^{s_2 t} + \cdots + K_n e^{s_n t}$$

for any numbers $K_1, K_2, \ldots, K_n$, satisfies the differential equation.

So long as the roots of the characteristic equation, $s_1, s_2, \ldots, s_n$, are distinct, the above solution contains n independent arbitrary constants, $K_1, K_2, \ldots, K_n$, and so is the general solution to the n(th)-order equation.

Consider the second-order equation

$$\frac{d^2y}{dt^2} + 5\frac{dy}{dt} + 6y = 0$$

Substituting $y(t) = Ke^{st}$ gives

$$s^2 Ke^{st} + 5sKe^{st} + 6Ke^{st} = 0$$
$$s^2 + 5s + 6 = 0$$

which is the characteristic equation. The characteristic equation factors as

$$(s + 2)(s + 3) = 0$$

and so has solutions

$$s_1 = -2$$

and

$$s_2 = -3$$

The general solution to the original second-order differential equation is then

$$y(t) = K_1 e^{-2t} + K_2 e^{-3t}$$

which has two independent arbitrary constants.

For the equation

$$2\frac{d^2y}{dt^2} + 4\frac{dy}{dt} + y = 0$$

the characteristic equation is

$$2s^2 + 4s + 1 = 0$$

which does not factor easily. Using the quadratic formula,

$$s_1, s_2 = \frac{4 \pm \sqrt{4^2 - 4 \cdot 2 \cdot 1}}{2 \cdot 2}$$

$$= \frac{-4 \pm \sqrt{8}}{4}$$

$$= -1 + \sqrt{2}, \ -1 - \sqrt{2}$$

The general solution is then

$$y(t) = K_1 e^{(-1+\sqrt{2})t} + K_2 e^{(-1-\sqrt{2})t}$$

The equation

$$\frac{d^2y}{dt^2} + 5\frac{dy}{dt} = 0$$

has characteristic equation

$$s^2 + 5s = 0$$
$$s(s + 5) = 0$$

and general solution

$$y(t) = K_1 e^{0t} + K_2 e^{-5t} = K_1 + K_2 e^{-5t}$$

5.7.4 Repeated Roots

A special situation occurs when two or more of the roots of the characteristic equation are the same number. For example,

$$\frac{d^2 y}{dt^2} + 2 \frac{dy}{dt} + y = 0$$

has characteristic equation

$$s^2 + 2s + 1 = (s + 1)(s + 1) = 0$$

giving

$$s_1 = -1, \qquad s_2 = -1$$

Now

$$K_1 e^{-t} + K_2 e^{-t} = (K_1 + K_2)e^{-t} = Ke^{-t}$$

cannot be the general solution to this second-order differential equation since the arbitrary constants K_1 and K_2 are not independent.

It can be shown that the general solution in this case is

$$y(t) = K_1 e^{-t} + K_2 t e^{-t}$$

If a root s_i is repeated three times, the corresponding terms in the solution of the homogeneous equation are

$$K_1 e^{s_i t} + K_2 t e^{s_i t} + K_3 t^2 e^{s_i t},$$

and so on.

The case of repeated roots is of limited practical importance, since the numbers in the equation have to be "just right" for repeated roots to occur.

5.7.5 Complex Roots

It is possible that some of the roots of the characteristic equation will be complex numbers. The solution to the homogeneous equation is still of the form

$$y(t) = K_1 e^{s_1 t} + K_2 e^{s_2 t} + \cdots + K_n e^{s_n t}$$

but some algebraic manipulation is then expedient so that the solution is in a more easily visualized form.

Each set of complex conjugate roots, for example

$$s_1, s_2 = \alpha \pm j\beta$$

gives rise to terms in the homogeneous solution

$$y(t) = K_1 e^{(\alpha + j\beta)t} + K_2 e^{(\alpha - j\beta)t}$$
$$= e^{\alpha t}(K_1 e^{j\beta t} + K_2 e^{-j\beta t})$$

Algebraic manipulations using Euler's relation (discussed in detail in Chapter Eight) may be used to convert the above to the equivalent form

$$y(t) = e^{\alpha t}(D \cos \beta t + E \sin \beta t)$$

where D and E are the arbitrary constants. This trigonometric form is generally easier to deal with and to visualize for equations of relatively low order.

D5-12

Find the general solutions to the following homogeneous differential equations:

(a) $2\dfrac{dy}{dt} + y = 0$ *ans.* $Ke^{-(1/2)t}$

(b) $-\dfrac{dy}{dt} - 3y = 0$ *ans.* Ke^{-3t}

(c) $\dfrac{d^2 y}{dt^2} + 6\dfrac{dy}{dt} + 8y = 0$ *ans.* $K_1 e^{-2t} + K_2 e^{-4t}$

(d) $2\dfrac{d^2 y}{dt^2} + \dfrac{dy}{dt} = 0$ *ans.* $K_1 + K_2 e^{-(1/2)t}$

(e) $3\dfrac{d^2 y}{dt^2} + 5\dfrac{dy}{dt} + y = 0$ *ans.* $K_1 e^{s_1 t} + K_2 e^{s_2 t}$,
                                               where $s_1, s_2 = (-5 \pm \sqrt{13})/6$

(f) $\dfrac{d^3 y}{dt^3} + 2\dfrac{d^2 y}{dt^2} + \dfrac{dy}{dt} = 0$ *ans.* $K_1 + K_2 e^{-t} + K_3 t e^{-t}$

(g) $\dfrac{d^4 y}{dt^4} = 0$ *ans.* $K_1 + K_2 t + K_3 t^2 + K_4 t^3$

(h) $2\dfrac{d^3 y}{dt^3} + 8\dfrac{d^2 y}{dt^2} + \dfrac{dy}{dt} = 0$ *ans.* $K_1 + K_2 e^{s_1 t} + K_3 e^{s_2 t}$,
                                               where $s_1, s_2 = (-4 \pm \sqrt{14})/2$

General Solutions to Homogeneous Differential Equations

An (n)th-order homogeneous linear, constant-coefficient differential equation is of the form

$$a_n \frac{d^n y}{dt^n} + a_{n-1} \frac{d^{n-1} y}{dt^{n-1}} + \cdots + a_1 \frac{dy}{dt} + a_0 y = 0$$

The general solution to this equation is

$$y(t) = K_1 e^{s_1 t} + K_2 e^{s_2 t} + \cdots + K_n e^{s_n t}$$

where $s_1, s_2, \ldots, s_n$ are the roots of the characteristic equation,

$$a_n s^n + a_{n-1} s^{n-1} + \cdots + a_1 s + a_0 = 0$$

providing the roots are distinct.

If there are repeated roots, s_1, the terms corresponding to these roots appear in the general solution as

$$K_1 e^{s_1 t} + K_2 t e^{s_1 t} + K_3 t^2 e^{s_1 t} + \cdots$$

5.8 General Solutions to Driven Equations

5.8.1 Forced and Natural Components of Solutions

For an equation with a nonzero driving function, the corresponding homogeneous equation is the original equation with the driving function replaced by zero. The general solution to the entire equation is the *general* solution to the homogeneous equation plus any *one* solution to the entire equation.

For the equation

$$a_n \frac{d^n y}{dt^n} + a_{n-1} \frac{d^{n-1} y}{dt^{n-1}} + \cdots + a_1 \frac{dy}{dt} + a_0 y = f(t)$$

let the general solution to the homogeneous equation be

$$y_n(t) = K_1 e^{s_1 t} + K_2 e^{s_2 t} + \cdots$$

so that

$$\left[a_n \frac{d^n}{dt^n} + a_{n-1} \frac{d^{n-1}}{dt^{n-1}} + \cdots + a_1 \frac{d}{dt} + a_0 \right] y_n(t) = 0$$

Let $y_f(t)$ be any one solution to the original equation:

$$\left[a_n \frac{d^n}{dt^n} + a_{n-1} \frac{d^{n-1}}{dt^{n-1}} + \cdots + a_1 \frac{d}{dt} + a_0 \right] y_f(t) = f(t)$$

Then

$$y(t) - y_n(t) + y_J(t)$$

satisfies the driven equation,

$$\left[a_n \frac{d^n}{dt^n} + a_{n-1} \frac{d^{n-1}}{dt^{n-1}} + \cdots + a_1 \frac{d}{dt} + a_0 \right] [y_n(t) + y_J(t)] - f(t)$$

Since the function

$$y(t) = y_n(t) + y_f(t)$$

satisfies the driven (n)th-order equation and contains n independent, arbitrary constants [in $y_n(t)$], it is the general solution.

The solution to the homogeneous equation is called the *natural* (or *homogeneous* or *transient*) component of the solution. Physically, this is the solution of the network with the fixed sources set to zero. It is how the network would respond "naturally." Usually this component of the solution dies out in time; energy that is initially stored in inductors and capacitors is eventually dissipated by the network's resistors.

The single solution to the whole equation is called the *forced* (or *particular* or *steady state*) component of the solution.

5.8.2   Forced Constant Response

Finding a single solution to the entire equation, a forced solution, is rather complicated in general. There are some types of driving functions, however, for which the solution is easy.

Except in a special case, if the driving function of the equation is a constant, the forced solution is also a constant. For example, for

$$\frac{d^3 y}{dt^3} + 8 \frac{d^2 y}{dt^2} + 7 \frac{dy}{dt} + 6y = 10$$

$$y_f = \frac{10}{6}$$

a constant, satisfies the equation.

To find the forced solution due to a constant driving function, simply substitute a constant $y_f = A$ into the equation:

$$a_n \frac{d^n A}{dt^n} + a_{n-1} \frac{d^{n-1} A}{dt^{n-1}} + \cdots + a_1 \frac{dA}{dt} + a_0 A = f$$

Since all derivatives of a constant are zero, this gives

$$a_0 A = f$$

$$y_f = A = \frac{f}{a_0}$$

For the equation

$$\frac{d^2y}{dt^2} + 5\frac{dy}{dt} + 6y = -7$$

the homogeneous equation is

$$\frac{d^2y_n}{dt^2} + 5\frac{dy_n}{dt} + 6y_n = 0$$

and the characteristic equation is

$$s^2 + 5s + 6 = (s+2)(s+3) = 0$$
$$s_1 = -2, \qquad s_2 = -3$$

The natural part of the solution is then

$$y_n(t) = K_1 e^{-2t} + K_2 e^{-3t}.$$

Substituting a trial forced solution,

$$y_f = A$$

a constant, into the entire equation gives

$$\frac{d^2 A}{dt^2} + 5\frac{dA}{dt} + 6A = -7$$

$$6A = -7$$

$$y_f = A = -\frac{7}{6}$$

The general solution to the equation is thus

$$y(t) = y_n(t) + y_f$$

$$= K_1 e^{-2t} + K_2 e^{-3t} - \frac{7}{6}$$

A special case occurs when the differential equation has no term proportional to the function itself, that is, when $a_0 = 0$. The equation

$$\frac{d^2y}{dt^2} + 8\frac{dy}{dt} = 10$$

is of this type. Substitution of a constant $y_f = A$ into the equation gives

$$0 = 10$$

so this cannot be the correct forced solution. The solution is actually a constant times t.

This special situation occurs whenever the equation driving function has the same shape as one of the terms in the natural component of the solution.

For the example problem, the characteristic equation is

$$s^2 + 8s - 0$$

giving

$$y_n(f) = K_1 e^{-8t} + K_2$$

The constant K_2 in the natural part of the solution and the constant driving function 10 are indicative of this special case, which will be examined in terms of the network in Chapter Seven.

Except for that special case, if the driving function of a differential equation is a constant, the forced component of the solution is a constant.

5.8.3   Other Forced Responses and Superposition

There are several other driving functions for which determination of a forced solution to a linear, constant-coefficient differential equation is rather easy. Exponential and sinusoidal driving functions will be of great importance in later chapters.

Except in a special case, if the driving function is exponential, the forced solution is exponential with the same exponential constant as the driving function.

Except in a special case, if the driving function is sinusoidal, the forced solution is sinusoidal with the same frequency as the driving function.

As with linear algebraic equations, driving functions of linear differential equations may be superimposed. The solution due to a sum of two driving function component terms is the sum of the solutions for the individual terms.

D5-13

Find the general solutions to the following constant driving function differential equations:

(a) $3\dfrac{dy}{dt} + 4y = -2$ ans. $(-\frac{1}{2}) + Ke^{-(4/3)t}$

(b) $\dfrac{dy}{dt} - 3y = 4$ ans. $(-\frac{4}{3}) + Ke^{3t}$

(c) $\dfrac{d^2y}{dt^2} + 5\dfrac{dy}{dt} + 6y = 10$ ans. $(\frac{5}{3}) + K_1 e^{-2t} + K_2 e^{-3t}$

(d) $\dfrac{d^2y}{dt^2} + 2\dfrac{dy}{dt} + y = 3$ ans. $3 + K_1 e^{-t} + K_2 te^{-t}$

(e) $\dfrac{d^2y}{dt^2} - 9y = 8$ ans. $(-\frac{8}{9}) + K_1 e^{3t} + K_2 e^{-3t}$

(f) $\quad -\dfrac{dy}{dt} = 3$ $\qquad\qquad$ *ans.* $-3t + K$

(g) $\quad \dfrac{d^2y}{dt^2} + 8\dfrac{dy}{dt} + 3y = 2$ $\qquad$ *ans.* $(\frac{2}{3}) + K_1 e^{s_1 t} + K_2 e^{s_2 t}$,
$\qquad\qquad\qquad\qquad\qquad\qquad\qquad$ where $s_1, s_2 = (-4 \pm \sqrt{13})$

(h) $\quad \dfrac{d^3y}{dt^3} + 8\dfrac{d^2y}{dt^2} = 10$ $\qquad$ *ans.* $(\frac{5}{8})t^2 + K_1 + K_2 t + K_3 e^{-8t}$

General Solutions to Driven Differential Equations

The general solution to a linear, constant-coefficient differential equation consists of the sum of the general solution to the homogeneous equation plus a single solution to the entire equation.

The general solution of the homogeneous equation is called the *natural* component of the solution. It contains the arbitrary constants. The single solution to the entire equation is called the *forced* component of the solution.

Except in the special case in which one of the terms in the natural component of the solution is a constant, if the driving function of an equation is constant, the forced solution of the equation is constant.

Driving function components may be superimposed, as in linear algebraic equations.

5.9 Specific Solutions and Boundary Conditions

5.9.1 The Meaning of the General Solution

The general solutions to the differential equations describing networks indicate that there is a whole family of possible network solutions, different possibilities for each different set of arbitrary constants. Of course, only one of these possibilities applies to any specific situation. Which of all the possible solutions applies depends on when and how the elements of the network were connected together.

For example, the solution for $v(t)$ in the network of Figure 5-28 depends on the time at which the network is connected together and on the amount of charge on the capacitor plates (or, equivalently, the capacitor voltage) when the network was first connected. After the network is connected, $v(t)$ satisfies the same differential equation, no matter what the circumstances of the network connection; but different initial capacitor charges will result in different possible solutions for $v(t)$ applying.

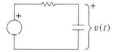

Figure 5-28   Simple network involving a capacitor

5.9.2   Boundary Conditions

Which of all of the possible solutions of a differential equation is the specific one that applies to a given situation may be described with boundary conditions.

Suppose it is known that of all the solutions of the form

$$y(t) = Ke^{-3t} + 7$$

the one for which the boundary condition

$$y(0) = 5$$

applies in a certain situation. Then the arbitrary constant K may be evaluated.

$$y(0) = Ke^0 + 7 = K + 7 = 5$$
$$K = -2$$

Of all the possible solutions,

$$y(t) = -2e^{-3t} + 7$$

is the one that satisfies the boundary condition.

Mathematically, the most convenient time t for a boundary condition is $t = 0$, since at $t = 0$ an exponential function Ke^{st} has value

$$Ke^0 = K$$

For this reason, the origin of the time scale in a problem is usually chosen so that boundary conditions occur at $t = 0$.

A second-order differential equation has two arbitrary constants in the general solution and requires two independent boundary conditions to select one specific solution from the general solution.

Consider the equation

$$\frac{d^2y}{dt^2} + 5\frac{dy}{dt} + 4y = 12$$

with the boundary conditions

$$y(0) = 2 \quad \text{and} \quad \left.\frac{dy}{dt}\right|_{t=0} = 7$$

The homogeneous equation is

$$\frac{d^2 y_n}{dt^2} + 5\frac{dy_n}{dt} + 4y_n = 0$$

the characteristic equation is

$$s^2 + 5s + 4 = (s + 1)(s + 4) = 0$$

and the natural component of the general solution is

$$y_n(t) = K_1 e^{-t} + K_2 e^{-4t}$$

The forced solution is given by

$$y_f = A$$

$$\frac{d^2 A}{dt^2} + 5\frac{dA}{dt} + 4A = 12$$

$$4A = 12$$

$$y_f = A = 3$$

The general solution is thus

$$\begin{aligned} y(t) &= y_n(t) + y_f \\ &= K_1 e^{-t} + K_2 e^{-4t} + 3 \end{aligned}$$

Applying the first boundary condition,

$$y(0) = K_1 + K_2 + 3 = 2$$
$$K_1 + K_2 = -1$$

The second boundary condition in this example involves the derivative of the function.

$$\frac{dy}{dt} = -K_1 e^{-t} - 4K_2 e^{-4t}$$

$$\left.\frac{dy}{dt}\right|_{t=0} = -K_1 - 4K_2 = 7$$

Collecting the conditions on K_1 and K_2,

$$\begin{cases} K_1 + K_2 = -1 \\ -K_1 - 4K_2 = 7 \end{cases}$$

$$K_1 = \frac{\begin{vmatrix} -1 & 1 \\ 7 & -4 \end{vmatrix}}{\begin{vmatrix} 1 & 1 \\ -1 & -4 \end{vmatrix}} = \frac{-3}{-3} = 1$$

$$K_2 = \frac{\begin{vmatrix} 1 & -1 \\ -1 & 1 \end{vmatrix}}{-3} = \frac{6}{-3} = -2$$

So of all the possible solutions of the form

$$y(t) = K_1 e^{-t} + K_2 e^{-4t} + 3$$

the specific solution

$$y(t) = e^{-t} - 2e^{-4t} + 3$$

is the one that applies in this situation.

D5-14

Find the specific solutions to the following differential equations with the indicated boundary conditions:

(a) $\dfrac{dy}{dt} + 3y = 6$

    $y(0) = 0$ *ans.* $2 - 2e^{-3t}$

(b) $\dfrac{dy}{dt} + y = 7$

    $y(0) = 5$ *ans.* $7 - 2e^{-t}$

(c) $2\dfrac{dy}{dt} + 5y = 12$

    $y(0) = -2$ *ans.* $(12/5) - (22/5)e^{-(5/2)t}$

(d) $4\dfrac{dy}{dt} + 3y = 12$

    $y(1) = 0$ *ans.* $4 - [3e^{(3/4)}]e^{-(3/4)t}$

(e) $2\dfrac{dy}{dt} + 5y = 0$

    $\left.\dfrac{dy}{dt}\right|_{t=0} = 3$ *ans.* $(-6/5)e^{-(5/2)t}$

(f) $\dfrac{dy}{dt} + 6y = 10$

    $\left.\dfrac{dy}{dt}\right|_{t=0} = 0$ *ans.* $5/3$

(g) $\dfrac{d^2y}{dt^2} + 3\dfrac{dy}{dt} + 2y = 10$

$y(0) = 5$

$\dfrac{dy}{dt}\bigg|_{t=0} = 0$ *ans.* 5

(h) $\dfrac{d^2y}{dt^2} + \dfrac{dy}{dt} = 4$

$y(0) = 0$

$\dfrac{dy}{dt}\bigg|_{t=0} = 2$ *ans.* $4t - 2 + 2e^{-t}$

Boundary Conditions

One specific solution to a differential equation, of all the possibilities given by the general solution, may be selected through the application of boundary conditions.

5.10   The Exponential Function

An exponential function is a function of the form

$f(t) = Ke^{st}$

where K and s are constants. The constant K is called the *amplitude* of the exponential function, and the constant s is called the *exponential constant* of the function.

If the exponential constant is positive (and a real number), the exponential function is said to be expanding since e^{st} gets larger and larger with time, t.

Figure 5-29 Expanding and decaying exponential functions

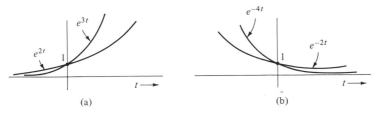

(a)   (b)

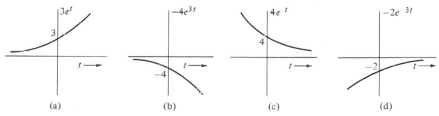

Figure 5-30   Examples of exponential functions with various amplitudes

The larger the value of s, the faster e^{st} "blows up." Of the two expanding exponential functions in Figure 5-29(a), e^{3t} expands more quickly than e^{2t}.

If the exponential constant is negative (and a real number), the exponential function is said to be decaying, since e^{st} approaches zero with time, t. The larger the negative value of s, the faster e^{st} "dies out." Examples of rapid and not-so-rapidly decaying exponential functions are given in Figure 5-29(b).

The effect of the amplitude K is illustrated in the examples of Figure 5-30.

The *time constant* of an exponential function (with a real value of s) is the inverse of the magnitude of the exponential constant. For

$$f(t) = 7e^{-2t}$$

the time constant is

$$\tau = \tfrac{1}{2}$$

In one time constant, a decaying exponential function decays by a factor of $(1/e)$, or to about 37% of its initial value.

The above function has value 7 at time $t = 0$. In a time of one time constant, it has value $7/e$. After another time-constant time interval has passed, the value of the function has decayed by another factor of $1/e$, or to $7/e^2$, as illustrated in Figure 5-31.

For

$$f(t) = 8e^{3t}$$

the time constant is

$$\tau = \tfrac{1}{3}$$

Figure 5-31   Time constant of a decaying exponential function

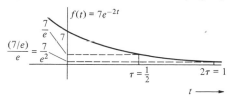

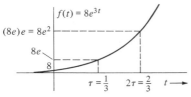

Figure 5-32   Time constant of an expanding exponential function

An expanding exponential expands by a factor of e in each time interval of one time constant, as illustrated for this function in Figure 5-32.

D5-15

Sketch the following exponential functions. Find the amplitude and time constant of each.

(a) $f(t) = -e^{3t}$ 　　　　　　　　　　　　*ans.* $-1, \frac{1}{3}$

(b) $f(t) = 7e^{-5t}$ 　　　　　　　　　　　　*ans.* $7, \frac{1}{5}$

(c) $f(t) = -10e^{-(4/3)t}$ 　　　　　　　　*ans.* $-10, \frac{3}{4}$

(d) $f(t) = 10^4 e^{-10^6 t}$ 　　　　　　　　*ans.* $10^4, 10^{-6}$

D5-16

Find the exponential functions (approximately) from the plots:

(a)

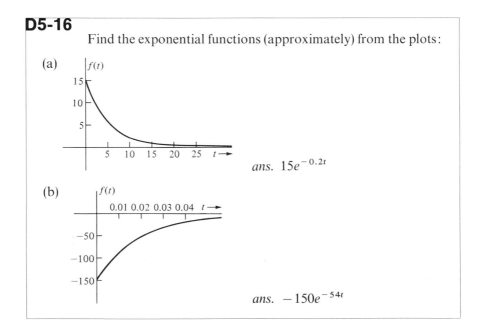

ans. $15e^{-0.2t}$

(b)

ans. $-150e^{-54t}$

The Exponential Function

Exponential functions have the form

$$f(t) = Ke^{st}$$

where the amplitude K and the exponential constant s are constants. For s a real number, the time constant of an exponential function is

$$\tau = \frac{1}{|s|}$$

It is the time interval over which a decaying exponential decays by a factor of $e^{-1} = 1/e$ and over which an expanding exponential function grows by a factor of $e^1 = e$.

Chapter Five Problems

Basic Problems

Capacitor Relations

1. Find $i(t)$, $p_{\text{into } C}(t)$, and $W_C(t)$:

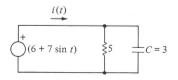

2. Find $v(t)$, $p_{\text{into } C}(t)$, and $W_C(t)$:

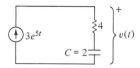

3. Find the equivalent capacitance:

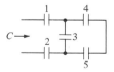

Inductor Relations

4.   Find $v(t)$, $p_{\text{into }L}(t)$, and $W_L(t)$:

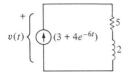

5.   Find $i(t)$, $p_{\text{into }L}(t)$, and $W_L(t)$:

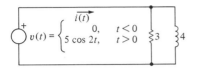

6.   Find the equivalent inductance:

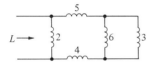

Mutually Coupled Inductor Relations

7.   Find the voltage-current relations:

(a)

(b)

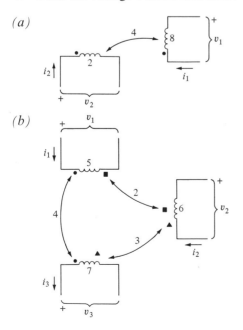

Controlled Source Equivalents for Inductive Coupling

8. Find the equivalent controlled voltage-source model for the following set of coupled inductors. Label the A, B, C, and D terminals of the equivalents and identify the controlling signals:

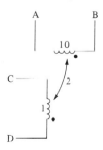

Systematic Simultaneous Equations

9. Write (but do not attempt to solve) systematic integrodifferential mesh equations for the following networks, in terms of the indicated currents:

(a)

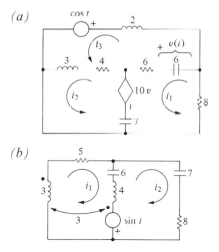

(b)

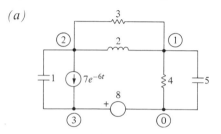

10. Write (but do not attempt to solve) systematic integrodifferential nodal equations for the following networks, in terms of the indicated variables:

(a)

(b)

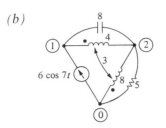

Differential Equations

11.   Find the general solutions to the following differential equations:

(a) $3\dfrac{dy}{dt} + y = 0$

(b) $4\dfrac{dy}{dt} + 3y = 5$

(c) $\dfrac{d^2y}{dt^2} + 6\dfrac{dy}{dt} = 0$

(d) $3\dfrac{d^2y}{dt^2} + 7\dfrac{dy}{dt} + 4y = -2$

12.   Find the specific solutions to the following differential equations with the indicated boundary conditions:

(a) $4\dfrac{dy}{dt} + 2y = 3$

   $y(0) = 5$

(b) $\dfrac{dy}{dt} + 4y = -2$

   $y(0) = -3$

(c) $\dfrac{d^2y}{dt^2} + 7\dfrac{dy}{dt} + 6y = 10$

   $y(0) = 5$

   $\dfrac{dy}{dt}\bigg|_{t=0} = 0$

(d) $3\dfrac{d^2y}{dt^2} + 8\dfrac{dy}{dt} + 2y = 6$

   $y(0) = 4$

   $\dfrac{dy}{dt}\bigg|_{t=0} = -3$

13. The following differential equations have complex characteristic roots. Find their general solutions:

(a) $\dfrac{d^2y}{dt^2} + 4\dfrac{dy}{dt} + 29y = -6$

(b) $\dfrac{d^3y}{dt^3} + 4\dfrac{d^2y}{dt^2} + 20\dfrac{dy}{dt} = 0$

Exponential Functions

14. Sketch the following exponential functions:
(a) $100e^{-10t}$
(b) $-0.01e^{0.02t}$
(c) $-10^6e^{-10^8t}$
(d) $8e^{-10^{-5}t}$

Practical Problems

Capacitors

1. The current through a certain $\frac{1}{2}$-F capacitor is graphed for a 7-s interval of time, in Figure 5-33. Draw graphs of three of the possible capacitor voltages during this time interval.

2. Physical capacitors will withstand only limited voltages and currents. The voltage limit is that beyond which the dielectric material between the capacitor plates will "break down" and become a conductor. Manufacturers generally rate capacitors in terms of a guaranteed maximum constant (or dc *working voltage*.

Capacitors with air, vacuum, oil, and similar dielectrics have the advantage of recovery of their properties if excessive voltage has caused conduction through the dielectric. Dielectric breakdown in other materials such as paper and plastic generally destroys the capacitor.

Exceeding the current limit in a physical capacitor results in excessive heating, due to small resistances, of the plates and the dielectric. The maximum safe current for a capacitor is usually only of concern in applications

Figure 5-33   Practical Problem 1

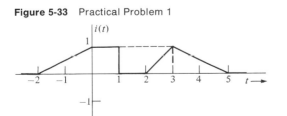

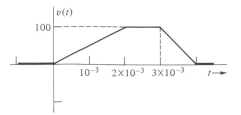

Figure 5-34 Practical Problem 2

such as power distribution and radio transmission, where relatively large power flows are involved.

In an *electrolytic* capacitor a large capacitance is formed by depositing an extremely thin layer of insulating material on one plate of an electrolytic cell. It is important that the voltage applied to such a capacitor (and other *polarized* types of capacitors) be only of a single, specified polarity. The wrong polarity of applied voltage will cause destruction of the deposited insulating material.

If a 10-μF capacitor has the applied voltage sketched in Figure 5-34, what is the largest capacitor current? Could an electrolytic capacitor be used in this situation?

3. The leakage of current through the dielectric material between the plates of a physical capacitor may be modeled by the capacitance in parallel with a leakage resistance, as indicated in Figure 5-35.

Capacitors designed for small leakage can have leakage resistances that cause a charged-capacitor voltage to decay only a few percent in a year's time. For such devices, the current leakage through contaminants on the surface of their package may be much greater than the leakage through the dielectric. More typical low-leakage capacitors have voltage decay rates of perhaps a few percent per minute.

In many applications resistors are deliberately connected across capacitors to increase the rate of decay of their voltages when equipment power is turned off, insuring that dangerous amounts of charge do not remain long on the plates. Capacitors present a danger if charged to lethal voltages and

Figure 5-35 Practical Problem 3

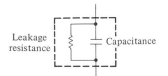

Model of physical capacitor

Small power transformers of the type used in power supplies for electronic instruments. These consist of coupled inductors and are used to develop several voltages larger or smaller than the power line voltage. *(Photo courtesy of Triad-Utrad.)*

by their capability of delivering extremely high currents if accidently short circuited. When large amounts of energy are stored, a capacitor's short-circuit current may be capable of vaporizing wires, screwdrivers, and the like.

Suppose a 0.1-μF capacitor has a 150-kΩ leakage resistance. For the capacitor voltage for Problem 2, carefully sketch the current through this physical capacitor.

Inductors

4. The current through a certain 3-H inductor is sketched in Figure 5-36. Draw a sketch of the inductor voltage for the same time interval.

5. Physical inductors are composed of a coil of wire. At normal temperatures, the resistance of that wire, called the *winding resistance*, may be

Figure 5-36 Practical Problem 4

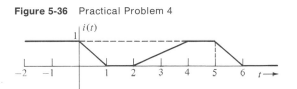

Model of physical inductor

Figure 5-37 Practical Problem 5

significant. A model for a physical inductor that takes its winding resistance into account is that of Figure 5-37.

Suppose the 3-H inductor of Problem 4 has a 10-Ω winding resistance. For the given current of Problem 4, carefully sketch the voltage across this inductor.

6. The inductance of a coil of wire may be greatly increased by channeling the magnetic flux through a core of ferromagnetic material. Most of the

Digital voltmeter is used to align a mobile amateur radio transceiver. (*Photo courtesy of Hewlett-Packard, Inc.*)

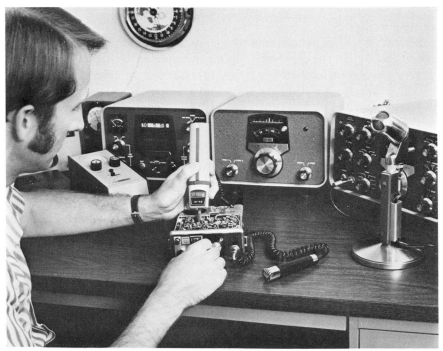

magnetic flux due to any loop of the coil may then be made to pass through all of the other loops.

Unfortunately, ferromagnetic materials have a nonlinear relation between current and flux, exhibiting permanent magnetism and saturation effects. Thus air gaps are commonly built into inductor cores to improve the linearity of the current–magnetic flux relation.

"Soft" ferromagnetic core material such as iron may respond rather sluggishly to a changed current with a changed flux, making the inductor effective only for relatively slowly varying currents.

To reduce the effects of induced currents in the core material itself, an iron inductor core may be *laminated*—that is, cut into thin strips that are electrically insulated from one another and placed side by side; or a conductive ferromagnetic material may be finely powdered and pressed or glued so that the conductive particles are insulated from one another.

Suppose that for a certain coil and core, the relation between coil current i and core flux Φ is

$$\Phi = 20i + i^3$$

Neglecting any resistance of the coil, what is the voltage-current relation of this nonlinear inductor?

7. The current limitation for a physical inductor is determined by the maximum heating of the conductor and core that may be sustained. An inductor's maximum voltage is the voltage at which breakdown of the insulation occurs. Safe voltage and current limits are termed "working voltages" and "working currents."

The current $i(t)$ through a 100-mH inductor with a 16-Ω winding resistance R_W is sketched in Figure 5-38.

(a) What is the maximum inductor voltage?

(b) What is the maximum amount of energy stored in the inductor? At what time is the maximum energy stored?

(c) Approximating the power loss in the inductor by

$$p_{loss}(t) = i^2(t)R_W$$

Figure 5-38   Practical.Problem 7

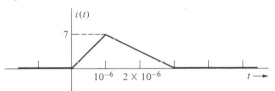

what is the total energy dissipated in the inductor in the time interval shown?

Exponential Functions

8. Radioactive decay is described by a decaying exponential function, and the half-life of a radioactive material is the time interval over which the decay is to one-half its initial value. Relate the half-life of an exponential to its time constant.

9. When exponential data are available only for a span of time that is small compared to a time constant, it is helpful to be able to identify the exponential function from its value and slope at a given time. Carefully describe how this may be done for the time $t = 0$. Repeat, assuming that the value and slope at some time $t = a$ are known, instead.

10. It is commonly asserted that for all practical purposes an exponential function has decayed to zero in a time span of five time constants. By what factor has such a function actually decayed in that time?

Automatic testing of integrated circuit assemblies. Advanced systems such as this one are capable of making thousands of measurements every second. (*Photo courtesy of GenRad, Inc.*)

Adanced Problems

Inductor and Capacitor Voltage-Current Relations

1.   *(a)*   Develop voltage-divider and current-divider rules for inductors.
     (b)   Develop voltage-divider and current-divider rules for capacitors.

Power and Stored Energy

2.   Show that the electrical power flow into a series connection of capacitors at any instant of time is equal to the power flow into the equivalent capacitor.

It is generally true that the power flow into any two-terminal combination of capacitors (or a two-terminal combination of inductors) is the same as the power flow into the equivalent capacitor (or inductor).

3.   Suppose a certain electrical element stores energy according to

$$W(t) = v^4(t)$$

Using $p_{into}(t) = dW/dt$, find the sink reference voltage-current relation for the element.

Mutual Inductance

4.   Show that the sink reference currents in terms of the voltages for two mutually coupled inductors are

$$
\begin{cases}
i_1 = \dfrac{L_2}{L_1 L_2 - M^2} \displaystyle\int_{\infty}^{t} v_1 \, dt \mp \dfrac{M}{L_1 L_2 - M^2} \displaystyle\int_{-\infty}^{t} v_2 \, dt \\[4mm]
i_2 = \mp \dfrac{M}{L_1 L_2 - M^2} \displaystyle\int_{-\infty}^{t} v_1 \, dt + \dfrac{L_1}{L_1 L_2 - M^2} \displaystyle\int_{-\infty}^{t} v_2 \, dt
\end{cases}
$$

5.   Find the equivalent inductances of the two series-connected coupled inductor arrangements shown in Figure 5-39.

Any two-terminal combination of inductors, including those that are mutually coupled, is equivalent to a single inductor.

6.   Find the equivalent inductances of the two parallel-connected coupled inductors shown in Figure 5-40.

Figure 5-39   Advanced Problem 5

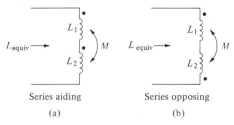

Series aiding	Series opposing
(a)	(b)

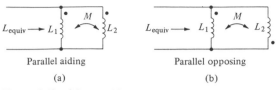

M
L_{equiv}
L_1
L_2
Parallel aiding
(a)
M
L_{equiv}
L_1
L_2
Parallel opposing
(b)

Figure 5-40 Advanced Problem 6

Simultaneous Equations

7. Find the equivalent capacitance of the two-terminal combination of capacitors of Figure 5-41.

Differential Equations

8. One method of differential equation solution is to find an *integrating factor.* For a first-order equation,

$$a_1 \frac{dy}{dt} + a_0 y = f(t)$$

multiplying by the integrating factor

$$p(t) = e^{(a_0/a_1)t}$$

gives

$$a_1 e^{(a_0/a_1)t} \frac{dy}{dt} + a_0 e^{(a_0/a_1)t} y = f(t)$$

$$a_1 \frac{d}{dt} (e^{(a_0/a_1)t} y) = f(t)$$

$$\frac{d}{dt} (e^{(a_0/a_1)t} y) = \frac{f(t)}{a_1}$$

which may be solved by integration.

Find the general solution to

$$-2 \frac{dy}{dt} + 3y = e^{-4t}$$

using the integrating factor method.

Figure 5-41 Advanced Problem 7

1
3
2
5
4

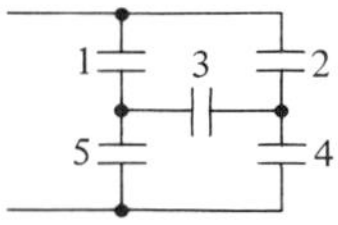

Unfortunately, finding an integrating factor for a second-order equation involves solving another second-order differential equation, and the situation is similar for equations of higher order.

9. Show that if

$$y(t) = K_1 e^{(\alpha + j\beta)t} + K_2 e^{(\alpha - j\beta)t}$$

is the general solution of a second-order differential equation, for arbitrary K_1 and K_2, then

$$y(t) = e^{\alpha t}(A \cos \beta t + B \sin \beta t)$$

is also the general solution, for arbitrary A and B.

10.   Find a differential equation for which

$$y(t) = K_1 e^{4t} + K_2 e^{-2t} + K_3 t e^{-2t} + 6$$

is the general solution, where K_1, K_2, and K_3 are arbitrary constants.

CHAPTER SIX
Switched Networks

6.1 Introduction

The complete solution of first-order networks, those containing a single energy-storing element, is now developed in detail.

The concluding sections of this chapter cover the important series and parallel second-order networks, together with discussion of the various types of second-order responses.

When you complete this chapter, you should know

1. How to solve first-order switched inductive networks and to graph any signal in the network.
2. What the unit step function is and how it may be used to describe source functions in switched networks.
3. How to solve first-order switched capacitive networks and to graph any signal in the network.
4. How to solve series and the parallel switched RLC networks and to graph their signals.

6.2   Switched First-Order Inductive Networks

6.2.1   Differential Equation for the Inductor Current

A first-order network is a network that contains sources, resistors, and one energy-storing element, either a capacitor or an inductor. If the energy-storing element is an inductor, the network is said to be an inductive first-order network.

So far as the inductor is concerned, the rest of a first-order inductive network consists of sources and resistors, and so may be represented by a Thévenin equivalent, as in the diagram of Figure 6-1. In terms of the Thévenin voltage and current, the mesh equation for the inductor current $i(t)$ is

$$L\frac{di}{dt} + R_1 i = v_T(t)$$

6.2.2   Equation Solution and Forced Constant Response

The solution of the differential equation for the inductor current in a first-order inductive network consists of the natural component plus the forced component.

The natural component of $i(t)$ is the solution to the homogeneous equation

$$L\frac{di_n}{dt} + R_1 i_n - 0$$

Substituting

$$i_n = Ke^{st}$$

$$sL + R_T = 0$$

$$s = -\frac{R_T}{L}$$

$$i_n(t) = Ke^{-(R_T/L)t}$$

For the usual case of positive R_T and L, this part of the solution dies out in time, with time constant L/R_T.

Figure 6-1   Thévenin equivalent used for first-order inductive network

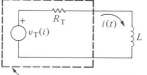

Thévenin equivalent of the rest of the network

This general result applies to any first-order inductive network: The natural component of the inductor current is as given above, where R_T is the Thévenin resistance of the rest of the network that is connected to the inductor.

The forced component of the inductor current depends on the driving function $v_T(t)$. If all of the network sources are constant, v_T will be constant, and finding the forced component of the solution is particularly easy. For a constant v_T,

$$i_f = A$$

is constant. Substituting into the differential equation,

$$L\frac{dA}{dt} + R_T A = v_T$$

$$R_T A = v_T$$

$$i_f = A = \frac{v_T}{R_T}$$

So far as the *forced* response to *constant* sources is concerned, the inductor voltage,

$$L\frac{di_f}{dt} = L\frac{dA}{dt} = 0$$

is zero; that is, the inductor behaves as a short circuit. To find the forced component of the inductor current due to *constant* sources, replace the inductor by a short circuit and solve for the corresponding current.

Consider the network of Figure 6-2(a). The natural component of the inductor current is given by

$$i_n(t) = Ke^{-(R_T/L)t} = Ke^{-(6/35)t}$$

where the Thévenin resistance is found by setting the sources to zero, as in Figure 6-2(b).

Since the network involves only constant sources, the forced component of the inductor current may be found by replacing the inductor by a short circuit, which is done in Figure 6-2(c).

Solving,

$$\begin{cases} 5i_1 - 2i_2 = -10 \\ -2i_1 + 2i_2 = 3 \end{cases}$$

$$i_f = i_2 = \frac{\begin{vmatrix} 5 & -10 \\ -2 & 3 \end{vmatrix}}{\begin{vmatrix} 5 & -2 \\ -2 & 2 \end{vmatrix}} = \frac{-5}{6}$$

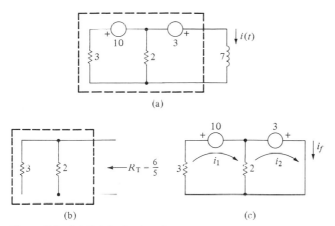

(a)

(b)   (c)

Figure 6-2   (a) Original network
(b) Finding Thévenin resistance of sources and resistors connected to inductor
(c) Finding forced behavior due to constant sources by replacing inductor
with a short circuit

The general solution for the inductor current is then

$$i(t) = i_n(t) + i_f$$
$$= Ke^{(6/35)t} - 5/6$$

6.2.3   Switched Networks and Continuity of Inductor Current

Changes in a network may be represented with switches, Figure 6-3, beside
which are indicated the times of openings and closings. In most cases the
origin of the time scale is best chosen so that a switching is made at time
$t = 0$ for numerical convenience.

Switched networks with a single switching are a type of problem in which
the forced network response before the switching is sought, and both forced
and natural components of the response are desired after the switching. It
is implied that the network has been connected a long time before the
switching and that any natural response arising from the original connection
has long since decayed, leaving the forced response.

The network, then, is changed by the switching. The response changes to
a new forced component and a new natural component, governed by the
equations for the new network after the switching.

Figure 6-3   (a) Connection made at time $t = 0$
(b) Connection opened at time $t = 5$
(c) Connection changed from one conductor to the other at time $t = 0$

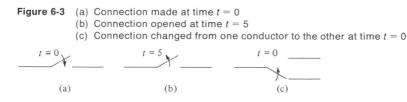

(a)   (b)   (c)

Only one of all the possible solutions given by the general solution after the switching applies in a particular situation. It is the one for which the inductor current is continuous. If the inductor current were discontinuous (if it "jumped" from one value to another), the inductor voltage,

$$v = L\frac{di}{dt}$$

would be infinite. Just the right amount of the natural component of inductor current is present after the switching so that the inductor current is continuous.

Consider the network in Figure 6-4(a). Since no information is given as to when or in what manner the network was connected before $t = 0$, it is implied that it has been connected, with the switch open, for a long time prior to $t = 0$. If there was any natural response when the network was first wired up, long ago, it has died out and is not of interest. So up to time $t = 0$, only the forced parts of the network signals are to be found.

Because the network source is disconnected from the network before $t = 0$, the Thévenin voltage and thus the forced component of $i(t)$ are zero before $t = 0$, as shown in Figure 6-4(b).

At time $t = 0$, the network is suddenly changed as the switch is closed. Now both the forced and the natural components of the inductor current are of interest. Because the source is constant, the forced component of the current may be found, as it was before $t = 0$, by replacing the inductor by a short circuit. The network is different now that the switch is closed, so i_f will differ from what it was before time $t = 0$. As indicated in Figure 6-4(c),

$$i_f = \frac{5}{2}$$

after $t = 0$.

Figure 6-4   (a) Complete problem
(b) Forced inductor current before $t = 0$
(c) Forced inductor current after $t = 0$
(d) Thévenin resistance after $t = 0$

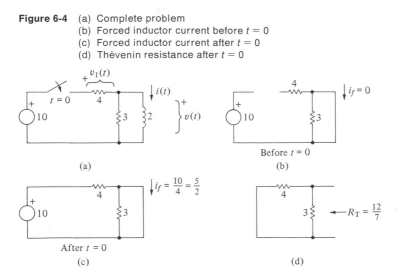

The natural component of $i(t)$ after $t - 0$ is

$$i_n(t) = Ke^{-(R_T/L)t} = Ke^{-(6/7)t}$$

R_T is the Thévenin resistance of the source-resistor network connected to the inductor after $t = 0$, after the switch has closed, Figure 6-4(d). The general solution for the inductor current after $t - 0$ is thus

$$i(t) = i_n(t) + i_f$$

$$= Ke^{-(6/7)t} + \frac{5}{2}$$

Of all the possible solutions, for each possible value of K in the general solution, only one is the solution to this specific problem. It is the solution for which the inductor current is continuous through the switching time, $t = 0$. Before $t = 0$, the inductor current was $i_f = 0$. Just after $t = 0$, when the general solution above applies, the current must be zero also:

$$i(0) = Ke^0 + \frac{5}{2} = K + \frac{5}{2} = 0$$

$$K - -\frac{5}{2}$$

So the solution for the inductor current in this problem is

$$i(t) = \begin{cases} 0, & t \le 0 \\ \dfrac{5}{2} - \dfrac{5}{2}e^{-(6/7)t}, & t \ge 0 \end{cases}$$

A sketch of $i(t)$ is shown in Figure 6-5. The inductor current changes from one forced value ($i_f = 0$) to another ($i_f = \frac{5}{2}$). It cannot change instantaneously from one value to the other, however, because the inductor current must be continuous. The current changes from one forced value to the other along the characteristic exponential curve for the network. Just the right amount of natural current added to the forced current after $t = 0$ makes the inductor current continuous.

6.2.4 Finding Other Network Signals

Once the inductor current is found, any other voltage or current in the network may be easily found. Suppose, in the previous network, it is also desired to find $v(t)$ and $v_1(t)$ in Figure 6-4(a).

Figure 6-5 Sketch of the inductor current

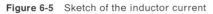
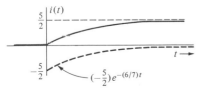

The voltage $v(t)$ is

$$v(t) = 2\frac{di(t)}{dt}$$

$$= \frac{d}{dt}\begin{cases} 0, & t \leq 0 \\ \dfrac{5}{2} - \dfrac{5}{2}e^{-(6/7)t}, & t \geq 0 \end{cases}$$

$$= \begin{cases} 0, & t < 0 \\ -\left(\dfrac{6}{7}\right)\left(-\dfrac{5}{2}\right)e^{-(6/7)t}, & t > 0 \end{cases}$$

$$= \begin{cases} 0, & t < 0 \\ \dfrac{15}{7}e^{-(6/7)t}, & t > 0 \end{cases}$$

A sketch of $v(t)$ is shown in Figure 6-6(a). The inductor current must be continuous, but other network voltages and currents, such as this one, need not be continuous.

Before $t = 0$, $v_1 = 0$ since, with the switch open, there is no current through the 4-Ω resistor. After $t = 0$,

$$v_1(t) = 10 - v(t) = 10 - \frac{15}{7}e^{-(6/7)t}$$

as sketched in Figure 6-6(b).

An alternative solution procedure to using voltage-current relations, as above, is to use the substitution theorem (Section 2.5): Replace the inductor with a current source with source function the known inductor current. All network voltages and currents are unchanged, and the resulting source-resistor network may be solved for any other network voltage or current.

6.2.5 Systematic Solutions

Consider the switched first-order inductive network of Figure 6-7. A step-by-step, systematic solution for the indicated signals v_1 and v_2 follows.

Figure 6-6 Other switched network signals

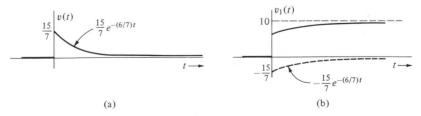

(a)   (b)

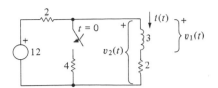

Figure 6-7   Original network

1. Before $t = 0$, find the forced inductor current. Since the network source is constant, the forced inductor current will be constant and may be found by replacing the inductor by a short circuit, Figure 6-8:

$$i_f = 3, \qquad t < 0$$

2. After $t = 0$, find the forced inductor current. The same procedure is used to find the forced inductor current after $t = 0$, Figure 6-9:

$$i_f = \frac{12}{5}, \qquad t > 0$$

Figure 6-8   Forced inductor current before $t = 0$

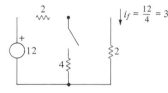

Figure 6-9   Forced inductor current after $t = 0$

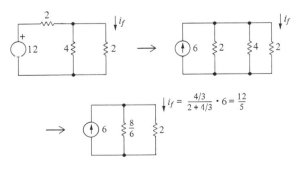

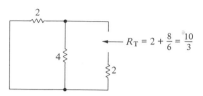

$$R_T = 2 + \frac{8}{6} = \frac{10}{3}$$

Figure 6-10 Thévenin resistance after $t = 0$

3. After $t = 0$, find the Thévenin resistance looking back from the inductor terminals. See Figure 6-10:

$$R_T = \frac{10}{3}$$

4. Form the general solution after $t = 0$ and apply the boundary condition of continuous inductor current. After $t = 0$, the general solution for the inductor current is

$$i(t) = \frac{12}{5} + Ke^{-(R_T/L)t}$$

$$= \frac{12}{5} + Ke^{-(10/9)t}$$

Before $t = 0$, $i = 3$ so that at time $t = 0$,

$$i(0) = \frac{12}{5} + K = 3$$

$$K = \frac{3}{5}$$

and

$$i(t) = \begin{cases} 3, & t \le 0 \\ \dfrac{12}{5} + \dfrac{3}{5}e^{-(10/9)t}, & t \ge 0 \end{cases}$$

A sketch of $i(t)$ is given in Figure 6-11.

5. Using the inductor current, find any other signals of interest in the network.

$$v_1(t) = L\frac{di}{dt} = \begin{cases} 0, & t < 0 \\ 3\left(-\dfrac{10}{9}\right)\left(\dfrac{3}{5}\right)e^{-(10/9)t} = -2e^{-(10/9)t}, & t > 0 \end{cases}$$

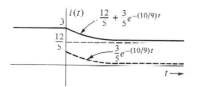

Figure 6-11 Sketch of inductor current

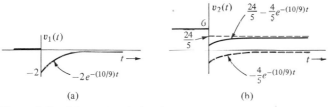

(a) (b)

Figure 6-12 Other network signals

This signal is sketched in Figure 6-12(a).

$$v_2(t) = 3\frac{di}{dt} + 2i$$

$$= \begin{cases} 6, & t < 0 \\ 2e^{-(10/9)t} + 2\left[\dfrac{12}{5} + \dfrac{3}{5}e^{-(10/9)t}\right] \end{cases}$$

$$= \frac{24}{5} - \frac{4}{5}e^{-(10/9)t}, \qquad t > 0$$

A sketch is shown in Figure 6-12(b).

D6-1

Find the Thévenin equivalents of the two-terminal networks connected to the inductor. Then write the single-loop differential equation in terms of the inductor current i.

(a)

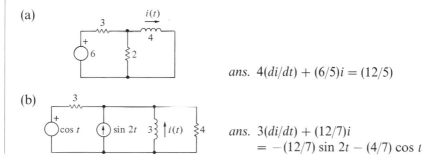

ans. $4(di/dt) + (6/5)i = (12/5)$

(b)

ans. $3(di/dt) + (12/7)i$
$\qquad = -(12/7)\sin 2t - (4/7)\cos t$

(c)

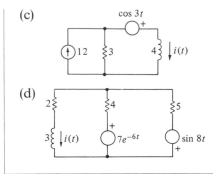

ans. $4(di/dt) + 3i = 36 + \cos 3t$

(d)

ans. $3(di/dt) + (38/9)i$
$= (35/9)e^{-6t} - (4/9)\sin 8t$

D6-2

Find the inductor current both before and after time $t = 0$ and sketch it:

(a)

ans. $0, (5/3) - (5/3)e^{-(3/4)t}$

(b)

ans. $7, 7e^{-(12/35)t}$

(c)

ans. $8/9, (8/5) - (32/45)e^{-(5/6)t}$

(d)

ans. $0, 5 - 5e^{-(3/2)t}$

D6-3

Find the indicated signals before and after time $t = 0$. Sketch both the inductor current and the signal of interest:

(a)

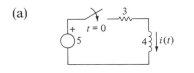

ans. $0, 6e^{-(2/5)t}$

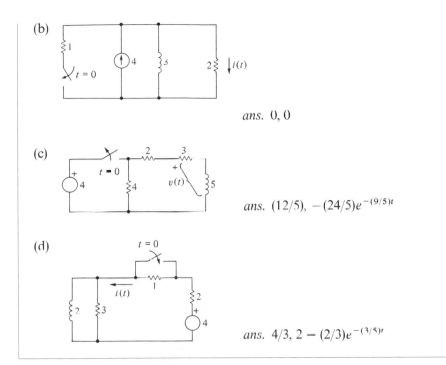

(b)

ans. $0, 0$

(c)

ans. $(12/5), -(24/5)e^{-(9/5)t}$

(d)

ans. $4/3, 2 - (2/3)e^{-(3/5)t}$

Switched Networks Containing a Single Inductor

The natural component of the inductor current in a first-order network is of the form

$$i_n(t) = Ke^{-(R_T/L)t}$$

where R_T is the Thévenin resistance of all of the source-resistor network connected to the inductor.

The forced component of the inductor current due to *constant* sources may be found by replacing the inductor by a short circuit and solving for the corresponding current.

In switched networks with a single switching time, it is desired to find the forced network response prior to the switching and both forced and natural components of the response after the switching.

The specific solution for the inductor current after the switching may may be found from the general solution by applying the boundary condition of continuous inductor current.

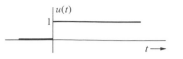

Figure 6-13 Graph of the unit step function

6.3 The Unit Step Function

6.3.1 The Function and Step Sources

The unit step function is defined as follows:

$$u(t) = \begin{cases} 0, & t < 0 \\ 1, & t > 0 \end{cases}$$

Its graph is shown in Figure 6-13. Since this function is discontinuous, one normally would not be interested in its value at $t = 0$, but if it were important, it could be defined.

The unit step function offers a convenient alternative to switches in switched network problems, particularly in complicated situations. A switched-on voltage source is easily represented with this notation, as is a switched-on current source, Figure 6-14.

6.3.2 Inductive Network Example

The following is a step-by-step example of the solution of a network where the switching is described by the step function, Figure 6-15(a).

1. Before $t = 0$, find the forced inductor current. The current source function is zero prior to $t = 0$, Figure 6-15(b), giving $i_f = 0$.
2. After $t = 0$, find the forced inductor current. Since the network source is constant, the forced inductor current may be found by replacing

Figure 6-14 (a) Switched voltage source
(b) Switched current source

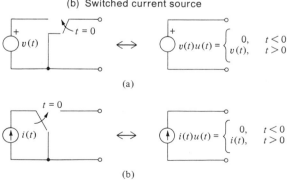

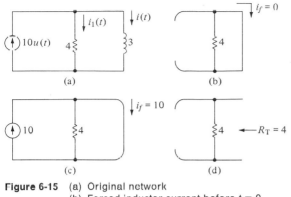

Figure 6-15   (a) Original network
                (b) Forced inductor current before $t = 0$
                (c) Forced inductor current after $t = 0$
                (d) Thévenin resistance after $t = 0$

the inductor by a short circuit, as in Figure 6-15(c), where it is seen that $i_f = 10$.

3. After $t = 0$, find the Thévenin resistance looking back from the inductor terminals. From Figure 6-15(d), $R_T = 4$.

4. Form the general solution after $t = 0$ and apply the boundary condition of continuous inductor current.

$$i(t) = 10 + Ke^{-(4/3)t}, \qquad t > 0$$
$$i(0) = 10 + K = 0$$
$$K = -10$$

so

$$i(t) = \begin{cases} 0, & t \leq 0 \\ 10 - 10e^{-(4/3)t}, & t \geq 0 \end{cases}$$

A sketch of $i(t)$ is given in Figure 6-16.

5. Using the inductor current, find any other signals of interest in the network. Before $t = 0$, $i_1 = 0$. After $t = 0$, the network is as in Figure 6-17, and

$$i_1(t) = 10 - i(t) = 10e^{-(4/3)t}$$

This current is sketched in Figure 6-18.

Figure 6-16   Sketch of the inductor current

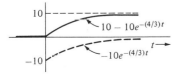

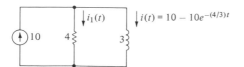

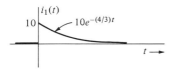

Figure 6-17 Network after $t = 0$

Figure 6-18 Another network signal

D6-4

Find and sketch the indicated signals before and after time $t = 0$. For comparison, sketch the inductor current also:

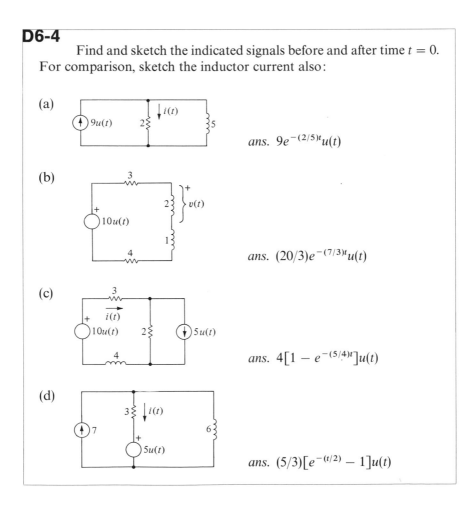

(a)

ans. $9e^{-(2/5)t}u(t)$

(b)

ans. $(20/3)e^{-(7/3)t}u(t)$

(c)

ans. $4[1 - e^{-(5/4)t}]u(t)$

(d)

ans. $(5/3)[e^{-(t/2)} - 1]u(t)$

The Unit Step Function

The unit step function is defined as

$$u(t) = \begin{cases} 0, & t < 0 \\ 1, & t > 0 \end{cases}$$

It is a convenient alternative to switches in describing switched networks.

6.4 Switched First-Order Capacitive Networks

6.4.1 Differential Equation for the Capacitor Voltage

If a first-order network contains a capacitor, the solution may be obtained in a manner similar to that for first-order inductive networks, except that the Norton equivalent should be used to give an equivalent two-node network and a differential equation in terms of the capacitor voltage.

For a network containing sources, resistors, and a single capacitor, construct the Norton equivalent of the source-resistor network connected to the capacitor, as indicated in Figure 6-19. The two networks are equivalents so far as the capacitor voltage is concerned, and that voltage satisfies the two-node differential equation

$$\left(C\frac{d}{dt} + \frac{1}{R_T} \right) v(t) = i_N(t)$$

Had a Thévenin equivalent been used to convert to an equivalent single-loop problem, the loop-current equation would have involved an integral. The equation should then be differentiated or a change of variables made to convert it to a differential equation. The problem is better formulated in terms of the capacitor voltage because a differential equation results without further manipulation and because the boundary condition applies directly to the capacitor voltage.

Figure 6-19 Norton equivalent used for first-order capacitive network

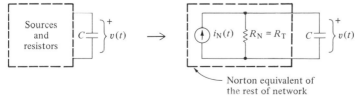

Norton equivalent of
the rest of network

6.4.2 Equation Solution and Forced Constant Response

The solution of the differential equation for the capacitor voltage in a first-order capacitive network consists of the natural component plus the forced component. The natural component of $v(t)$ is the solution to the homogeneous equation

$$C \frac{dv_n}{dt} + \frac{1}{R_T} v_n = 0$$

Substituting

$$v_n = K e^{st}$$

$$sC + \left(\frac{1}{R_T} \right) = 0$$

$$s = -\frac{1}{R_T C}$$

$$v_n(t) = K e^{-(1/R_T C)t}$$

If, as usual, R_T and C are positive, this part of the solution dies out in time, with time constant $R_T C$.

This is a general result that applies to any first-order capacitive network. The natural component of the capacitor voltage is as given above, where R_T is the Thévenin (or Norton) resistance of the rest of the network that is connected to the capacitor.

The forced component of the capacitor voltage depends on the driving function $i_N(t)$. If all of the network sources are constant, i_N will be a constant, and

$$v_f = A$$

is constant. Substituting into the differential equation,

$$C \frac{dA}{dt} + \frac{1}{R_T} A = i_N$$

$$\frac{A}{R_T} = i_N$$

$$v_f = A = i_N R_T$$

So far as the *forced* response to *constant* sources is concerned, the capacitor current,

$$C \frac{dv_f}{dt} = C \frac{dA}{dt} = 0$$

is zero. The capacitor behaves as an open circuit. To find the forced component of the capacitor voltage due to *constant* sources, replace the capacitor by an open circuit and solve for the corresponding voltage.

6.4.3   Switched Networks and Continuity of Capacitor Voltage

For a switched capacitive network, it is the capacitor voltage that must remain continuous since if it were discontinuous, the capacitor current,

$$i = C \frac{dv}{dt}$$

would be infinite.

6.4.4   Systematic Solutions

Consider the first-order switched capacitive network of Figure 6-20(a). A step-by-step, systematic solution follows.

1. Before $t = 0$, find the forced capacitor voltage. Since the network source is constant, the forced capacitor voltage will be constant and may be found by replacing the capacitor by an open circuit, Figure 6-20(b).

    $$v_f = \frac{15}{7}$$

    before $t = 0$.
2. After $t = 0$, find the forced capacitor voltage. As shown in Figure 6-20(c),

    $$v_f = 5$$

    after $t = 0$.

Figure 6-20   (a) Original network
(b) Forced capacitor voltage before $t = 0$
(c) Forced capacitor voltage after $t = 0$
(d) Thévenin resistance after $t = 0$

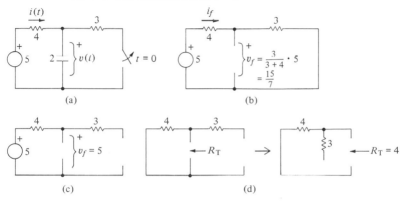

3. After $t = 0$, find the Thévenin resistance looking back from the capacitor terminals.

$$R_T = 4$$

as in Figure 6-20(d).

4. Form the general solution after $t = 0$ and apply the boundary condition of continuous capacitor voltage. After $t = 0$, the general solution for the capacitor voltage is

$$v(t) = 5 + Ke^{-(1/R_TC)t}$$
$$= 5 + Ke^{-(1/8)t}$$

Before $t = 0$, $v = 15/7$, so that at time $t = 0$,

$$v(0) = 5 + K = \frac{15}{7}$$

$$K = -\frac{20}{7}$$

and

$$v(t) = \begin{cases} \dfrac{15}{7}, & t \leq 0 \\ 5 - \left(\dfrac{20}{7}\right)e^{-(1/8)t}, & t \geq 0 \end{cases}$$

A sketch of $v(t)$ is shown in Figure 6-21.

5. Using the capacitor voltage, find any other signals of interest in the network. Before $t = 0$,

$$i_f = \frac{5}{7}$$

from Figure 6-20(b).

After $t = 0$, the network is as shown in Figure 6-22, and

$$i(t) = \frac{5 - 5 + (20/7)e^{-(1/8)t}}{4} = \left(\frac{5}{7}\right)e^{-(1/8)t}$$

A sketch of $i(t)$ is shown in Figure 6-23.

Figure 6-21   Sketch of the capacitor voltage

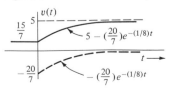

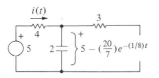

$$5 - (\tfrac{20}{7})e^{-(1/8)t}$$

Figure 6-22 Network after $t = 0$

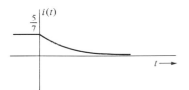

Figure 6-23 Another network signal

Consider the first-order switched capacitive network of Figure 6-24(a), which is described in terms of the unit step function. A systematic solution follows:

1. Before $t = 0$, find the forced capacitor voltage. This calculation is done in Figure 6-24(b), where it is seen that

$$v_f = 6$$

Figure 6-24 (a) Original network
            (b) Forced capacitor voltage before $t = 0$
            (c) Forced capacitor voltage after $t = 0$
            (d) Thévenin resistance after $t = 0$

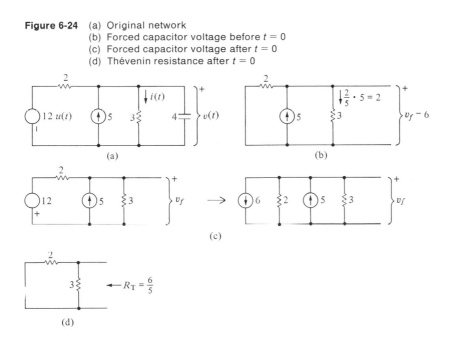

2. After $t = 0$, find the forced capacitor voltage. Using equivalent circuits, Figure 6-24(c),

$$(\tfrac{1}{2} + \tfrac{1}{3})v_f = 5 - 6$$

$$v_f = -\frac{6}{5}$$

3. After $t = 0$, find the Thévenin resistance looking back from the capacitor terminals. From Figure 6-24(d),

$$R_T = \frac{6}{5}$$

4. Form the general solution after $t = 0$ and apply the boundary condition of continuous capacitor voltage.

$$v(t) = -\frac{6}{5} + Ke^{-(1/R_T C)t}$$

$$= -\frac{6}{5} + Ke^{-(5/24)t}$$

$$v(0) = -\frac{6}{5} + K = 6$$

$$K = \frac{36}{5}$$

$$v(t) = \begin{cases} 6, & t \le 0 \\ \left(-\frac{6}{5}\right) + \left(\frac{36}{5}\right)e^{-(5/24)t}, & t \ge 0. \end{cases}$$

A sketch of the capacitor voltage is shown in Figure 6-25.
5. Using the capacitor voltage, find any other signals of interest in the network.

$$i(t) = \frac{v(t)}{3} = \begin{cases} 2, & t \le 0 \\ \left(-\frac{2}{5}\right) + \left(\frac{12}{5}\right)e^{-(5/24)t}, & t \ge 0 \end{cases}$$

Figure 6-25 Sketch of capacitor voltage

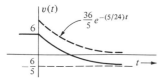

D6-5

Find the Norton equivalents of the two-terminal networks connected to the capacitor. Then write the two-node differential equation
in terms of the capacitor voltage $v(t)$.

(a)

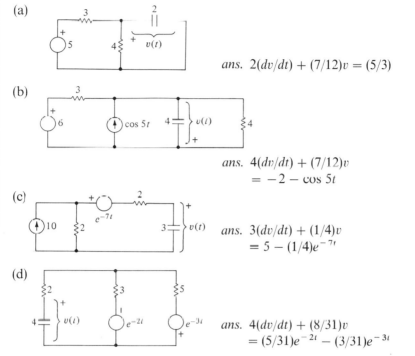

ans. $2(dv/dt) + (7/12)v = (5/3)$

(b)

ans. $4(dv/dt) + (7/12)v$
$= -2 - \cos 5t$

(c)

ans. $3(dv/dt) + (1/4)v$
$= 5 - (1/4)e^{-7t}$

(d)

ans. $4(dv/dt) + (8/31)v$
$= (5/31)e^{-2t} - (3/31)e^{-3t}$

D6-6

Find the capacitor voltage both before and after time $t - 0$ and
sketch it:

(a)

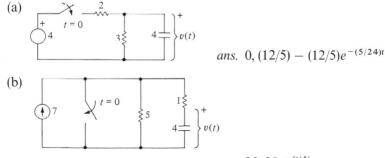

ans. $0, (12/5) - (12/5)e^{-(5/24)t}$

(b)

ans. $35, 35e^{-(t/4)}$

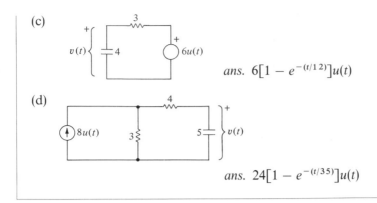

(c)

ans. $6[1 - e^{-(t/12)}]u(t)$

(d)

ans. $24[1 - e^{-(t/35)}]u(t)$

D6-7

Find and sketch the indicated signals before and after time $t = 0$. For comparison, sketch the capacitor current also:

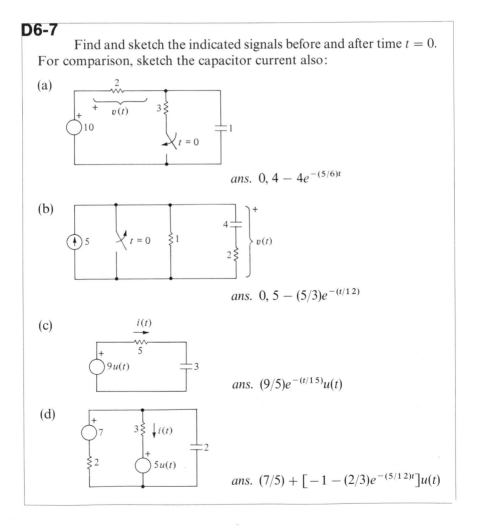

(a)

ans. $0, 4 - 4e^{-(5/6)t}$

(b)

ans. $0, 5 - (5/3)e^{-(t/12)}$

(c)

ans. $(9/5)e^{-(t/15)}u(t)$

(d)

ans. $(7/5) + [-1 - (2/3)e^{-(5/12)t}]u(t)$

Switched Networks Containing a Single Capacitor

The natural component of the capacitor voltage in a first-order network is of the form

$$v_n(t) = Ke^{-(1/R_TC)t}$$

where R_T is the Thévenin resistance of all of the source-resistor network connected to the capacitor.

The forced component of the capacitor voltage due to *constant* sources may be found by replacing the capacitor by an open circuit and solving for the corresponding voltage.

In switched networks, the specific solution for the capacitor voltage after the switching may be found from the general solution by applying the boundary condition of continuous capacitor voltage.

6.5   Series *RLC* Networks

6.5.1   Differential Equation for the Capacitor Voltage

The series *RLC* circuit is diagramed in Figure 6-26. It contains two energy-storing elements, the inductor and the capacitor, and so is described by a second-order integrodifferential equation.

The loop equation for $i(t)$ is

$$L\frac{di}{dt} + Ri + \frac{1}{C}\int_{-\infty}^{t} i\,dt = v(t)$$

If this equation is rewritten in terms of the capacitor voltage,

$$v_C(t) = \frac{1}{C}\int_{-\infty}^{t} i\,dt$$

a differential equation, without integrals, results; and the other network voltages and currents are easily expressed in terms of the capacitor voltage,

Figure 6-26   Series *RLC* network

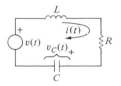

since

$$i(t) = C \frac{dv_C}{dt}$$

In terms of $v_C(t)$, the equation is

$$LC \frac{d^2 v_C}{dt^2} + RC \frac{dv_C}{dt} + v_C = v(t)$$

The boundary conditions for this network are that the capacitor voltage, v_C, and the inductor current,

$$i(t) = C \frac{dv_C}{dt}$$

thus dv_C/dt, must be continuous.

Consider the specific network of Figure 6-27. Before time $t = 0$, the source voltage is zero, and all voltages and currents are zero.

After $t = 0$, the network is governed by

$$4 \frac{di}{dt} + 3i + \frac{1}{2} \int_{-\infty}^{t} i \, dt = 5$$

Eliminating the running integral by dealing with the capacitor voltage,

$$8 \frac{d^2 v_C}{dt^2} + 6 \frac{dv_C}{dt} + v_C = 5$$

The solution of this network problem is thus formulated mathematically as the solution of the above differential equation with the boundary conditions

$$v_C(0) = 0$$

$$\frac{dv_C}{dt}\bigg|_{t=0} = 0$$

6.5.2   Solution with Real Characteristic Roots

For the example RLC network, Figure 6-27, the homogeneous equation is

$$8 \frac{d^2 v_{C_n}}{dt^2} + 6 \frac{dv_{C_n}}{dt} + v_{C_n} = 0$$

Figure 6-27   Example series RLC network

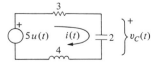

for which the characteristic equation is

$$8s^2 + 6s + 1 = 0$$

$$s_1, s_2 = \frac{-6 \pm \sqrt{36 - 32}}{16}$$

$$= -\tfrac{1}{4}, -\tfrac{1}{2}$$

The natural component of the capacitor voltage is therefore of the form

$$v_{C_n}(t) = K_1 e^{-(1/4)t} + K_2 e^{-(1/2)t}$$

Substituting a constant forced solution

$$v_{C_f} = A$$

gives

$$8\frac{d^2}{dt^2}(A) + 6\frac{d}{dt}(A) + (A) = 5$$

$$A = 5 = v_{C_f}$$

so that the general solution for the capacitor voltage is

$$v_C(t) = v_{C_f} + v_{C_n}(t)$$
$$= 5 + K_1 e^{-(1/4)t} + K_2 e^{-(1/2)t}$$

Applying the boundary conditions gives

$$v_C(0) = 5 + K_1 + K_2 = 0$$

and

$$\left.\frac{dv_C}{dt}\right|_{t=0} = -\frac{1}{4}K_1 - \frac{1}{2}K_2 = 0$$

which may be solved for K_1 and K_2:

$$\begin{cases} K_1 + K_2 = -5 \\ -\tfrac{1}{4}K_1 - \tfrac{1}{2}K_2 = 0 \end{cases}$$

$$K_1 = \frac{\begin{vmatrix} -5 & 1 \\ 0 & -\tfrac{1}{2} \end{vmatrix}}{\begin{vmatrix} 1 & 1 \\ -\tfrac{1}{4} & -\tfrac{1}{2} \end{vmatrix}} = \frac{5/2}{-1/4} = -10$$

$$K_2 = \frac{\begin{vmatrix} 1 & -5 \\ -\tfrac{1}{4} & 0 \end{vmatrix}}{-\tfrac{1}{4}} = 5$$

Hence the solution for the capacitor voltage after $t = 0$ in this network is

$$v_C(t) = 5 - 10e^{-(1/4)t} + 5e^{-(1/2)t}, \qquad t \geq 0$$

which is sketched in Figure 6-28(a)

Other network signals are easily found from $v_C(t)$. For example,

$$i(t) = C\frac{dv_C}{dt} = 2\frac{dv_C}{dt}$$

$$= 5e^{-(1/4)t} - 5e^{-(1/2)t}$$

after $t = 0$. This signal is sketched in Figure 6-28(b).

6.5.3   Complex Roots and Oscillatory Response

The natural response of second- and higher-order networks may be governed by characteristic equations with complex roots, as they are in the event that

$$\frac{4L}{C} > R^2$$

in the series RLC network, for which

$$s_1, s_2 = \frac{-R \pm \sqrt{R^2 - (4L/C)}}{2L}$$

$$= \alpha \pm j\beta$$

where

$$j = \sqrt{-1}$$

is the imaginary unit.

With complex characteristic roots, the natural response component is of the form

$$K_1 e^{s_1 t} + K_2 e^{s_2 t} = K_1 e^{(\alpha + j\beta)t} + K_2 e^{(\alpha - j\beta)t}$$
$$= e^{\alpha t}[K_1 e^{j\beta t} + K_2 e^{-j\beta t}]$$

where K_1 and K_2 are conjugate complex numbers. The alternate, ex-

Figure 6-28   Sketches of signals in the series RLC network

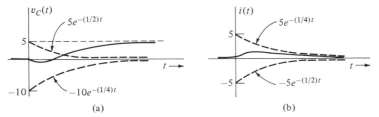

(a)   (b)

ponential times sinusoidal form

$$e^{\alpha t}[A \cos \beta t + B \sin \beta t]$$

where A and B are real numbers, is generally preferred because it is easier to deal with and to visualize.

Consider the example series *RLC* network of Figure 6-29(a). After time $t = 0$, the loop equation for $i(t)$ is

$$\frac{di}{dt} + 2i + 5 \int_{-\infty}^{t} i \, dt = 10$$

In terms of the capacitor voltage, $v_C(t)$, this equation is

$$\frac{1}{5}\frac{d^2 v_C}{dt^2} + \frac{2}{5}\frac{dv_C}{dt} + v_C = 10$$

Continuity of the capacitor voltage requires that

$$v_C(0) = 0$$

Before time $t = 0$, the network is as sketched in Figure 6-29(b). In view of the constant source, the forced inductor current before $t = 0$ may be found by replacing the inductor by a short circuit. It is

$$i = \frac{10}{2} = 5$$

Inductor current continuity then requires that the solution after $t = 0$ have

$$i(0) = 5$$

In terms of the capacitor voltage, this means that

$$i = C\frac{dv_C}{dt}$$

$$\frac{1}{5}\frac{dv_C}{dt}\Big|_{t=0} = 5$$

or

$$\frac{dv_C}{dt}\Big|_{t=0} = 25$$

Figure 6-29 (a) Original network
(b) Network before $t = 0$

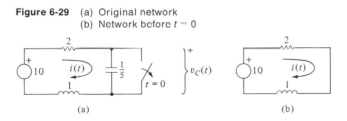

(a)   (b)

Proceeding now to the solution of this network's differential equation with the stated boundary conditions, the homogeneous equation is

$$\frac{1}{5}\frac{d^2 v_{C_n}}{dt^2} + \frac{2}{5}\frac{dv_{C_n}}{dt} + v_{C_n} = 0$$

which has characteristic equation

$$\tfrac{1}{5}s^2 + \tfrac{2}{5}s + 1 = 0,$$

$$s_1, s_2 = \frac{-(2/5) \pm \sqrt{(4/25) - (4/5)}}{2/5}$$

$$= -1 \pm \sqrt{-4}$$
$$= -1 \pm j2$$

The natural component of the capacitor voltage is of the form

$$
\begin{aligned}
v_{C_n} &= K_1 e^{s_1 t} + K_2 e^{s_2 t} \\
&= K_1 e^{(-1+j2)t} + K_2 e^{(-1-j2)t} \\
&= e^{-t}[K_1 e^{j2t} + K_2 e^{-j2t}] \\
&= e^{-t}[A \cos 2t + B \sin 2t]
\end{aligned}
$$

The forced component of the capacitor voltage, due to the constant source, is found by substituting a constant trial solution into the differential equation, obtaining

$$v_{C_f} = 10$$

The general solution of the equation is thus

$$v_C(t) = v_{C_f} + v_{C_n}(t) = 10 + e^{-t}[A \cos 2t + B \sin 2t]$$

after time $t = 0$.

Applying the boundary conditions,

$$v_C(0) = 10 + A = 0$$

$$A = -10$$

And

$$\frac{dv_C}{dt} = -e^{-t}[A \cos 2t + B \sin 2t] + e^{-t}[-2A \sin 2t + 2B \cos 2t]$$

$$\left.\frac{dv_C}{dt}\right|_{t=0} = -A + 2B = 25$$

$$B = \frac{25 - 10}{2} = \frac{15}{2}$$

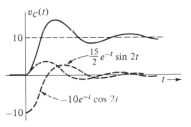

Figure 6-30   Oscillatory *RLC* network capacitor voltage

The solution to this problem,

$$
v_C(t) = \begin{cases} 0, & t \le 0 \\ 10 + e^{-t}\left(-10\cos 2t + \dfrac{15}{2}\sin 2t\right), & t \ge 0 \end{cases}
$$

$$
= \left[10 + e^{-t}\left(-10\cos 2t + \frac{15}{2}\sin 2t\right)\right]u(t)
$$

is sketched in Figure 6-30.

D6-8

Write (but do not solve) differential equations for the capacitor voltage, $v_C(t)$, after time $t = 0$. Also find the boundary conditions $v_C(0)$ and $(dv_C/dt)|_{T-0}$:

(a)

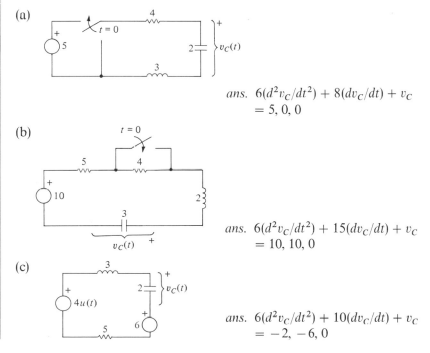

ans. $6(d^2 v_C/dt^2) + 8(dv_C/dt) + v_C$
      $= 5, 0, 0$

(b)

ans. $6(d^2 v_C/dt^2) + 15(dv_C/dt) + v_C$
      $= 10, 10, 0$

(c)

ans. $6(d^2 v_C/dt^2) + 10(dv_C/dt) + v_C$
      $= -2, -6, 0$

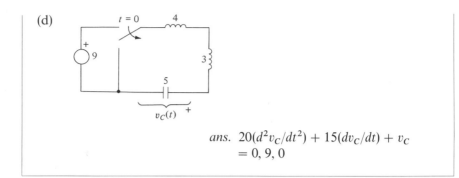

(d)

ans. $20(d^2v_C/dt^2) + 15(dv_C/dt) + v_C$
$= 0, 9, 0$

D6-9

Find and sketch the indicated signals before and after time $t = 0$. There are two signals of interest in each network:

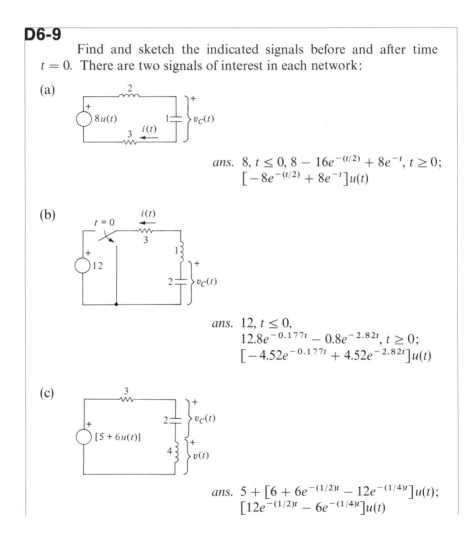

(a)

ans. $8, t \le 0, 8 - 16e^{-(t/2)} + 8e^{-t}, t \ge 0$;
$[-8e^{-(t/2)} + 8e^{-t}]u(t)$

(b)

ans. $12, t \le 0,$
$12.8e^{-0.177t} - 0.8e^{-2.82t}, t \ge 0$;
$[-4.52e^{-0.177t} + 4.52e^{-2.82t}]u(t)$

(c)

ans. $5 + [6 + 6e^{-(1/2)t} - 12e^{-(1/4)t}]u(t)$;
$[12e^{-(1/2)t} - 6e^{-(1/4)t}]u(t)$

(d)

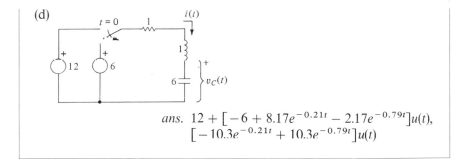

ans. $12 + [-6 + 8.17e^{-0.21t} - 2.17e^{-0.79t}]u(t),$
$[-10.3e^{-0.21t} + 10.3e^{-0.79t}]u(t)$

D6-10

The following series *RLC* network has oscillatory natural response. Find and sketch $v_C(t)$ before and after time $t = 0$:

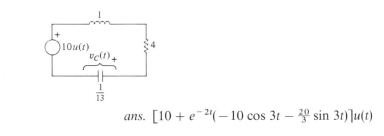

ans. $[10 + e^{-2t}(-10 \cos 3t - \frac{20}{3} \sin 3t)]u(t)$

The Series *RLC* Network

For the solution of switched second- and higher-order networks, it is expedient to first concentrate on obtaining a network differential equation and boundary conditions. Thereafter the problem is a mathematical one of solving the equation and applying the boundary conditions.

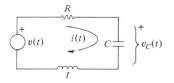

The series *RLC* network has loop equation

$$L \frac{di}{dt} + Ri + \frac{1}{C} \int_{-\infty}^{t} i \, dt = v(t)$$

Rewriting the loop equation in terms of the capacitor voltage,

$$v_C(t) = \frac{1}{C} \int_{-\infty}^{t} i\, dt, \qquad i = C\frac{dv_C}{dt}$$

gives

$$LC\frac{d^2v_C}{dt^2} + RC\frac{dv_C}{dt} + v_C = v(t)$$

which does not involve an integral.

The two boundary conditions are that the capacitor voltage v_C must be continuous and that the inductor current,

$$i(t) = C\frac{dv_C}{dt}$$

must be continuous. In terms of the capacitor voltage, continuous inductor current means dv_C/dt must be continuous.

6.6 Parallel *RLC* Networks

6.6.1 Differential Equation for the Inductor Current

The two-node equation for a parallel *RLC* network, Figure 6-31, is

$$C\frac{dv}{dt} + \frac{1}{R}v + \frac{1}{L}\int_{-\infty}^{t} v\, dt = i(t)$$

If this equation is rewritten in terms of the inductor current,

$$i_L(t) = \frac{1}{L}\int_{-\infty}^{t} v\, dt, \qquad v = L\frac{di_L}{dt}$$

a differential equation results:

$$LC\frac{d^2i_L}{dt^2} + \frac{L}{R}\frac{di_L}{dt} + i_L = i(t)$$

Figure 6-31 Parallel *RLC* network

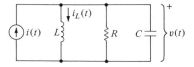

The boundary conditions for this network are that the inductor current i_L must be continuous and that the capacitor voltage,

$$v = L\frac{di_L}{dt}$$

must be continuous. In terms of the i_L, continuous capacitor voltage means that the slope of i_L, di_L/dt, must be continuous.

Consider the specific network of Figure 6-32. Before time $t = 0$, the network is governed by

$$\frac{1}{4}\frac{d^2 i_L}{dt^2} + \frac{4}{3}\frac{di_L}{dt} + i_L = 6$$

The forced solution to this equation is

$$i_L = 6$$

The forced capacitor voltage before $t = 0$ is

$$v = L\frac{di_L}{dt} = 0$$

After $t = 0$, the equation becomes

$$\frac{1}{4}\frac{d^2 i_L}{dt^2} + \frac{4}{3}\frac{di_L}{dt} + i_L = 13$$

with boundary conditions

$$i_L(0) = 6$$

and

$$v(0) = L\frac{di_L}{dt}\bigg|_{t=0} = 0$$

or

$$\frac{di_L}{dt}\bigg|_{t=0} = 0$$

Figure 6-32 Example parallel *RLC* network

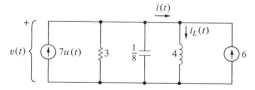

6.6.2   Solution

The example network equation has solution as follows: The homogeneous equation is

$$\frac{1}{4}\frac{d^2 i_{L_n}}{dt^2} + \frac{4}{3}\frac{d i_{L_n}}{dt} + i_{L_n} = 0$$

and has the characteristic equation

$$\frac{1}{4}s^2 + \frac{4}{3}s + 1 = 0$$

$$s_1, s_2 = \frac{-(4/3) \pm \sqrt{(16/9) - 1}}{\frac{1}{2}}$$

$$= -0.021, -5.31$$

So,

$$i_{L_n}(t) = K_1 e^{-0.021t} + K_2 e^{-5.31t}$$

The forced component of i_L after $t = 0$, due to the constant source, is easily found by substituting

$$i_{L_f} = A$$

into the entire equation, giving

$$i_{L_f} = A = 13$$

so that the general solution for $i_L(t)$ is

$$i_L(t) = 13 + K_1 e^{-0.021t} + K_2 e^{-5.31t}, \qquad t \geq 0$$

Applying the boundary conditions,

$$i_L(0) = 13 + K_1 + K_2 = 6$$

$$\frac{d i_L}{dt}\bigg|_{t=0} = -0.021 K_1 - 5.31 K_2 = 0$$

gives

$$K_1 = \frac{\begin{vmatrix} -7 & 1 \\ 0 & -5.31 \end{vmatrix}}{\begin{vmatrix} 1 & 1 \\ -0.021 & -5.31 \end{vmatrix}} = \frac{37.17}{-5.29} = -7.03$$

$$K_2 = \frac{\begin{vmatrix} 1 & -7 \\ -0.021 & 0 \end{vmatrix}}{-5.29} = 0.03$$

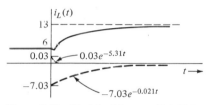

Figure 6-33 Sketch of the parallel *RLC* network inductor current

The solution is thus

$$i_L(t) = \begin{cases} 6, & t \geq 0 \\ 13 - 7.03e^{-0.021t} + 0.03e^{-5.31t}, & t \geq 0. \end{cases}$$

A sketch of $i_L(t)$ is shown in Figure 6-33.

Other signals of interest are easily found from $i_L(t)$. For example, the signal $i(t)$ in the network diagram is

$$i(t) = i_L(t) - 6$$

$$= \begin{cases} 0, & t \leq 0 \\ 7 - 7.03e^{-0.021t} + 0.03e^{-5.31t}, & t \geq 0 \end{cases}$$

Whereas the series *RLC* network has oscillatory response for sufficiently small R; in the parallel *RLC* network, large R gives oscillatory response.

D6-11

Write (but do not solve) differential equations for the inductor current $i_L(t)$ after time $t = 0$. Also find the boundary conditions $i_L(0)$ and $(di_L/dt)|_{t=0}$:

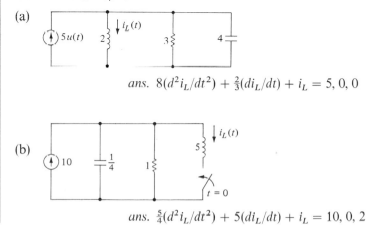

(a)

ans. $8(d^2i_L/dt^2) + \frac{2}{3}(di_L/dt) + i_L = 5, 0, 0$

(b)

ans. $\frac{5}{4}(d^2i_L/dt^2) + 5(di_L/dt) + i_L = 10, 0, 2$

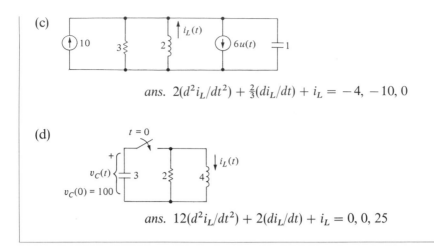

(c)

ans. $2(d^2 i_L/dt^2) + \frac{2}{3}(di_L/dt) + i_L = -4, -10, 0$

(d)

ans. $12(d^2 i_L/dt^2) + 2(di_L/dt) + i_L = 0, 0, 25$

D6-12

Find and sketch the indicated signals before and after time $t = 0$. There are two signals of interest in each network:

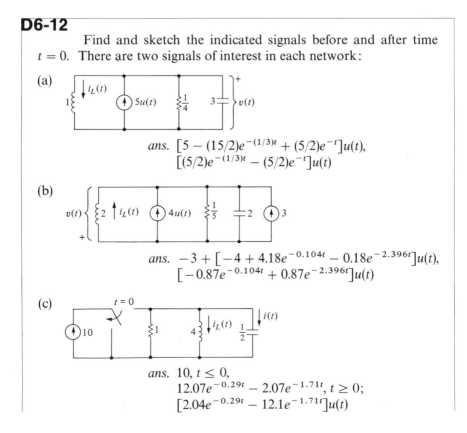

(a)

ans. $[5 - (15/2)e^{-(1/3)t} + (5/2)e^{-t}]u(t),$
$[(5/2)e^{-(1/3)t} - (5/2)e^{-t}]u(t)$

(b)

ans. $-3 + [-4 + 4.18e^{-0.104t} - 0.18e^{-2.396t}]u(t),$
$[-0.87e^{-0.104t} + 0.87e^{-2.396t}]u(t)$

(c)

ans. $10, t \le 0,$
$12.07e^{-0.29t} - 2.07e^{-1.71t}, t \ge 0;$
$[2.04e^{-0.29t} - 12.1e^{-1.71t}]u(t)$

(d)

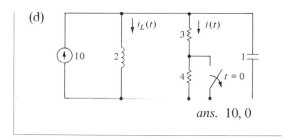

ans. 10, 0

D6-13

The parallel *RLC* network below has oscillatory natural response. Find and sketch $i_L(t)$ before and after time $t = 0$:

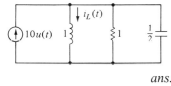

ans. $[10 + e^{-t}(-10 \cos t - 10 \sin t)]u(t)$

The Parallel *RLC* Network

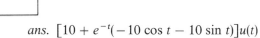

The two-node equation for $v(t)$ is

$$C \frac{dv}{dt} + \frac{1}{R} v + \frac{1}{L} \int_{-\infty}^{t} v \, dt = i(t)$$

Rewriting the nodal equation in terms of the inductor current,

$$i_L(t) = \frac{1}{L} \int_{-\infty}^{t} v(t) \, dt, \qquad v = L \frac{di_L}{dt}$$

gives

$$LC \frac{d^2 i_L}{dt^2} + \frac{L}{R} \frac{di_L}{dt} + i_L = i(t)$$

which does not involve an integral.

The two boundary conditions are that the inductor current i_L must be continuous and that the capacitor voltage,

$$v(t) = L \frac{di_L}{dt}$$

must be continuous. In terms of the inductor current, continuous capacitor voltage means di_L/dt must be continuous.

Chapter Six Problems

Basic Problems

Switched First-Order Inductive Networks

1.   Find $i(t)$ and $v(t)$ both before and after time $t = 0$ and sketch them:

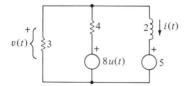

2.   Find $i(t)$ and $i_1(t)$ both before and after time $t = 0$ and sketch them:

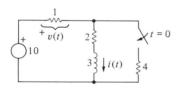

Find $i(t)$ and $v(t)$ both before and after time $t = 0$ and sketch them:

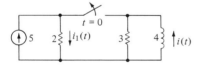

Switched First-Order Capacitive Networks

4.   Find $v(t)$ and $i(t)$ both before and after time $t = 0$ and sketch them:

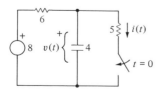

5. Find $v(t)$ and $v_1(t)$ both before and after time $t = 0$ and sketch them:

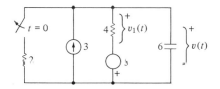

6. Find $v(t)$ and $i(t)$ both before and after time $t = 0$ and sketch them:

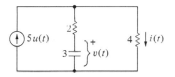

Series RLC Networks

7. Find the differential equation for the capacitor voltage $v_C(t)$ after $t = 0$, and the boundary conditions $v_C(0)$ and $(dv_C/dt)|_{t=0}$:

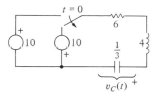

8. Find and sketch the indicated signals before and after time $t = 0$:

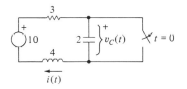

Parallel RLC Networks

9. Find the differential equation for the inductor current $i_L(t)$ after $t = 0$, and the boundary conditions $i_L(0)$ and $(di_L/dt)|_{t=0}$:

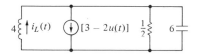

10. Find and sketch the indicated signals before and after time $t = 0$:

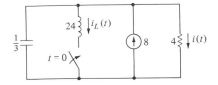

Practical Problems

Inductors

1. Inductors may be used to produce very high voltages by rapidly changing the current through them. This method is used in most television receivers to develop the 15 to 30-kV accelerating voltage for the cathode-ray tube.

If the current through a 90-mH inductor is "triangular" shaped, as in the sketch of Figure 6-34, what must the amplitude A be in order to generate an inductor voltage with pulses that have an amplitude 10 kV?

2. Suppose one attempts to interrupt an inductor current with a switch, as indicated in Figure 6-35(a). If the inductor current were changed instantly, an infinite voltage would be produced across the switch terminals. In practice, the voltage developed at the switch will rise rapidly, as the switch disconnects, until the air surrounding the switch contacts becomes ionized. An electric arc at the switch will then develop.

If the switch contacts are moved far enough apart sufficiently rapidly, the voltage developed across the inductor will cause the insulation around the

Figure 6-34 Practical Problem 1

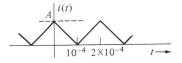

Figure 6-35 (a) Interrupting an inductor current
(b) Simplified ignition circuit

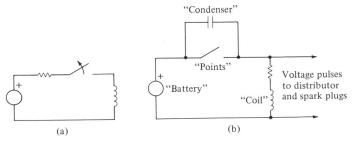

Assortment of switches. A wide variety of types and contact arrangements are shown. (*Photo courtesy of Cutler-Hammer Co.*)

inductor wires to break down, causing arcing between turns in the inductor winding itself. The inductor current *will* be continuous, even if it must complete its flow outside of the conductors.

Automobile ignition circuits, which are capable of generating 5 to 30-kV voltage pulses, interrupt an inductor current produced by the storage battery The "points" are a set of cam-driven switch contacts which open each time a voltage pulse is desired. A simplified ignition circuit is shown in Figure 6-35(b).

Rather than to allow an arc to occur at the points, a capacitor (often called a "condenser" in this application) is placed across the points to slow slightly the rise and fall of the point voltage so that the arcing will, instead, take place at the spark plug. A faulty "condenser" will allow arcing at the points and cause their rapid destruction.

In electronic ignition systems, the switching action is done by a transistor rather than the mechanical "points."

Performance of the system is improved by using coupled coils to magnify the voltage still further.

If the current through the coupled inductor L_1 in Figure 6-36 is $i(t)$ given in Problem 1 with $A = 1$, sketch the open-circuit voltages $v_1(t)$, $v_2(t)$, and $v_3(t)$.

Capacitors

3. One method of measuring the capacitance of large-valued capacitors is to charge them to some convenient voltage, then allow them to discharge

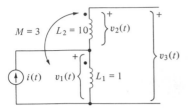

Figure 6-36 Practical Problem 2

through a voltmeter, as in the drawing of Figure 6-37. With a known voltmeter internal resistance, the time constant of the capacitor discharge $R_{meter}C$, which may be obtained by timing the voltage decay, is used to calculate C.

For a voltmeter with 20,000-Ω internal resistance, a certain capacitor voltage decays from 100 V to 84.5 V in 10 s. What is the value of the capacitor?

4. When a capacitor is discharged by short circuiting it with a switch, Figure 6-38, the resistance of the connecting wires and of the switch contacts must be taken into account for accurate results. For an initial capacitor voltage V, the current that flows at the instant the switch is closed is V/R, since the capacitor voltage is continuous. This initial current may be very large if R is sufficiently small.

If 1 J of energy is stored in a 10-μF capacitor, what will the magnitude of the initial current be if the capacitor is discharged through 0.3 Ω?

5. Low voltages may be momentarily converted to higher voltages by charging several capacitors in parallel, then connecting the capacitors in

Figure 6-37 Practical Problem 3

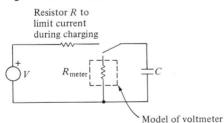

Figure 6-38 Practical Problem 4

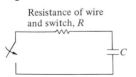

A simple open-frame relay. When the relay coil (the dark cylinder in the picture) is energized, the soft iron bar at the top of the frame is attracted to the coil, moving the metal arm at the right from one contact to the other. (*Photo courtesy of Potter and Brumfield Division of AMF, Inc.*)

series for their discharge. This technique is often used in battery-powered geiger counters and photoflash units, where high voltage pulses are needed which would be inconvenient to obtain directly from a battery.

Five initially uncharged 1-μF capacitors are charged for 1 s as shown in Figure 6-39(a). What is the voltage across the capacitors?

The capacitors are then connected as in Figure 6-39(b). What is the voltage v after $\frac{1}{10}$ s of discharge?

Periodic Switching

6. The repetitive square-wave voltage sketched in Figure 6-40(a) is the source voltage for the RL network of (b). Sketch $i(t)$. [It is not required to find $i(t)$ analytically.]

Figure 6-39 Practical Problem 5

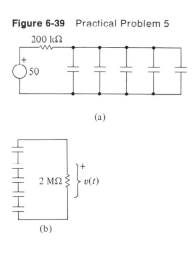

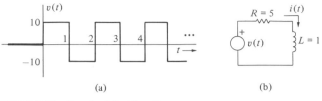

Figure 6-40 Practical Problem 6

When, as in this case, the network time constant is much smaller than the time between switchings of the square wave, the response becomes essentially a switched RL network response repeated over and over. This repetition is suitable for display on an oscilloscope.

Telephone circuit dispatching center. Long-distance audio, video, and data communications signals are routed worldwide. Many of the relay switching circuits are expected to soon be replaced by electronic systems. (*Photo courtesy of A. T. & T. Long Lines*)

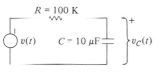

Figure 6-41 Practical Problem 8

7. Sketch $i(t)$ for the source voltage and network of Problem 6, but with $R = 5, L = 10$.

8. The source voltage in the network of Figure 6-41 is $v(t)$ in Problem 6. Sketch the voltage $v_C(t)$.

9. The source voltage in the network of Problem 8 is, instead, the one sketched in Figure 6-42. Sketch $v_C(t)$.

Advanced Problems

Switched Inductive Networks

1. At time $t = 0$, 20 J of energy are stored in the inductor of Figure 6-43. Find the two possible inductor currents after $t - 0$.

2. For the network of Figure 6-44, the first switching occurs at $t = 0$. Before the natural component of the response dies out, a second switching occurs at time $t = 1$. Find the inductor current $i(t)$.

Figure 6-42 Practical Problem 9

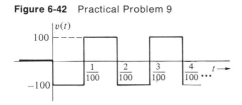

Figure 6-43 Advanced Problem 1

Figure 6-44 Advanced Problem 2

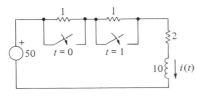

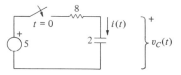

Figure 6-45 Advanced Problem 4

Unit Step Function

3. Sketch a graph of the function

$$f(t) = 3u(-t) + 4u(t) - 3u(t-2) - u(t-3)$$

Switched Capacitive Networks

4. The voltage across an ideal capacitor, when nothing is connected across it, can be anything. Thus in the network of Figure 6-45, it is necessary to specify the capacitor voltage prior to $t = 0$ to be able to solve the network. Find $i(t)$ if $v_C(0) = -5$.

5. A capacitor is to be used to store energy supplied by a voltage generator that is modeled by a Thévenin equivalent. The energy storage might be used in place of storage batteries for an electric vehicle. The capacitor voltage v will eventually reach 100 V in the circuit of Figure 6-46.

Show that the energy eventually stored in the capacitor equals the net energy that is dissipated in (that is, flows into) the resistance. This result holds regardless of the (positive) values of R and C.

6. The problem of finding how a constant voltage divides across two capacitors, Figure 6-47(a), is indeterminant. A more accurate model of the physical capacitor includes a large parallel resistance representing the leakage of current between the capacitor plates. Show that with this model, a net constant voltage is *eventually* distributed as

$$v_1 = \frac{R_1}{R_1 + R_2} v$$

Figure 6-46 Advanced Problem 5

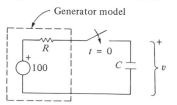

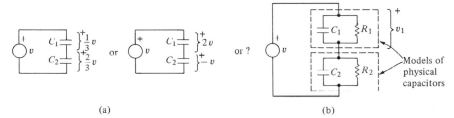

(a)   (b)

Figure 6-47   Advanced Problem 6

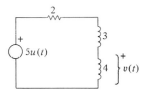

Figure 6-48   Advanced Problem 8

RLC Networks

7. Find two different switched first-order inductive networks in which the inductor current after $t = 0$ satisfies

$$3\frac{di}{dt} + 4i = 5$$

with the boundary condition

$$i(0) = -3$$

Repeat, finding two different switched capacitive networks for which the capacitor voltage satisfies

$$3\frac{dv}{dt} + 4v = 5$$

after $t = 0$, with the boundary condition

$$v(0) = -3$$

8. The network of Figure 6-48 is first order, even though it contains two inductors, because the two inductors in series are equivalent to a single inductor. Find $v(t)$.

CHAPTER SEVEN
The Impedance Concept

7.1 Introduction

Exponential functions are certainly a key in the natural behavior of differential equations and hence to networks described by these equations. Exponential functions are also fundamental so far as forced behavior is concerned, as will be seen in this chapter.

When a network is driven by an exponential source, the forced part of every network voltage and current is also exponential, proportional to the source. The ratio of forced voltage to current in any element is then a constant, just as if the element were a resistor.

In this chapter simple methods are developed for finding the forced response of networks due to exponential sources. Generalized impedance is discussed in connection with transfer functions, then these same methods are used to introduce a general method of finding network natural response.

When you complete this chapter, you should know

1. How to find the general solution to linear, time-invariant differential equations with exponential driving functions.
2. What impedance is and the impedances of capacitors, inductors, and coupled inductors.

3. How to find forced exponential network response using the impedance method.
4. How to superimpose sources to accommodate exponential functions with different time constants.
5. How to solve simple switched networks with exponential sources.
6. What network transfer functions are.
7. The relationship between impedance and network natural behavior.

7.2   Solutions to Differential Equations with Exponential Driving Functions

Except in a very special case, if the driving function to a differential equation is an exponential function, varying as e^{st}, the forced component of the solution of the differential equation is also exponential, with the same value of the exponential constant, s.

For example, consider

$$\frac{d^2 y}{dt^2} + 5\frac{dy}{dt} + 4y = 10e^{-2t}$$

Substituting,

$$y(t) - Ae^{-2t}$$
$$(-2)^2 Ae^{-2t} + 5(-2)Ae^{-2t} + 4Ae^{-2t} = 10e^{-2t}$$
$$-2Ae^{-2t} = 10e^{-2t}$$
$$A = -5$$

and

$$y_f(t) = -5e^{-2t}$$

satisfies this equation.

The characteristic equation is

$$s^2 + 5s + 4 = (s + 1)(s + 4) = 0$$

so that the natural component of the solution is

$$y_n(t) = K_1 e^{-t} + K_2 e^{-4t}$$

The general solution of the equation is thus

$$y(t) = y_n(t) + y_f(t)$$
$$= K_1 e^{-t} + K_2 e^{4t} - 5e^{2t}$$

A second example is as follows. The equation

$$\frac{d^3y}{dt^3} + 5\frac{d^2y}{dt^2} + 6\frac{dy}{dt} = 7e^{4t}$$

has characteristic equation

$$s^3 + 5s^2 + 6s = s(s + 2)(s + 3) = 0$$

so

$$y_n(t) = K_1 + K_2 e^{-2t} + K_3 e^{-3t}$$

Substituting a trial forced solution

$$y(t) = Ae^{4t}$$

into the entire equation gives

$$4^3 Ae^{4t} + 5 \cdot 4^2 Ae^{4t} + 6 \cdot 4Ae^{4t} = 7e^{4t}$$

$$168 Ae^{4t} = 7e^{4t}$$

$$A = \frac{7}{168} = \frac{1}{24}$$

so

$$y_f(t) = Ae^{4t} = \left(\frac{1}{24}\right)e^{4t}$$

and the general solution is

$$y(t) = y_n(t) + y_f(t) = K_1 + K_2 e^{-2t} + K_3 e^{-3t} + \left(\frac{1}{24}\right)e^{4t}$$

A special case occurs when the driving function has the same shape as one of the terms in the natural component of the solution. For example, for the equation

$$\frac{d^2y}{dt^2} + 3\frac{dy}{dt} + 2y = 4e^{-t}$$

the characteristic equation is

$$s^2 + 3s + 2 = (s + 1)(s + 2) = 0$$

giving

$$y_n(t) = K_1 e^{-t} + K_2 e^{-2t}$$

Substituting a trial forced solution

$$y = Ae^{-t}$$

into the original equation in an attempt to find a forced solution gives

$$0 = 4e^{-t}$$

A forced solution of the form Ae^{-t} cannot be found in this case. Actually the forced solution is of the form Ate^{-t}. This special case will be examined later from the standpoint of the network.

Except in this special case, then, if the driving function of a differential equation varies as e^{st}, the forced solution is of the form Ae^{st}.

D7-1

Find the general solution of the following differential equations:

(a) $\dfrac{dy}{dt} + 3y - 4e^{5t}$ $\qquad$ ans. $(\frac{1}{2})e^{5t} + Ke^{-3t}$

(b) $3\dfrac{dy}{dt} + 2y = e^{-4t}$ $\qquad$ ans. $(-\frac{1}{10})e^{-4t} + Ke^{-(2/3)t}$

(c) $\dfrac{d^2y}{dt^2} + 3\dfrac{dy}{dt} + 2y = 5e^{4t}$ $\qquad$ ans. $(\frac{1}{6})e^{4t} + K_1e^{-t} + K_2e^{-2t}$

(d) $\dfrac{d^2y}{dt^2} + 5\dfrac{dy}{dt} + 6y = -2e^{-3t}$ $\qquad$ ans. $2te^{-3t} + K_1e^{-2t} + K_2e^{-3t}$

Differential Equations with Exponential Driving Equations

Except when the driving function is of the same form as one of the natural behavior terms, the forced solution to a differential equation with an exponential driving function, varying as e^{st}, is exponential, with the same exponential constant, s.

The forced solution, Ae^{st}, may be found by substituting into the equation and solving for A.

7.3 Impedance

7.3.1 The Impedance Concept

When the network sources are exponential, all with the same exponential constant, the differential equations for each network voltage and current have exponential driving functions. The forced components of every voltage and every current are then exponential, with the same exponential constant as the sources.

Since the forced component of every voltage and every current varies as e^{st}, the ratio of forced voltage to forced current in any element is a constant, just as it is for a resistor. Generally the forced exponential voltage-current ratio in an element depends on the exponential constant, s.

7.3.2 Impedances of the Basic Elements

The impedance of a two-terminal element is defined as follows:

$$Z(s) = \left. \frac{\text{voltage across the element } v(t)}{\text{current through the element } i(t)} \right|_{\substack{\text{when } v(t) \text{ and } i(t) \text{ have the sink reference relation} \\ \text{and when } v(t) \text{ and } i(t) \text{ each vary as } e^{st}}}$$

The unit of impedance is the same as that of resistance, the ohm. The inverse of impedance, $1/Z(s)$, is called *admittance*.

The impedance of a resistor is just its resistance, R:

$$Z_R = \frac{v(t)}{i(t)} = R$$

With the inductor and capacitor, the ratio of sink reference voltage to current is *not* a constant unless both the voltage and current are exponential.

For the inductor, Figure 7-1(a), if the inductor current is

$$i_L(t) = Ae^{st}$$

the sink reference inductor voltage is also exponential and is

$$v_L(t) = L\frac{di_L}{dt} = sLAe^{st}$$

so the impedance of an inductance is

$$Z_L = \frac{v_L(t)}{i_L(t)} = \frac{sLAe^{st}}{Ae^{st}} = sL$$

For the capacitor, Figure 7-1(b), if the capacitor voltage is exponential,

$$v_C(t) = Ae^{st}$$

Figure 7-1 Exponential voltages and currents for the inductor and the capacitor

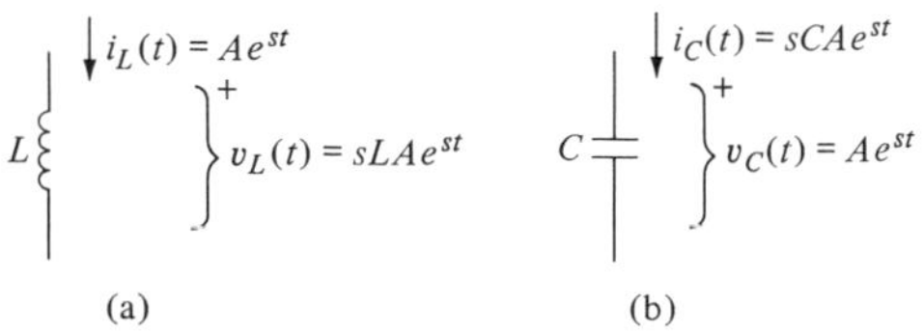

(a)   (b)

the sink reference capacitor current is also exponential,

$$i_C(t) = C\frac{dv_C(t)}{dt} = sCAe^{st}$$

The impedance of a capacitor is thus

$$Z_C = \frac{v_C(t)}{i_C(t)} = \frac{Ae^{st}}{sCAe^{st}} = \frac{1}{sC}$$

D7-2

The impedance of a certain element, as a function of the exponential constant s is

$$Z(s) = \frac{s}{2s+3}$$

If the element voltage and current are both exponential and the current through the element is

$$i(t) = 4e^{-5t}$$

what is the sink reference element voltage?   *ans.* $(20/7)e^{5t}$

Impedance

When the sources in a network are all exponential, with the same exponential constant s, the forced components of each voltage and current vary as e^{st}.

The ratio of forced exponential voltage to current in any element is a constant. Thus for forced exponential voltages and currents, every network element has the same voltage-current relation as, and so is equivalent to, a resistor.

This equivalent resistance, the sink reference voltage-current ratio when both the voltage and current vary as e^{st}, is called the impedance of the element. The impedance of an inductor is

$$Z_L = sL$$

and the impedance of a capacitor is

$$Z_C = \frac{1}{sC}$$

The inverse of an impedance is termed admittance.

7.4 Solutions for Forced Exponential Response

7.4.1 The Impedance Equivalent

To find the forced component of any voltage or current in a network driven by one or more exponential sources with the same exponential constant s, simply convert the original network to an equivalent network in which the inductors and capacitors are replaced by equivalent resistances that are those element's impedances. The exponential constants of the sources must all be the same and each of the impedances is evaluated at the value of s of the sources to obtain the equivalent resistance of the element.

Consider the network of Figure 7-2. The impedance of the inductor is $sL = s \cdot 3$ which, for exponential signals of the form of the source e^{2t}, is $sL = 2 \cdot 3$. For signals which vary as e^{2t}, the inductor is equivalent to a 6-Ω resistor. Similarly, the capacitor is equivalent to a resistor of $1/sC = 1/2 \cdot 4 = \frac{1}{8}\,\Omega$. The inductor and the capacitor are replaced by their impedances in the equivalent network.

All of the solution methods for source-resistor networks now apply to this problem.

$$i(t) = \frac{7e^{2t}}{2 + 6 + \frac{1}{8}} = \frac{56e^{2t}}{65}$$

$$v(t) = \frac{42}{2 + 6 + \frac{1}{8}} e^{2t} = \frac{336}{65} e^{2t}$$

Instead of redrawing the network, these equivalent resistances, the element impedances, may be indicated on the original diagram. It is helpful to circle the impedances to distinguish them from the element values, as in the network of Figure 7-3, for which

$$(\tfrac{1}{6} + \tfrac{1}{4} + \tfrac{1}{2})v(t) = e^{3t} - 5e^{3t}$$

$$v(t) = -\frac{48}{11} e^{3t}$$

Figure 7-4 is an example in which the network has mutual inductive coupling. Replacing the coupling by the controlled voltage-source equivalent, the equivalent network results. For exponential currents i_1 and i_2,

Figure 7-2 Impedance equivalent for network solution

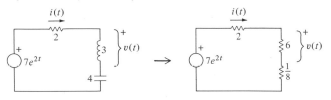

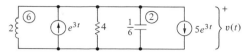

Figure 7-3 Example of using impedance equivalent

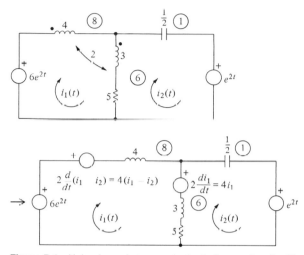

Figure 7-4 Using impedance equivalents for a network with mutual inductive coupling

each varying as e^{2t},

$$\frac{di_1}{dt} = 2i_1 \quad \text{and} \quad \frac{di_2}{dt} = 2i_2$$

so the mutual coupling term derivatives may also be eliminated.

The network equations are as follows:

$$\begin{cases} (6 + 8)i_1 \quad 6i_2 = 6e^{2t} - 4i_1 + 4i_2 - 4i_1 \\ -6i_1 + (6 + 1)i_2 = -e^{2t} + 4i_1 \end{cases}$$

$$\begin{cases} 22i_1 - 10i_2 - 6e^{2t} \\ -10i_1 + 7i_2 = -e^{2t} \end{cases}$$

$$i_1 = \frac{\begin{vmatrix} 6e^{2t} & -8 \\ -e^{2t} & 7 \end{vmatrix}}{\begin{vmatrix} 18 & -8 \\ -8 & 7 \end{vmatrix}} = \frac{34e^{2t}}{62}$$

$$i_2 = \frac{\begin{vmatrix} 18 & 6e^{2t} \\ -8 & -e^{2t} \end{vmatrix}}{62} - \frac{30e^{2t}}{62}$$

7.4.2 The Case of Constant Sources

Constant sources are of the form

$$Ae^{0t} = A$$

for which the exponential constant s is zero, so forced exponential response includes forced constant response as a special case.

For constant sources, $s = 0$, the impedance of an inductor is $sL = 0$, and the impedance of a capacitor is $1/sC = \infty$. This is to say that the forced response of a network with constant sources may be found by replacing all inductors by short circuits (zero impedance) and all capacitors by open circuits (infinite impedance). This result, which is a special case of the impedance equivalent, was introduced in Chapter Six in connection with switched networks.

The example network of Figure 7-5 has constant sources. For it,

$$8i = 20 - 12$$
$$i = 1$$

7.4.3 Negative Element Impedances

For sources with negative exponential constants, s, the impedances of inductors and capacitors are negative, but they are handled just as are positive resistors so far as the mathematics is concerned. For the network

Figure 7-5 Impedance methods for a network with constant sources

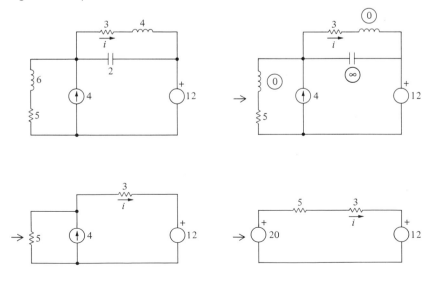

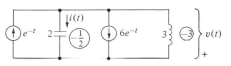

Figure 7-6 Network with negative impedances

of Figure 7-6,

$$(-2 - \tfrac{1}{3})v(t) = 6e^{-t} - e^{-t}$$

$$v(t) = -\frac{15}{7} e^{-t}$$

$$i(t) = \frac{-v(t)}{-\frac{1}{2}} = -\frac{30}{7} e^{-t}$$

Figure 7-7 is an example of impedance solution in which equivalent circuits are used. When different types of elements, each of which is equivalent to a resistance, are combined to form an overall impedance, the combination element is denoted by the general symbol for an element, a "box." For this network,

$$\left(-4 - \frac{15}{2}\right) i(t) = 3e^{-2t} - 4e^{-2t}$$

$$i(t) = \frac{2}{23} e^{-2t}$$

$$v(t) = -\frac{15}{2} i(t) = -\frac{15}{23} e^{-2t}$$

Figure 7-7 Equivalent circuit solution resulting in combined impedances

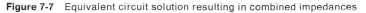

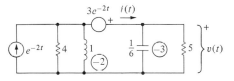

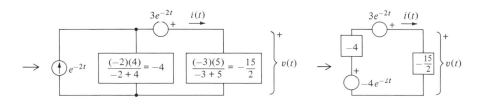

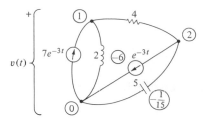

Figure 7-8 Example of using impedance and nodal equations

An example in which nodal equations are used in the solution is given in Figure 7-8. For this network,

$$\begin{cases} (-\frac{1}{6} + \frac{1}{4})v_1 - (\frac{1}{4})v_2 = 7e^{-3t} \\ -(\frac{1}{4})v_1 + (\frac{1}{4} - 15)v_2 = -e^{-3t} \end{cases}$$

$$\begin{cases} v_1 - 3v_2 = 84e^{-3t} \\ -v_1 - 59v_2 = -4e^{-3t} \end{cases}$$

$$v(t) = v_1 = \frac{\begin{vmatrix} 84e^{-3t} & -3 \\ -4e^{-3t} & -59 \end{vmatrix}}{\begin{vmatrix} 1 & -3 \\ -1 & -59 \end{vmatrix}}$$

$$= -\frac{4944}{56}e^{-3t}$$

7.4.4 Singular Cases

In the network of Figure 7-9, a difficulty occurs. This situation is indicative of a special case of a network differential equation in which the forced network response is not exponential.

The current $i(t)$ satisfies

$$2\frac{di}{dt} + 6i = e^{-3t}$$

Figure 7-9 Special case in which the forced network response is not exponential

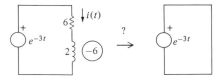

The homogeneous equation is

$$2\frac{di_n}{dt} + 6i_n = 0$$

and the characteristic equation is

$$2s + 6 = 0$$
$$s = -3$$

The natural component of $i(t)$ is

$$i_n(t) = Ke^{-3t}$$

which is of the same form as the driving function.

An inconsistency such as this in the equivalent impedance network will always occur in these special cases. For the example, a forced current of the form

$$i(t) = Ate^{-3t}$$

will satisfy the differential equation:

$$2(Ae^{-3t} - 3Ate^{-3t}) + 6Ate^{-3t} = e^{-3t}$$
$$A = \tfrac{1}{2}$$
$$i(t) - (1/2)te^{-3t}$$

These special cases are of limited practical importance because of the difficulty of obtaining exactly the right element values in a network for the situation to occur.

D7-3

Using impedance methods, find the indicated signals:

(a)

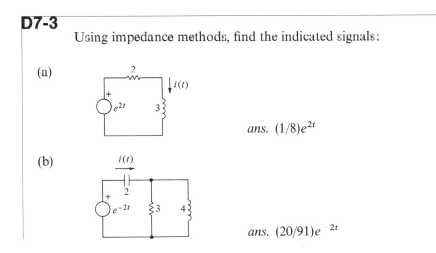

ans. $(1/8)e^{2t}$

(b)

ans. $(20/91)e^{-2t}$

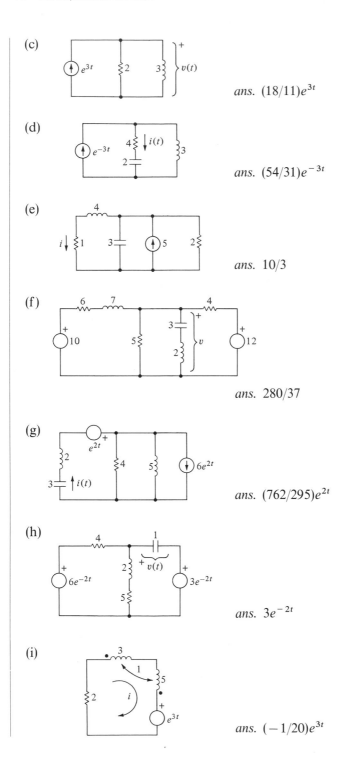

(c)

ans. $(18/11)e^{3t}$

(d)

ans. $(54/31)e^{-3t}$

(e)

ans. $10/3$

(f)

ans. $280/37$

(g)

ans. $(762/295)e^{2t}$

(h)

ans. $3e^{-2t}$

(i)

ans. $(-1/20)e^{3t}$

(j)

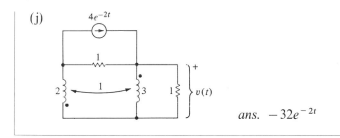

$$ans. \ -32e^{-2t}$$

Finding Forced Exponential Response

When all sources in a network are exponential, Ae^{st}, with the same exponential constant s, all network forced voltages and currents vary as e^{st}. The ratio of forced voltage to current in an inductor or capacitor under these circumstances is a constant, the impedance of the element.

In terms of the exponential constant s, the impedance of an inductance L is

$$Z_L = sL$$

and the impedance of a capacitance C is

$$Z_C = \frac{1}{sC}$$

To solve for any forced voltage or current in such a network, replace the inductors and capacitors by their impedances and solve the corresponding source-resistor network.

7.5   Superposition of Sources

If a network has two or more sources that are exponential, but with different exponential constants s, the sources may be superimposed so that in each subproblem all voltages and currents are exponential with a single value of s.

An example is shown in Figure 7-10. For this network,

$$(-5 - 2 + 4)i_1 = 3e^{-t}$$
$$i_1(t) = -e^{-t}$$

and

$$i_2(t) = \frac{6}{6+5}e^t = \frac{6}{11}e^t$$

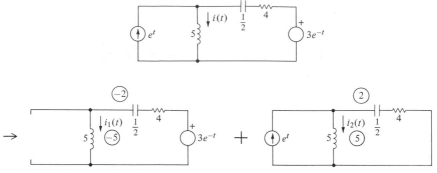

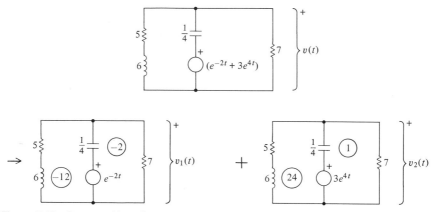

Figure 7-10   Superposition of sources

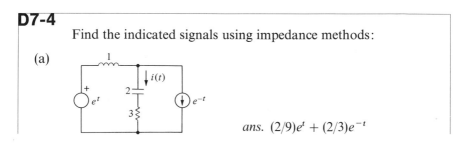

Figure 7-11   Superposition of source components

Then

$$i(t) = i_1(t) + i_2(t) = \frac{6}{11} e^t - e^t$$

Figure 7-11 shows an example of superposition for which the source has more than one exponential component.

D7-4

Find the indicated signals using impedance methods:

(a)

ans. $(2/9)e^t + (2/3)e^{-t}$

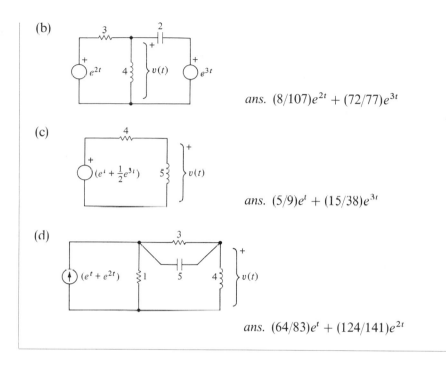

(b)

ans. $(8/107)e^{2t} + (72/77)e^{3t}$

(c)

ans. $(5/9)e^{t} + (15/38)e^{3t}$

(d)

ans. $(64/83)e^{t} + (124/141)e^{2t}$

Solving Networks with Different Exponential Source Functions

If there are several exponential sources in a network that have different exponential constants s, they may be superimposed to give a set of problems each of which involves only a single exponential constant at a time.

Source functions consisting of the sum of several different exponential terms may also be accommodated by superposition.

7.6 Switched Networks with Exponential Sources

The same development could now be carried through for switched networks with exponential sources as was done previously for the switched networks with constant sources. The same methods of solution apply, of course, except that the calculation of the forced response differs.

An example is given in Figure 7-12. Since the network source is disconnected before $t = 0$, the forced inductor current is zero prior to $t = 0$.

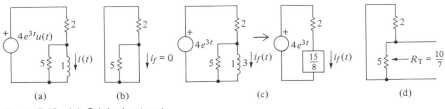

Figure 7-12 (a) Original network
(b) Forced inductor current before $t = 0$
(c) Forced inductor current after $t = 0$
(d) Thévenin resistance after $t = 0$

After time $t = 0$, the source is exponential, so the impedance method with $s = 3$ applies:

$$\left(2 + \frac{15}{8}\right) i_f(t) = 4e^{3t}$$

$$i_f(t) = \frac{32}{31} e^{3t}$$

After $t = 0$, the general solution for the inductor current is thus

$$i(t) = \frac{32}{31} e^{3t} + Ke^{-(R_T/L)t}$$

$$= \frac{32}{31} e^{3t} + Ke^{-(10/7)t}$$

Before $t = 0$, $i = 0$, so

$$i(0) = \frac{32}{31} + K = 0$$

$$K = -\frac{32}{31}$$

$$i(t) = \begin{cases} 0, & t \le 0 \\ \dfrac{32}{31} e^{3t} - \dfrac{32}{31} e^{-(10/7)t}, & t \ge 0 \end{cases}$$

D7-5

Find the indicated signal and roughly sketch it. Show the signal slightly before $t = 0$ on the sketch:

(a)

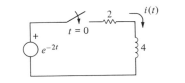

ans. $(1/6)[e^{-(1/2)t} - e^{-2t}]$, $t \ge 0$

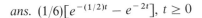

(b)

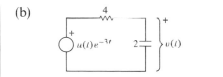

$$ans. \ (1/23)\left[e^{-(1/8)t} - e^{-3t}\right]u(t)$$

Switched Networks with Exponential Sources

When exponential sources are involved in switched networks, the forced exponential response may be found using impedance methods.

7.7 Transfer Functions

Every forced voltage and current due to a single exponential source in a network varies as does the source, as e^{st}. The ratio of *any* forced voltages or currents in a single-source network is thus constant.

Impedance is an exponential signal ratio of this kind, that of the sink reference exponential voltage to current in one element. Other ratios of one forced exponential voltage or current to another are called *transfer functions* and are often given the symbol $T(s)$.

To calculate network transfer functions, as functions of the exponential constant s, let the source be the general exponential function Ae^{st}. All elements then behave as impedances so far as the forced network signals are concerned. Solve for the exponential signals involved, in terms of s, and form the desired ratio.

For the network of Figure 7-13, the source $v(t)$ has been taken to be a general exponential function, and the network elements have been represented by their impedances, as functions of s.

Suppose that it is desired to find the transfer function

$$T_1(s) = \left. \frac{v_1(t)}{v(t)} \right|_{\text{when all signals vary as } e^{st}}$$

Figure 7-13 Network with a single exponential source

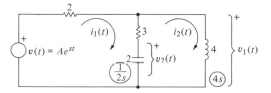

The systematic loop equations for the network are

$$\begin{cases} \left(5 + \dfrac{1}{2s}\right)i_1 - \left(3 + \dfrac{1}{2s}\right)i_2 = v \\[2mm] -\left(3 + \dfrac{1}{2s}\right)i_1 + \left(3 + \dfrac{1}{2s} + 4s\right)i_2 = 0 \end{cases}$$

Solving for i_2 in terms of v,

$$i_2 = \frac{\begin{vmatrix} \left(5 + \dfrac{1}{2s}\right) & v \\[3mm] -\left(3 + \dfrac{1}{2s}\right) & 0 \end{vmatrix}}{\begin{vmatrix} \left(5 + \dfrac{1}{2s}\right) & -\left(3 + \dfrac{1}{2s}\right) \\[3mm] -\left(3 + \dfrac{1}{2s}\right) & \left(3 + \dfrac{1}{2s} + 4s\right) \end{vmatrix}}$$

$$= \frac{\left(3 + \dfrac{1}{2s}\right)v}{15 + \dfrac{5}{2s} + 20s + \dfrac{3}{2s} + \dfrac{1}{4s^2} + 2 - 9 - \dfrac{3}{s} - \dfrac{1}{4s^2}}$$

$$= \frac{(6s + 1)v}{40s^2 + 16s + 2}$$

The voltage v_1 is then, for exponential signals,

$$v_1 = (4s)i_2 = \frac{(24s^2 + 4s)v}{40s^2 + 16s + 2}$$

and the desired transfer function is

$$T_1(s) = \left.\frac{v_1(t)}{v(t)}\right|_{\substack{\text{when all signals} \\ \text{vary as } e^{st}}} = \frac{12s^2 + 2s}{20s^2 + 8s + 1}$$

To find the transfer function

$$T_2(s) = \left.\frac{v_2(t)}{v_1(t)}\right|_{\substack{\text{when all signals} \\ \text{vary as } e^{st}}}$$

the voltage divider rule may be used. For exponential signals, the ratio of

$v_2(t)$ to $v_1(t)$ is

$$\frac{v_2(t)}{v_1(t)} = \frac{\dfrac{1}{2s}}{3 + \dfrac{1}{2s}} = \frac{1}{6s + 1} = T_2(s)$$

Other network transfer functions may be similarly found.

D7-6

For the given network, find the following transfer functions, as a function of the exponential constant s:

(a) $T_a(s) = \dfrac{i_1(t)}{v(t)}\bigg|_{\text{when all signals vary as } e^{st}}$

(b) $T_b(s) = \dfrac{v_1(t)}{v(t)}\bigg|_{\text{when all signals vary as } e^{st}}$

(c) $T_c(s) = \dfrac{i_2(t)}{v(t)}\bigg|_{\text{when all signals vary as } e^{st}}$

(d) $T_d(s) = \dfrac{i_2(t)}{i_1(t)}\bigg|_{\text{when all signals vary as } e^{st}}$

(e) $T_e(s) = \dfrac{i_2(t)}{v_1(t)}\bigg|_{\text{when all signals vary as } e^{st}}$

ans. $1/(2 + 3s)$, $3/(60s^2 + 5)$,
$12s/(60s^2 + 5)$,
$(36s^2 + 24s)/(60s^2 + 5)$, $4s$

Transfer Functions

In a network with a single exponential source, with source function Ae^{st}, the forced component of every voltage and every current is also exponential, varying as e^{st}.

A network transfer function is the ratio of one forced exponential voltage or current to another, as a function of the exponential constant s.

7.8   Using Impedance to Find Natural Behavior

7.8.1   Loop Impedance Method

The natural component of a voltage or current in any network is of the form

$$y_n(t) = K_1 e^{s_1 t} + K_2 e^{s_2 t} + \cdots + K_n e^{s_n t}$$

where $K_1, K_2, \ldots, K_n$ are arbitrary constants that are dependent on the network initial conditions. The exponential constants $s_1, s_2, \ldots, s_n$ are called the "characteristic roots" of the natural behavior.

The natural component of any network voltage or current is the general solution of the homogeneous differential equations describing the signal. In terms of the network, forming homogeneous equations is to set all fixed sources to zero. Thus the natural components of network voltages and currents are those signals that can exist in the network without fixed sources.

Since each natural behavior term, $K_i e^{s_i t}$, is exponential, element impedances may be used in determining which exponential signals may exist in the source-free network. For example, for the single-loop network in Figure 7-14, setting the source to zero results in the source-free network of Figure 7-14(b). For an exponential natural current

$$i_n(t) = K e^{st}$$

the element impedances may be used, and the loop equation is

$$\left(2s + 10 + \frac{12}{s}\right) i_n(t) = 0$$

The current $i_n(t)$ can be nonzero only if

$$\left(2s + 10 + \frac{12}{s}\right) = \frac{2s^2 + 10s + 12}{s} = 0$$

$$2(s + 2)(s + 3) = 0$$

$$s = -2, -3$$

That is, only exponential loop currents of the shape e^{-2t} and e^{-3t} can exist

Figure 7-14   Exponential signals in a source-free network

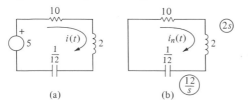

(a)   (b)

in the source-free network. The most general such current is

$$i_n(t) = K_1 e^{-2t} + K_2 e^{-3t}$$

and this is the form of the natural part of the loop current.

Equating the loop impedance, through which a current flows, to zero to find the roots of the natural component of that current is to ask if a source-free loop with impedance $Z(s)$ can have a nonzero current in it. Generally it can, for exponential currents with those values of s for which $Z(s) = 0$.

The drawings of Figure 7-15 illustrate this concept for a more complicated network. The original network in Figure 7-15(a) has all sources set to zero in Figure 7-15(b). There being no sources present, there are only natural voltages and currents in Figure 7-15(b), which are the natural components of the corresponding signals in the original network.

The conductor through which the natural current of interest, $i_n(t)$, flows is imagined pulled out from the rest of the network, and the entire impedance looking back into the rest of the network from this loop, $Z(s)$ in Figure 7-15(c), is calculated, as a function of the exponential constant s.

In Figure 7-15(d) the current $i_n(t)$ is seen to be nonzero only if

$$Z(s)i_n(t) = 0$$

that is, for those values of s for which

$$Z(s) = 0$$

Figure 7-15 Using a loop impedance to find network natural response

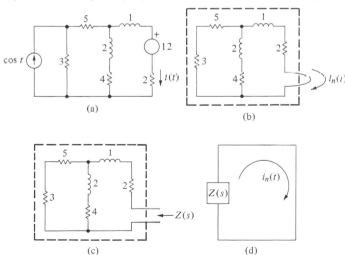

The loop impedance $Z(s)$ in the example is

$$Z(s) = s + 2 + \frac{(2s + 4)(8)}{2s + 4 + 8}$$

$$= \frac{2s^2 + 32s + 56}{2s + 12} = \frac{s^2 + 16s + 28}{s + 6}$$

Equating $Z(s)$ to zero,

$$Z(s) = \frac{(s + 2)(s + 14)}{s + 6} = 0$$

$$s_1, s_2 = -2, -14$$

So the natural component of this current $i(t)$ is of the form

$$i_n(t) = K_1 e^{-2t} + K_2 e^{-14t}$$

7.8.2 Characteristic Roots of the Natural Behavior

The natural component of each network voltage and current consists of a sum of exponential terms of the form

$$y_n(t) = K_1 e^{s_1 t} + K_2 e^{s_2 t} + \cdots + K_n e^{s_n t}$$

The operations involved in expressing one source-free network voltage or current in terms of another are differentiation, integration, multiplication by a constant, and addition, all of which only modify the amplitudes of each exponential term.

Hence the natural component of every network signal consists of the *same* exponential terms, in varying amounts. Every voltage and current is of the above form, but with generally different amplitudes $K_1, K_2, \ldots, K_n$. The characteristic roots of the natural behavior of a network, $s_1, s_2, \ldots, s_n$, are characteristic of the network; the same roots are involved in the natural component of every voltage and every current in the network. Except in special cases, every network loop impedance should yield the same characteristic roots. For example, for the network of Figure 7-16, in which all

Figure 7-16 Finding the same characteristic roots using any loop impedance

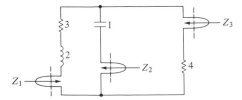

sources have been set to zero, each of the indicated loop impedances have the same roots, which are the characteristic roots of the network:

$$Z_1(s) = 2s + 3 + \cfrac{\dfrac{4}{s}}{4 + \dfrac{1}{s}} = \frac{8s^2 + 14s + 7}{4s + 1}$$

$$Z_2(s) = \frac{1}{s} + \frac{(2s + 3)4}{2s + 3 + 4} = \frac{8s^2 + 14s + 7}{2s^2 + 7s}$$

$$Z_3(s) = 4 + \cfrac{\dfrac{2s + 3}{s}}{2s + 3 + \dfrac{1}{s}} = \frac{8s^2 + 14s + 7}{2s^2 + 3s + 1}$$

7.8.3   Number of Natural Behavior Roots

The number of characteristic natural behavior roots n for a network is equal to the number of independent energy-storing elements in it; n is the number of independent initial energy storages that may be made. Since each inductor current and each capacitor voltage must be continuous, n is also the number of boundary conditions that apply to the network.

Mutual inductances are not counted in addition to self-inductances as independent energy-storing elements because the amount of energy stored in a mutual inductance is fixed once the initial currents in the associated self-inductances are given.

Ordinarily one needs only to total the number of inductors and capacitors in the network to determine the number of characteristic roots expected. However, there are some special circumstances for which some inductors and capacitors may not contribute to the number of characteristic roots.

After setting all fixed network sources to zero, if there are several inductors that may be replaced by a single equivalent inductor, as in Figure 7-17(a), only the single equivalent inductor should be counted as an independent energy-storing element. A similar consideration holds for equivalent capacitors.

If an inductor current is fixed by a current source or a capacitor voltage is fixed by a voltage source, the initial energy storage in that element is not arbitrary, and the element does not contribute to the number of characteristic roots. When the network fixed sources are set to zero, an inductor in series with a current source will be left hanging and a capacitor in parallel with a voltage source will be left short circuited, so these situations are quite obvious, as in the example network of Figure 7-17(b).

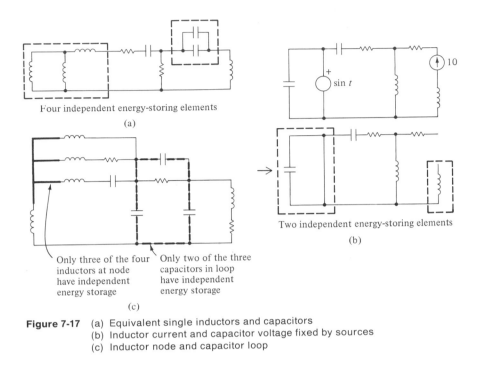

Four independent energy-storing elements

(a)

Two independent energy-storing elements

(b)

Only three of the four inductors at node have independent energy storage

Only two of the three capacitors in loop have independent energy storage

(c)

Figure 7-17 (a) Equivalent single inductors and capacitors
(b) Inductor current and capacitor voltage fixed by sources
(c) Inductor node and capacitor loop

Less obvious situations where the energies stored in the network energy-storing elements are not independent of one another involve a node joining only inductors or a loop involving only capacitors, as shown in Figure 7-17(c). In the case of the inductors, one inductor current is expressible in terms of the others, so the number of independent energy storages in inductors joined exclusively at a node is one less than the number of inductors.

Similarly, one of the capacitor voltages in the loop of capacitors is dependent on the others, so the energy stored in one of these capacitors is not independent of that stored in the others.

7.8.4 Terminal-Pair Impedance Method

The characteristic roots of the natural behavior of a network may also be found by considering the nonzero exponential network *voltages* that may exist with the fixed sources set to zero.

Consider a network such as the one in Figure 7-18(a) for which the fixed sources have been set to zero. The natural voltage $v_n(t)$, for example, may be considered to be the voltage across an impedance $Z'(s)$, as in Figure 7-18(b). A terminal pair has been drawn connected to the network to aid in visualizing $Z'(s)$.

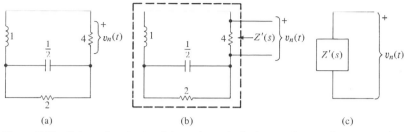

Figure 7-18   Using a terminal-pair impedance to find network natural response

In Figure 7-18(c), the natural voltage $v_n(t)$ is seen to be nonzero only if it satisfies the source-free two-node equation

$$\left(\frac{1}{Z'(s)}\right) v_n(t) = 0$$

or

$$\frac{1}{Z'(s)} = 0$$

For the example network

$$
Z'(s) = \frac{4\left[s + \dfrac{(4/s)}{2 + (2/s)}\right]}{4 + s + \dfrac{(4/s)}{2 + (2/s)}} = \frac{4\left(s + \dfrac{2}{s+1}\right)}{4 + s + \dfrac{2}{s+1}}
$$

$$
= \frac{4s^2 + 4s + 2}{s^2 + 5s + 6}
$$

The inverse of this terminal-pair impedance is

$$
\frac{1}{Z'(s)} = \frac{s^2 + 5s + 6}{4s^2 + 4s + 2} = \frac{(s+2)(s+3)}{4s^2 + 4s + 2}
$$

and equating to zero gives the natural behavior roots -2 and -3. Then

$$v_n(t) = K_1 e^{-2t} + K_2 e^{-3t}$$

and the natural components of all other network voltages and currents also consist of an e^{-2t} term plus an e^{-3t} term.

Thus an alternate method of finding the characteristic roots of a network is to set the fixed sources to zero, connect a terminal pair to the network, and calculate the impedance looking into the terminal pair. The roots of the *inverse* of a terminal-pair impedance are the characteristic roots since they describe the exponential terminal-pair voltages that may exist in the source-free network.

Except in special cases, the inverses of other terminal-pair impedances will yield the same characteristic roots, as will the loop impedances. For the example network, redrawn in Figure 7-19, the terminal-pair impedance Z_2 is

$$Z_2(s) = \cfrac{1}{\cfrac{s}{2} + \cfrac{1}{2} + \cfrac{1}{s+4}}$$

Its inverse is

$$\frac{1}{Z_2(s)} = \frac{s}{2} + \frac{1}{2} + \frac{1}{s+4} = \frac{s^2 + 5s + 6}{2s + 8}$$

which has the same roots as found previously, $s = -2$ and $s = -3$.

The loop impedance,

$$Z_1(s) = 4 + s + \cfrac{\cfrac{4}{s}}{2 + \cfrac{2}{s}} = 4 + s + \frac{1}{s+1}$$

$$= \frac{s^2 + 5s + 6}{s+1}$$

also gives the same roots, as do the other terminal-pair and loop impedances.

7.8.5 Impedance Poles and Zeros

The *zeros* of an impedance $Z(s)$ are the values of s for which

$$Z(s) = 0$$

Figure 7-19 Loop and terminal-pair impedances of the example network

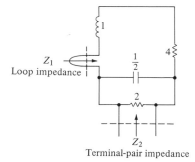

Z_2
Terminal-pair impedance

Its *poles* are the values of s for which

$$\frac{1}{Z(s)} = 0$$

that is, those values of s that make $Z(s)$ infinite.

When a network impedance function is rationalized, expressed as the ratio of two polynomials in s, its zeros are the roots of the numerator polynomial and its poles are the roots of the denominator polynomial. For example, the impedance indicated in Figure 7-20 is

$$Z(s) = 2s + \frac{\left(4 + \dfrac{1}{3s}\right)(1)}{4 + \dfrac{1}{3s} + 1} = 2s + \frac{12s + 1}{15s + 1}$$

$$= \frac{30s^2 + 14s + 1}{15s + 1}$$

The zeros of this impedance are given by

$$30s^2 + 14s + 1 = 0$$

$$s_1, s_2 = \frac{-14 \pm \sqrt{196 - 120}}{60}$$

$$= -0.38, \, -0.089$$

Its pole is given by

$$15s + 1 = 0$$

$$s = -\frac{1}{15}$$

The characteristic roots of the natural behavior of a network are thus the zeros of the loop impedances and they are the poles of the terminal-pair impedances. Ordinarily all loop impedances in a network have the same zeros and these zeros are the same as the poles of all terminal-pair impedances.

Figure 7-20   Impedance poles and zeros

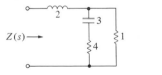

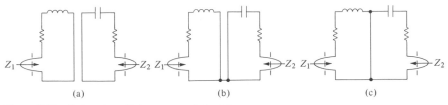

Figure 7-21 Networks with two uncoupled parts

7.8.6 Roots Unobservable from a Loop

It could happen that all of the network characteristic roots cannot be obtained from one of the loops or one of the terminal-pair impedances. That obviously will be the case if the network consists of two or more parts that are not coupled to one another, as in the examples of Figure 7-21.

In very special cases, quite ordinary-looking networks might also have loop or terminal-pair impedances from which not all of the characteristic roots are obtainable. The network of Figure 7-22 is an example.

The loop impedance Z_1 is

$$Z_1(s) = 2s + 6 + \frac{4(s + 3)}{s + 7}$$

$$= \frac{2s^2 + 24s + 54}{s + 7} = \frac{2(s + 3)(s + 9)}{s + 7}$$

which has roots $s = -3$ and $s = -9$. The loop impedance Z_2 is

$$Z_2(s) = 4 + \frac{2s + 6}{3} = \frac{2s + 18}{3}$$

which displays only the root $s = -9$.

This property of this special network is easily understood when its behavior is analyzed for exponential signals of the form Ke^{-3t}. Evaluating the impedances at $s = -3$, it is seen (Figure 7-23) that the Z_2 loop is decoupled from the other loop for $s = -3$.

Figure 7-22 Network where a natural behavior root is not observable from one of the loops

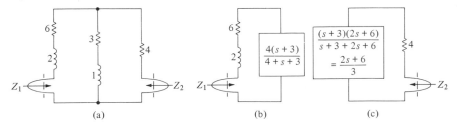

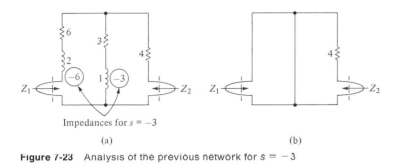

Impedances for $s = -3$

(a)   (b)

Figure 7-23   Analysis of the previous network for $s = -3$

Occasionally networks are deliberately designed in this way so that some of the characteristic roots are not observable in parts of the network. Thus one may see in practice, for example, networks involving three or more energy-storing elements, designed for second-order behavior so far as a certain part of the network is concerned.

If a network loop or terminal-pair impedance yields a lesser number of roots than the network has independent energy-storing elements, other impedances must be examined to obtain the remaining roots.

D7-7

Find the characteristic roots of the natural behavior of the following networks using the loop-impedance method, then using the terminal-pair method:

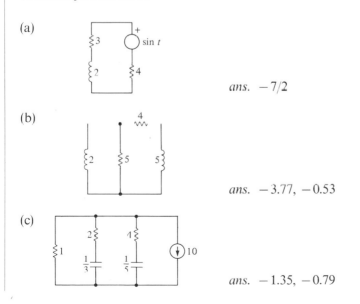

(a)

ans. $-7/2$

(b)

ans. $-3.77, -0.53$

(c)

ans. $-1.35, -0.79$

(d)

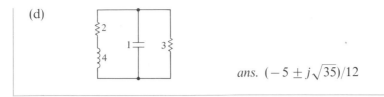

ans. $(-5 \pm j\sqrt{35})/12$

D7-8

Find the impedance of each of the following elements, as a function of the exponential constant s, and express as the ratio of two polynomials in s. Then find the poles and the zeros of each impedance:

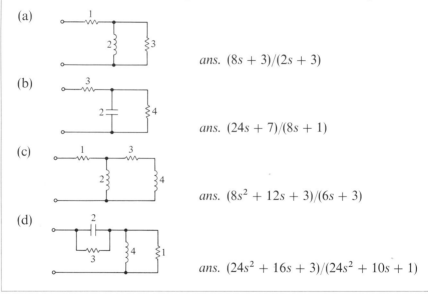

(a)

ans. $(8s + 3)/(2s + 3)$

(b)

ans. $(24s + 7)/(8s + 1)$

(c)

ans. $(8s^2 + 12s + 3)/(6s + 3)$

(d)

ans. $(24s^2 + 16s + 3)/(24s^2 + 10s + 1)$

Using Impedance to Find Natural Response

The natural component of each voltage and current in a network consists of a sum of exponential terms,

$$y_n(t) = K_1 e^{s_1 t} + K_2 e^{s_2 t} + \cdots + K_n e^{s_n t}$$

where n is the number of independent energy-storing elements in the network.

The set of exponential constants $s_1, s_2, \ldots, s_n$ (the *characteristic roots* of the natural behavior) are the same for every network voltage and current, although the constants $K_1, K_2, \ldots, K_n$ are generally different for each signal.

The characteristic roots of a network may be found by setting a loop impedance to zero or by setting the inverse of a network terminal-pair impedance to zero. These characteristic natural behavior roots are the zeros of the loop impedances and they are the poles of the terminal-pair impedances.

Chapter Seven Problems

Basic Problems

Differential Equations

1. Find the solution of

$$\frac{d^2y}{dt^2} + 7\frac{dy}{dt} + 12y = 5e^{-5t}$$

with the boundary conditions

$$y(0) = -3$$

$$\frac{dy}{dt}\bigg|_{t=0} = 2$$

Impedance Concept

2. The impedance of a certain element, as a function of the exponential constant s is

$$Z(s) = \frac{s-2}{3s+1}$$

If the element voltage and current are both exponential and the element voltage is

$$v(t) = -5e^{-4t}$$

what is the sink reference element current?

Networks with Real Exponential Sources

3. Using impedance methods with equivalent circuits, find $v(t)$ and $i(t)$:

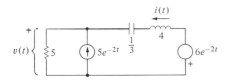

4. Using impedance methods and systematic mesh equations, find the indicated signals:

(a)

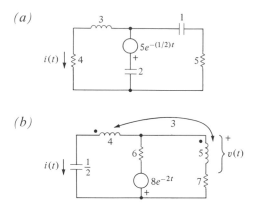

(b)

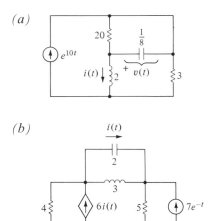

5. Using impedance methods and systematic nodal equations, find the indicated signals:

(a)

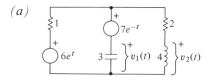

(b)

Superposition of Exponential Sources

6. Using impedance methods, find the indicated signals:

(a)

(b)

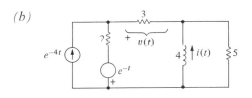

Switched Networks

7. Find $i(t)$ and $v(t)$ and roughly sketch them. Show the signals both before and after $t = 0$:

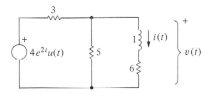

Transfer Functions

8. Find the transfer function

$$T(s) = \frac{i_1(t)}{i_2(t)}\bigg|_{\text{when all signals vary as } e^{st}}$$

for the network:

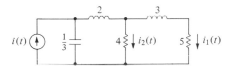

Natural Behavior

9. Find the equations satisfied by the roots of the natural behavior of the following networks using the loop-impedance method, then using the terminal-pair method:

(a)

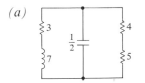

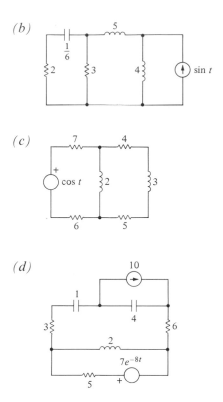

(b)

(c)

(d)

10. A certain two-terminal network has impedance

$$Z(s) = \frac{s^2 + 3s + 2}{s^2 + 5s + 6}$$

Find the characteristic roots of the natural behavior if
(a) The terminals are left unconnected.
(b) The terminals are short circuited.
(c) The terminals are connected to a voltage source.
(d) The terminals are connected to a current source.
(e) The terminals are connected to a 4-Ω resistor.

Practical Problems

Adjustable Inductors

1. Adjustable inductors may be constructed by using an adjustable slider contact that selects the number of coil turns used, or by using a movable magnetic-core material, the position of which controls the degree of magnetic-flux coupling from turn to turn of the coil.

Figure 7-24   Practical Problem 1

To what value of inductance should the inductor L in the network of Figure 7-24 be adjusted to produce a voltage $v_2(t)$ that is five times the voltage $v_1(t)$? What value of L will produce v_2 that is ten times v_1?

The inductor symbol with an arrow through it denotes an adjustable inductor.

An oscilloscope is used to display the relations between voltage and current for semiconductor devices such as transistors.   (*Photo courtesy of Tektronix, Inc.*)

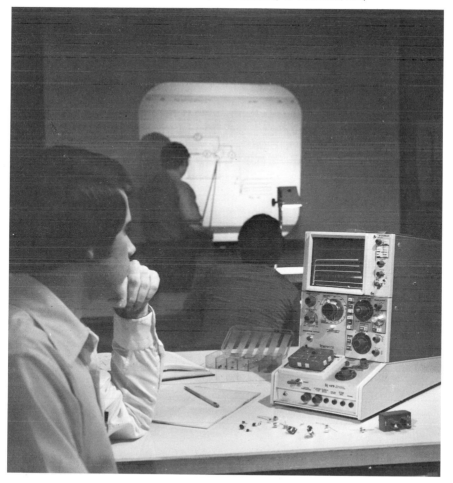

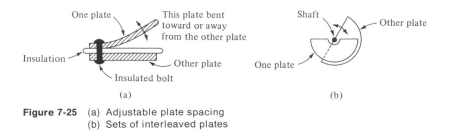

Figure 7-25   (a) Adjustable plate spacing
              (b) Sets of interleaved plates

2. When an inductance is changed with time, the sink reference voltage v is related to the inductor current i by

$$v = \frac{d}{dt}(Li)$$

For a constant inductor current

$$i = 5$$

find an inductance as a function of time $L(t)$ to produce the constant inductor voltage,

$$v = 100$$

Adjustable Capacitors

3. Adjustable capacitors may be constructed with movable sets of plates for which the plate separation or degree of overlap of the plates is changeable, as indicated in Figure 7-25.

What value of capacitance C will produce precisely zero voltage v_2 for an exponential voltage v_1 with exponential constant $a = -3$ in the network of Figure 7-26?

What value of C will produce zero voltage v_2 when $a = -4$?

The capacitor symbol with an arrow through it denotes an adjustable capacitor.

Figure 7-26   Practical Problem 3

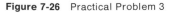
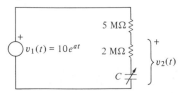

4.   When a capacitance is changed with time, the sink reference capacitor voltage v and current i are related by

$$i = \frac{d}{dt}(Cv)$$

Suppose that over an interval of time an open-circuited capacitance is increased according to

$$C(t) = 10 + 2t$$

If the capacitor voltage is 100 V at time $t = 0$, what is the capacitor voltage as a function of time thereafter?

Natural Behavior

5.   Design a two-terminal network that has natural behavior of the form $K_1 e^{-2t}$ when the terminals are open circuited and $K_2 e^{-3t}$ when the terminals are short circuited.

Digital system circuit cards are tested, and critical voltages are displayed on the oscilloscope screen to the right.   (*Photo courtesy of Tektronix, Inc*)

A television station master control room. The large-screen display devices in the top two rows are television monitors, which are similar to ordinary television sets. Below the monitors are oscilloscopes that display the television signals as functions of time. (*Photo courtesy of Radio Corporation of America*)

Advanced Problems

Exponentially Driven Differential Equations

1. Consider the system described by

$$\frac{dy}{dt} + 2y = e^{at}$$

with $y(0) = 0$. Find the solution $y(t)$ in terms of a for $a \neq -2$.

Impedance

2. Find the impedance of an element for which the sink reference voltage $v(t)$ and current $i(t)$ are related by

$$2\frac{dv}{dt} = 3i - 4\int_{-\infty}^{t} i\,dt$$

What is this element's admittance?

3. The impedance of a certain element, as a function of the exponential constant s, is

$$Z(s) = \frac{3s - 10}{s + 4}$$

Find a differential equation relating the sink reference voltage and current in this element.

4. The impedance of a certain combination of elements is known to be of the form

$$Z(s) = \frac{a_1 s}{a_2 s + 3}$$

where a_1 and a_2 are unknown constants. Find $Z(s)$ if it is known that a current $i(t) = 3e^{-t}$ produces the voltage $v(t) - 3e^{-t}$ and that a current $i(t) = -4e^t$ produces the voltage $v(t) = -2e^t$.

5. Using R's, L's, and C's as needed, design two-terminal networks with the following impedances:

(a) $\quad Z(s) = \dfrac{1}{s + 2}$

(b) $\quad Z(s) - 2s + \dfrac{3}{s + 4} = \dfrac{2s^2 + 8s + 3}{s + 4}$

(c) $\quad Z(s) = \dfrac{4s}{s + 3}$

(d) $\quad Z(s) = \dfrac{s + 3}{4s}$

Switched Exponential-Source Networks

6. For the network of Figure 7-27, write and solve the single loop differential equation for the current $i(t)$ after $t = 0$. Repeat using impedance methods and compare the two solution methods.

Figure 7-27   Advanced Problem 6

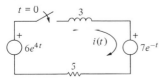

Figure 7-28 Advanced Problem 7

7. Find $v(t)$ in the network of Figure 7-28.

Natural Behavior

8. By considering the form of its open circuit and short-circuited natural response, show that it is not possible to construct a two-terminal network involving positive $L's$, $R's$, and $C's$ with an impedance which has poles or zeros that are positive numbers.

From these it is perceived how imaginary exponential quantities are reduced to the sine and cosine of real arc. For there is

$$e^{+v\sqrt{-1}} = \cos v + \sqrt{-1} \sin v$$

$$e^{-v\sqrt{-1}} = \cos v - \sqrt{-1} \sin v$$

Leonard Euler
Introduction in Analsyin infinitorum,
Lusannae, 1748

While alternating waves can be, and frequently are, represented graphically in rectangular coordinates, with the time as abscissas, and the instantaneous values of the wave as ordinates, the best insight with regard to the mutual relation of different alternating waves is given by their representation in polar coordinates, with the time as the angle or the amplitude one complete period being represented by one revolution and the instantaneous values as radius vectors.

Charles Proteus Steinmetz
*Theory and Calculation of Alternating
Current Phenomena, Fourth Edition,* **1908**

These next three chapters are devoted to networks with sinusoidal sources. Not only are sinusoidal sources of great practical importance in themselves, but sinusoidal network response is the basis of general solution methods for networks with arbitrary source functions.

In the coming chapter it is shown how impedance methods may also be applied to networks with sinusoidal sources by representing sinusoids as complex exponential functions. Source-resistor network methods continue to apply, but with impedances that are complex numbers.

Resonance and related topics are developed in the following chapter, and the final chapter is devoted to sinusoidal power.

PART THREE
Sinusoidal Network Response

To the Instructor

These remaining three chapters concern forced sinusoidal response. Properties of sinusoidal functions and complex algebra are first reviewed carefully.

Sinusoidal response calculations. Having introduced impedance for exponential sources in the previous chapter, the three basic techniques of sinusoidal response calculation easily follow.

Expanding the sinusoid into Euler components, their superposition and solution using impedance is the first, obvious, method. This *method of components* is very straightforward and is fundamental to any later work with the exponential form of the Fourier series.

The redundancy of the two Euler components leads quickly to the *method of sinors* (or rotating phasors) in which it is recognized that only one of the two component problems must be solved. The alternative interpretation of adding imaginary parts to the sources that contribute purely imaginary parts to the solution is also explained. It is pointed out that the sinor method is the usual solution technique in fields other than networks.

The *method of phasors* is presented as being simply a "snapshot" of the sinor solution at time $t = 0$. The phasor method, which is hereafter used exclusively, is then emphasized by drill which includes careful acquaintance with graphical solutions.

Frequency response and resonance. Series and parallel networks are used to introduce the concept of resonance. There then follows discussion of frequency response and of resonance in general, with more involved examples. Both the magnitude extrema and the zero imaginary part definitions of resonance are included and discussed, and attention is given to bandwidth and quality factor, particularly in high-Q networks.

Sinusoidal power. The material on real and reactive power in sinusoidally driven networks is conventional and includes balanced three-phase systems. Power factor correction follows naturally from the discussion of resonance in the previous chapter.

Special emphasis is given to the interpretation and solution of problems of the type found on engineer-in-training examinations, in which such details as polarities, rms values, and element models must often be inferred.

CHAPTER EIGHT
Response to Sinusoidal Sources

8.1   Introduction

This chapter begins with a discussion of sinusoidal functions and the relations between the various ways in which they are commonly represented. Then differential equations with sinusoidal driving functions are examined.

Because complex algebra plays a key role in all of the analysis to follow, considerable effort is devoted initially to a careful review of this topic.

Three impedance methods for solving for forced sinusoidal network response are then developed. Each of the three methods is closely related and each is commonly used in one or another branch of electrical engineering. The method of components is the most straightforward but involves much redundancy, which is exploited with the method of sinors. Additional simplification is obtained in the method of phasors.

A great deal of drill, particularly in phasor methods, is especially needed at this time, and these methods are further reinforced with graphical solutions.

When you complete this chapter, you should know

1. How sinusoidal functions are represented and the meaning of the terms *amplitude*, *frequency*, and *phase*.
2. How to solve differential equations that have sinusoidal driving functions.

3. How to perform algebraic operations with complex numbers, including Euler's relation and conversions between rectangular and polar forms.
4. How to use the method of components to solve networks with sinusoidal sources.
5. How to solve sinusoidally driven networks using the simplification known as the sinor method.
6. How to further streamline sinusoidal network solution with the phasor method.
7. Graphical methods for networks with a single sinusoidal source and how to use them.

8.2 Sinusoidal Functions

8.2.1 Amplitude, Frequency, and Phase

A sinusoidal function (or sinusoid) is a function that may be expressed in the form

$$f(t) = A \cos (\omega t + \theta)$$

where A is a positive constant called the *amplitude* of the sinusoidal function, ω is a positive constant called its *radian frequency* (in radians per second), and θ is a constant called the *phase angle*.

The function, Figure 8-1, repeats, over and over again, every

$$T = \frac{2\pi}{\omega} \quad \text{s}$$

This is to say that it has a *frequency*

$$f = \frac{1}{T} = \frac{\omega}{2\pi} \quad \text{cycles per second (cps) or hertz (Hz)}$$

The argument of the cosine function goes through 2π rad every cycle.

Figure 8-1 Sinusoidal function

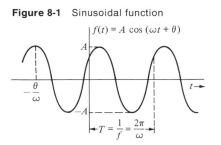

Since the maximum and minimum values of the cosine function are $+1$ and -1, respectively, $f(t)$ has maximum and minimum values of $+A$ and $-A$.

The position of the wave with respect to the origin is fixed by the phase angle θ, which may be expressed within a 2π rad (or $360°$) range. When

$$\omega t + \theta = 0$$

the argument of the cosine function is zero and the value of the function $f(t)$ is maximum. This occurs at the time

$$t = -\frac{\theta}{\omega}$$

and periodically, at multiples of $2\pi/\omega$ s before and after that time.

Although, strictly speaking, the argument of the cosine function is in radians, there is no reason why θ cannot be described in terms of degrees, for example,

$$f(t) = 10 \cos (3t + 45°)$$

so long as the use of degrees is clearly indicated. Then the angle in degrees can be converted to radians according to

$$\text{(angle in radians)} = \frac{2\pi}{360} \text{(angle in degrees)} \cong 0.0175 \text{(angle in degrees)}$$

if and when it is necessary to do so.

The radian frequency ω of a sinusoid is always expressed as a positive number. The trigonometric identity that cosine is an even function,

$$\cos (-x) = \cos x$$

allows conversion of any argument so that the multiplier of t is positive. For example,

$$8 \cos (-3t + 39°) = 8 \cos (3t - 39°)$$

The amplitude A of a sinusoid may also always be expressed as a non-negative number. Since

$$\cos (x + \pi) = -\cos x$$

a negative amplitude may always be converted to a positive one by adding π rad, $180°$, to the phase angle. For example,

$$-6 \cos (4t + 73°) = 6 \cos (4t + 253°)$$

Of course, any integer multiple of 2π rad, $360°$, may be added to, or subtracted from, the argument of a sinusoid without changing the function. So

$$6 \cos (4t + 253°) = 6 \cos (4t + 253° - 360°)$$
$$= 6 \cos (4t - 107°)$$

for example. The phase angle of a sinusoid may thus be expressed within a 360° range, although common practice is to express it within ±360° rather than to bother with deciding which 360° range (±180°, 0–360° or some other) should be the standard.

8.2.2   Representation in Terms of the Sine Function

Sinusoidal functions may also be expressed in terms of the sine function. As indicated by the trigonometric identities of Figure 8-2,

$$\sin x = \cos (x - 90°)$$
$$\cos x = \sin (x + 90°)$$

the only distinction between sine and cosine is a 90° difference in phase angle.

To convert a sinusoidal function expressed in terms of sine to an expression in terms of cosine, subtract 90° from the argument. For example,

$$7 \sin (8t - 105°) = 7 \cos (8t - 195°)$$

To convert a sinusoidal representation in terms of cosine to one in terms of sine, add 90° to the argument. For example,

$$10 \cos (20t - 140°) = 10 \sin (20t - 50°)$$

Sine is an odd function, so

$$\sin (-x) = -\sin x$$

As with cosine,

$$\sin (x + \pi) = -\sin x$$

And, adding or subtracting any integer multiple of 2π, 360°, to the phase angle leaves the sine unchanged.

8.2.3.   Quadrature Representation

Another useful way to represent a sinusoidal function is as the sum of a cosine and a sine, as

$$A \cos (\omega t + \theta) = D \cos \omega t + E \sin \omega t$$

Figure 8-2   Relation between sine and cosine

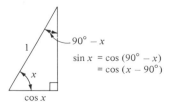

$90° - x$

$\sin x = \cos (90° - x)$
$\quad\quad = \cos (x - 90°)$

x

$\cos x$

This is known as the *quadrature* form for a sinusoid because it is in terms of cosine and sine which differ in phase by 90°, or are in quadrature to one another.

The constants D and E are related to A and θ by

$$D = A \cos \theta$$
$$E = -A \sin \theta$$

as is evident from the trigonometric identity

$$\cos (x + y) = \cos x \cos y - \sin x \sin y$$

That is,

$$A \cos (\omega t + \theta) = (A \cos \theta) \cos \omega t + (-A \sin \theta) \sin \omega t$$

An example of the conversion of a cosine and angle form to the quadrature form is the following:

$$10 \cos (3t + 60°) = (10 \cos 60°) \cos 3t - (10 \sin 60°) \sin 3t$$

$$= \left(\frac{10}{2}\right) \cos 3t + \left(\frac{-10\sqrt{3}}{2}\right) \sin 3t$$

In terms of D and E, the amplitude and phase angle of the quadrature sinusoid are

$$A = \sqrt{D^2 + E^2}$$

$$\theta = -\tan^{-1} \left(\frac{E}{D}\right)$$

To derive this identity, multiply and divide by $\sqrt{D^2 + E^2}$, giving

$$D \cos \omega t + E \sin \omega t = \sqrt{D^2 + E^2} \left[\left(\frac{D}{\sqrt{D^2 + E^2}}\right) \cos \omega t + \left(\frac{E}{\sqrt{D^2 + E^2}}\right) \sin \omega t\right]$$

Defining the angle ϕ as in the sketch of Figure 8-3, and using the identity

$$\cos (x - y) = \cos x \cos y + \sin x \sin y$$

there results

$$D \cos \omega t + E \sin \omega t = \sqrt{D^2 + E^2} [\cos \phi \cos \omega t + \sin \phi \sin \omega t]$$
$$= \sqrt{D^2 + E^2} \cos (\omega t - \phi)$$

where

$$\phi = -\theta = \tan^{-1} \left(\frac{E}{D}\right)$$

Figure 8-3   Angle ϕ in terms of D and E

The following is an example of conversion of a sinusoidal function from quadrature to the cosine and angle form:

$$3 \cos 7t + 3 \sin 7t = \sqrt{9 + 9} \cos (7t - \tan^{-1} 1)$$
$$= 3\sqrt{2} \cos (7t - 45°)$$

By converting to the quadrature form and back again, it is straightforward to convert sums of sinusoidal functions with the same frequency to a single sinusoidal function. For example,

$$4 \cos (2t - 60°) + 3 \cos (2t + 45°)$$

$$= \left(\frac{4}{2}\right) \cos 2t + \left(\frac{4\sqrt{3}}{2}\right) \sin 2t + \left(\frac{3}{\sqrt{2}}\right) \cos 2t - \left(\frac{3}{\sqrt{2}}\right) \sin 2t$$

$$= \left(2 + \frac{3}{\sqrt{2}}\right) \cos 2t + \left(2\sqrt{3} - \frac{3}{\sqrt{2}}\right) \sin 2t$$

$$= (4.12) \cos 2t + (1.63) \sin 2t$$

$$= \sqrt{(4.12)^2 + (1.63)^2} \cos \left(2t - \tan^{-1} \frac{1.63}{4.12}\right)$$

$$= (4.43) \cos (2t - 23°)$$

D8-1

Find the amplitude, radian frequency, and phase angle (in terms of the cosine function) for each of the following sinusoids:

(a) $-10 \cos \left(377t - \frac{5\pi}{6}\right)$ *ans.* $10 \cos (377t + \pi/6)$

(b) $10^5 \cos (-2\pi t + 700°)$ *ans.* $10^5 \cos (2\pi t + 20°)$

(c) $50 \sin \left(\frac{3\pi}{5} - 377t\right)$ *ans.* $50 \cos (377t - \pi/10)$

(d) $-100 \sin (3t + 40°)$ *ans.* $100 \cos (3t + 130°)$

D8-2

Sketch plots of the following sinusoidal functions, indicating amplitude, period, and time to a maximum, minimum, or zero crossing:

(a) $f(t) = 3 \cos\left(8t - \dfrac{\pi}{4}\right)$ *ans.* $3, 2\pi/8, \pi/32$

(b) $f(t) = -10 \cos(10^6 t + 30°)$ *ans.* $10, 2\pi \times 10^{-6}, 2.6 \times 10^{-6}$

(c) $f(t) = 7 \sin(0.3t + 120°)$ *ans.* $7, 20\pi/3, 19.2$

(d) $f(t) = -10^4 \sin\left(t - \dfrac{3\pi}{5}\right)$ *ans.* $10^4, 2\pi, \pi/10$

D8-3

Find, approximately, the sinusoidal functions from the plots. Express in the form $A \cos(\omega t + \theta)$ with $A \geq 0$, $\omega \geq 0$ and $-360° < \theta < 360°$:

(a)

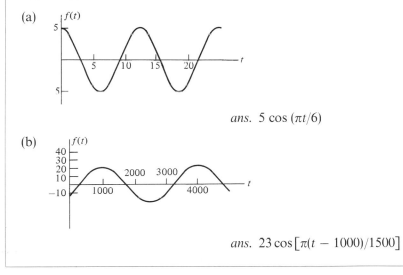

ans. $5 \cos(\pi t/6)$

(b)

ans. $23 \cos[\pi(t - 1000)/1500]$

D8-4

Convert to the form $A \cos(\omega t + \theta)$ with $A \geq 0$, $\omega > 0$ and $-360° < \theta < 360°$:

(a) $f(t) = -2 \cos 5t + \sin 5t$

ans. $\sqrt{5} \cos(5t + 206.5°)$

(b) $f(t) = -6 \cos 10^7 t - 3 \sin(-10^7 t)$

ans. $6.7 \cos(10^7 t + 153.5°)$

(c) $f(t) = -7 \cos 10t + 6 \cos (10t - 45°)$

> *ans.* $5.06 \cos (10t + 237°)$

(d) $f(t) = -3 \cos (2t - 90°) - 4 \cos (2t + 120°)$

> *ans.* $2.05 \cos (2t - 13°)$

Sinusoidal Functions

A sinusoidal function may be expressed in the form

$$f(t) = A \cos (\omega t + \theta)$$

with amplitude A nonnegative, radian frequency ω positive, and phase angle θ in a 360° range.

The following identities allow conversion of sinusoidal functions in other forms to this standard form:

$$\cos (-x) = \cos x, \quad \sin (-x) = -\sin x$$
$$\cos (x + 180°) = -\cos x, \quad \sin (x + 180°) = -\sin x$$
$$\sin x = \cos (x - 90°), \quad \cos x = \sin (x + 90°)$$

and

$$D \cos \omega t + E \sin \omega t = A \cos (\omega t + \theta)$$

where

$$A = \sqrt{D^2 + E^2}, \quad \theta = -\tan^{-1} \frac{E}{D}$$

$$D = A \cos \theta, \quad E = -A \sin \theta$$

8.3 Solutions to Differential Equations with Sinusoidal Driving Functions

For a linear, time-invariant differential equation with sinusoidal driving function,

$$a_n \frac{d^n y}{dt^n} + a_{n-1} \frac{d^{n-1} y}{dt^{n-1}} + \cdots + a_1 \frac{dy}{dt} + a_0 y = A \cos (\omega t + \theta)$$

the forced component of the solution is, except in a special case, sinusoidal with the same radian frequency, ω, as the driving function.

After all, the time derivative of a sinusoid,

$$\frac{d}{dt}[A \cos (\omega t + \theta)] = -A\omega \sin (\omega t + \theta)$$

is another sinusoid with the same frequency. Sums of sinusoids of the same frequency are sinusoidal with that frequency. So if the operations

$$a_n \frac{d^n}{dt^n} + a_{n-1} \frac{d^{n-1}}{dt^{n-1}} + \cdots + a_1 \frac{d}{dt} + a_0$$

are performed on a sinusoid with frequency ω, the result is a sinusoid with frequency ω. The problem is then to find the amplitude and phase angle of the solution that yields the correct driving function amplitude A and phase angle θ.

As the following example illustrates, it is easiest, in directly solving differential equations with sinusoidal driving functions, to deal with the sinusoids in quadrature form.

Consider the equation,

$$\frac{dy}{dt} + 2y = 4 \cos 3t + 2 \sin 3t$$

where the driving function has been expressed in quadrature form. Substituting a trial forced solution with the same frequency as the driving function,

$$y_f(t) = D \cos 3t + E \sin 3t$$

gives

$$\frac{dy_f}{dt} + 2y_f = -3D \sin 3t + 3E \cos 3t + 2D \cos 3t + 2E \sin 3t$$

$$= (2D + 3E) \cos 3t + (-3D + 2E) \sin 3t$$

$$= 4 \cos 3t + 2 \sin 3t$$

Equating coefficients,

$$\begin{cases} 2D + 3E = 4 \\ -3D + 2E = 2 \end{cases}$$

$$D = \frac{\begin{vmatrix} 4 & 3 \\ 2 & 2 \end{vmatrix}}{\begin{vmatrix} 2 & 3 \\ -3 & 2 \end{vmatrix}} = \frac{2}{13}$$

$$E = \frac{\begin{vmatrix} 2 & 4 \\ -3 & 2 \end{vmatrix}}{13} - \frac{16}{13}$$

So

$$y_f(t) = \left(\frac{2}{13}\right)\cos 3t + \left(\frac{16}{13}\right)\sin 3t$$

which could, if desired, be expressed in the cosine and angle form.

The form of the natural component of the solution is found, as usual, from the homogeneous equation,

$$\frac{dy_n}{dt} + 2y_n = 0$$

Subsituting

$$y_n(t) = Ke^{st}$$
$$(s + 2)Ke^{st} = 0$$
$$s = -2$$

and

$$y_n(t) = Ke^{-2t}$$

The general solution to the equation is then

$$y(t) = y_f(t) + y_n(t)$$

$$= \left(\frac{2}{13}\right)\cos 3t + \left(\frac{16}{13}\right)\sin 3t + K_e^{-2t}$$

which involves the one arbitrary constant.

The singular case, similar to that with exponential driving functions, occurs if the equation is such that the natural component of its solution is sinusoidal with the same frequency as the driving function. Then it will not be possible to equate coefficients for the forced component of the solution. Similar to the exponential case, the forced solution in this special circumstance then involves t times the sinusoid.

D8-5

Find the general solutions of each of the following differential equations:

(a) $\dfrac{dy}{dt} = 3\cos 2t$ *ans.* $K + (3/2)\sin 2t$

(b) $\dfrac{dy}{dt} + 2y = 6\sin 3t$ *ans.* $Ke^{-2t} + (-18/13)\cos 3t +$
                                                     $(12/13)\sin 3t$

(c) $2\dfrac{dy}{dt} + 3y = 4 \cos (2t - 45°)$ *ans.* $Ke^{-(3/2)t} - 0.113 \cos 2t +$
$0.8 \sin 2t$

(d) $\dfrac{d^2y}{dt^2} + 3\dfrac{dy}{dt} + 2y = 4 \cos 5t$ *ans.* $K_1e^{-t} + K_2e^{-2t} + (-46/$
$377) \cos 5t + (30/377) \sin 5t$

Solutions to Sinusoidally Driven Differential Equations

The sum of two sinusoids of the same frequency is another sinusoid of that frequency. The derivative of a sinusoid is a sinusoid of the same frequency.

The forced component of the solution of a linear, time-invariant differential equation with a sinusoidal driving function is (except in the singular case) another sinusoid with the same frequency as the driving function.

8.4   The Algebra of Complex Numbers

8.4.1   Rectangular Form

The rather large number of trigonometric identities involved in the solution of the differential equations describing networks may be avoided, or at least simplified, through the use of complex numbers. The principles of complex algebra are now summarized, as preparation for the use of complex numbers throughout the remainder of the text.

A complex number $m = a + jb$ consists of a real part,

$\text{Re}\,[m] = a$

plus $j = \sqrt{-1}$ times an imaginary part,

$\text{Im}\,[m] = b$

Mathematicians often use the symbol i for the imaginary unit, but because of the conflict with the symbol i for electric current (which was chosen before the usefulness of complex numbers in network analysis was generally recognized), engineers and most scientists prefer to use the symbol j.

The algebra and calculus of complex numbers are the same as real number algebra and calculus, where j is just treated as the constant it is. Powers of j are

$$j^2 = (\sqrt{-1})^2 = -1$$
$$j^3 = (\sqrt{-1})^3 = -1\sqrt{-1} = -j$$
$$j^4 = (\sqrt{-1})^4 = (-1)^2 = 1$$
$$j^5 = j$$

and so on.

A real number k times a complex number m gives

$$km = (ka) + j(kb)$$

The *complex conjugate* of a complex number m is denoted by m^* and is defined to be

$$m^* = (a + jb)^* = a - jb = a + j(-b)$$

The complex conjugate has the same real part as the original number, but the negative of its imaginary part. The solutions given by the quadratic formula, when they are complex, are complex conjugates.

If

$$m_1 = a_1 + jb_1 \qquad \text{and} \qquad m_2 = a_2 + jb_2$$

the sum of the complex numbers is given by

$$m_1 + m_2 = a_1 + jb_1 + a_2 + jb_2 = (a_1 + a_2) + j(b_1 + b_2)$$

The sum of a complex number and its complex conjugate is twice the real part of either number:

$$m + m^* = a + jb + a - jb = 2a = 2\,\mathrm{Re}\,[m] = 2\,\mathrm{Re}\,[m^*]$$

The product and quotient of two complex numbers have real and imaginary parts which also may be found using the ordinary rules of algebra. For the product,

$$m_1 m_2 = (a_1 + jb_1)(a_2 + jb_2) = a_1 a_2 + j^2 b_1 b_2 + jb_1 a_2 + jb_2 a_1$$
$$= (a_1 a_2 - b_1 b_2) + j(b_1 a_2 + a_1 b_2)$$

The product of a complex number with its complex conjugate is a real number:

$$mm^* = (a + jb)(a - jb) = a^2 + jab - jab + b^2$$
$$= a^2 + b^2$$

The quotient of two complex numbers requires a little more manipulation. Multiply numerator and denominator by the complex conjugate of the denominator, leaving j's only in the numerator:

$$\frac{m_1}{m_2} = \frac{a_1 + jb_1}{a_2 + jb_2} = \frac{(a_1 + jb_1)(a_2 - jb_2)}{(a_2 + jb_2)(a_2 - jb_2)} = \frac{(a_1a_2 + b_1b_2) + j(b_1a_2 - a_1b_2)}{a_2^2 + b_2^2}$$

For example,

$$\frac{3 - j2}{4 + j5} = \frac{(3 - j2)(-4 - j5)}{(-4 + j5)(-4 - j5)}$$

$$= \frac{-12 - j15 + j8 - 10}{16 + 25}$$

$$= -\frac{22}{41} - j\frac{7}{41}$$

8.4.2 Polar Coordinates

It is often convenient to plot the real and imaginary parts of complex numbers on a two-axis plot which is called the *complex plane*, Figure 8-4. An arrow is drawn from the origin to the point representing a complex number. Every complex number is represented by a unique point on the complex plane and every point on the plane represents a complex number.

The distance from the origin of the complex plane to the point on the plane representing a complex number is called the *magnitude* of a complex number m and is denoted by $|m|$. In terms of the number's real and imaginary parts,

$$|m| = |a + jb| = \sqrt{a^2 + b^2}$$

The *angle* of a complex number, denoted by $\angle m$, is the angle a line from the origin to the point representing the complex number makes with the positive real axis:

$$\angle m = \tan^{-1}\frac{b}{a}$$

The algebraic signs of both b and a must be examined to determine the quadrant of $\angle m$; the ratio alone, (b/a), is not sufficient to determine the angle.

Figure 8-4 Complex number plotted on the complex plane

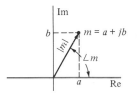

For example, if

$$m = -3 + j3$$
$$|m| = \sqrt{3^2 + 3^2} = 3\sqrt{2}$$

and

$$\angle m = \tan^{-1} \frac{3}{-3} = 135°$$

as shown in Figure 8-5.

In terms of its magnitude and angle, the real and imaginary parts of a complex number are

$$\text{Re}\,[m] = |m|\cos(\angle m)$$
$$\text{Im}\,[m] = |m|\sin(\angle m)$$

If

$$|m| = 4 \qquad \text{and} \qquad \angle m = 60°$$

$$m = a + jb = |m|\cos(\angle m) + j|m|\sin(\angle m)$$

$$= 4\left(\frac{1}{2}\right) + j4\left(\frac{\sqrt{3}}{2}\right)$$

$$= 2 + j2\sqrt{3}$$

Multiplication of a complex number m by a positive real number k multiplies the magnitude by k and leaves the angle unchanged:

$$\begin{cases} |km| = k|m| \\ \angle km = \angle m, \qquad \text{if } k \text{ is positive} \end{cases}$$

If the real number is negative, then the magnitude of the product is changed by the factor $|k|$, and the direction of the product is reversed, that is, the angle is changed by $180°$:

$$\begin{cases} |km| = |k|\,|m| \\ \angle km = \angle m + 180°, \qquad \text{if } k \text{ is negative} \end{cases}$$

Figure 8-6 shows these relationships.

Figure 8-5 Example of magnitude and angle of a complex number

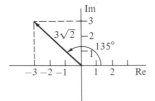

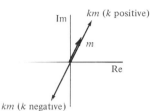

Figure 8-6 Graphical multiplication of a complex number by a real number

The addition of two complex numbers is easy to perform graphically on the complex plane. If a parallelogram is constructed from the sides m_1 and m_2, the diagonal of the parallelogram is the arrow for the sum $(m_1 + m_2)$.

Subtraction of one complex number from another,

$$m_1 - m_2$$

may be easily done graphically on the complex plane by first constructing $(-m_2)$ from m_2, by reversing the direction of the arrow for m_2, and then adding m_1 and $(-m_2)$ by constructing the addition parallelogram.

If

$$m_1 = 2 - j3 \quad \text{and} \quad m_2 = -1 + j4,$$

the sum

$$m_1 + m_2 - 1 + j$$

constructed graphically is shown in Figure 8-7(a). The difference,

$$m_1 - m_2 = 3 - j7$$

is constructed graphically in Figure 8.7(b).

Figure 8-7 Graphical addition and subtraction of complex numbers

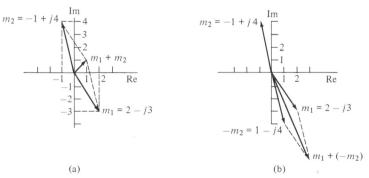

(a) (b)

8.4.3 Euler's Relation

Specifying the magnitude and angle of a complex number is equivalent to specifying its real and imaginary parts. There is a difference, though, in that up to now $m = a + jb$ is the complex number, and $(|m|, \angle m)$ are only a set of coordinates for the number. Leonard Euler (1707–1783), a Swiss mathematical genius, changed this viewpoint considerably. He showed that any complex number $m = a + jb$ could also be expressed in the polar form

$$m = a + jb = K e^{j\theta}$$

where

$$K = |m| \qquad \text{and} \qquad \theta = \angle m$$

Consider the power series for the following functions:

$$e^x = 1 + \frac{x}{1!} + \frac{x^2}{2!} + \frac{x^3}{3!} + \cdots$$

$$\cos x = 1 - \frac{x^2}{2!} + \frac{x^4}{4!} - \frac{x^6}{6!} + \cdots$$

$$\sin x = \frac{x}{1!} - \frac{x^3}{3!} + \frac{x^5}{5!} - \frac{x^7}{7!} + \cdots$$

The angle x must be expressed in *radians* for the series for $\cos x$ and $\sin x$ to be in this simple form.

If in the series for e^x, x is replaced in the series by $j\alpha$, there results

$$e^{j\alpha} = 1 + \frac{j\alpha}{1!} + \frac{j^2\alpha^2}{2!} + \frac{j^3\alpha^3}{3!} + \frac{j^4\alpha^4}{4!} + \frac{j^5\alpha^5}{5!} + \frac{j^6\alpha^6}{6!} + \frac{j^7\alpha^7}{7!} + \cdots$$

Now $j^2 = (\sqrt{-1})^2 = -1$, $j^3 = j(j^2) = -j$, $j^4 = j(j^3) = 1$, $j^5 = j(j^4) = j$, and so on, so

$$e^{j\alpha} = \left(1 - \frac{\alpha^2}{2!} + \frac{\alpha^4}{4!} - \frac{\alpha^6}{6!} + \cdots\right) + j\left(\frac{\alpha}{1!} - \frac{\alpha^3}{3!} + \frac{\alpha^5}{5!} - \frac{\alpha^7}{7!} + \cdots\right)$$

$$= \cos \alpha + j \sin \alpha$$

This result is known as Euler's relation.

A complex number may thus be expressed in terms of its magnitude

$$K = |m| = \sqrt{a^2 + b^2}$$

and angle

$$\theta = \angle m = \tan^{-1}\frac{b}{a}$$

as

$$m = a + jb = K \cos \theta + jK \sin \theta$$
$$= Ke^{j\theta}$$

For example,

$$-3 - j2 = 3.61e^{j214°} = 3.61e^{-j146°}$$

as shown in Figure 8-8.

Some identities which will be very useful later in this chapter may be obtained as follows. Substituting $-\alpha$ for α in Euler's relation,

$$e^{-j\alpha} = \cos(-\alpha) + j \sin(-\alpha)$$
$$= \cos \alpha - j \sin \alpha$$

Adding $e^{j\alpha}$ and $e^{-j\alpha}$ gives

$$e^{j\alpha} + e^{-j\alpha} = 2 \cos \alpha$$

or

$$\cos \alpha = \frac{e^{j\alpha} + e^{-j\alpha}}{2}$$

Similarly,

$$\sin \alpha = \frac{e^{j\alpha} - e^{-j\alpha}}{2j}$$

8.4.4 Multiplication and Division in Polar Form

Using the exponential properties

$$e^x e^y = e^{(x+y)} \qquad \text{and} \qquad \frac{e^x}{e^y} = e^{(x-y)}$$

multiplication and division of complex numbers is seen to be very easy when performed in the polar form.

The product of two complex numbers,

$$(K_1 e^{j\theta_1})(K_2 e^{j\theta_2}) = K_1 K_2 e^{j(\theta_1 + \theta_2)}$$

Figure 8-8 Rectangular to polar conversion

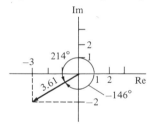

has magnitude the product of the individual magnitudes and angle that is the sum of the individual angles. For example,

$$(3e^{j36°})(4e^{-j120°}) = 12e^{-j84°}$$

The quotient of two complex numbers,

$$\frac{(K_1 e^{j\theta_1})}{(K_2 e^{j\theta_2})} = \frac{K_1}{K_2} e^{j(\theta_1 - \theta_2)}$$

has magnitude the quotient of the individual magnitudes and angle equal to the numerator angle minus the denominator angle. For example,

$$\frac{7e^{-j100°}}{10e^{j70°}} = \frac{7}{10} e^{-j170°}$$

More complicated products and quotients are similarly formed:

$$\frac{(11e^{j30°})(5e^{-j50°})}{(4e^{-j120°})(6e^{j70°})} = \frac{(11)(5)}{(4)(6)} e^{j(30° - 50° + 120° - 70°)}$$

$$= \frac{55}{24} e^{j30°}$$

D8-6

Express the following complex numbers in the form $a + jb$:

(a) $(2 - j3)(j4)$ *ans.* $12 + j8$

(b) $\dfrac{(1 - j2)}{(5 + j6)}$ *ans.* $(-7/61) - j(16/61)$

(c) $\dfrac{(-2 - j8)(4 + j1)}{(3 - j5)}$ *ans.* $5 - j3$

(d) $\dfrac{(-4 + j3)}{(-3 + j2)(5 + j9)}$ *ans.* $(81/1378) + j(-167/1378)$

D8-7

Sketch plots of the following complex numbers on the complex plane. Show the axes and scales carefully:

(a) $2 + j2$

(b) $2 - j2$

(c) $-2 + j2$

(d) $-2 - j2$

D8-8

Calculate the magnitudes and angles of each of the following complex numbers:

(a)	$1 + j2$	*ans.* $\sqrt{5}, 63.5°$
(b)	$3 - j4$	*ans.* $5, -53.1°$
(c)	$-4 + j3$	*ans.* $5, 143.1°$
(d)	$-2 - j1$	*ans.* $\sqrt{5}, 206.5°$

D8-9

Find approximate graphical solutions for the magnitudes and angles of the following complex numbers using a sketch of the number on the complex plane:

(a)	$7 - j8$	*ans.* $10.6, -48.7°$
(b)	$-10^4 + j10^5$	*ans.* $1.005 \times 10^5, 95.7°$
(c)	$-0.13 - j0.52$	*ans.* $0.54, -104°$
(d)	$1.7 + j4.6$	*ans.* $4.9, 69.4°$

D8-10

Express the following complex numbers in polar form, using the Euler identity. Do not use graphical methods:

(a)	$-4 + j4$	*ans.* $4\sqrt{2}e^{j135°}$
(b)	$3 - j4$	*ans.* $5e^{-j53.1°}$
(c)	$-4 - j3$	*ans.* $5e^{j216.8°}$
(d)	$j4(2 - j2)$	*ans.* $8\sqrt{2}e^{j45°}$

D8-11

Express the following complex numbers in rectangular form. Do not use graphical methods:

(a)	$6e^{-j45°}$	*ans.* $(6/\sqrt{2}) - j(6/\sqrt{2})$
(b)	$3e^{j30°}$	*ans.* $2.6 + j1.5$
(c)	$10^6e^{-j60°}$	*ans.* $5 \times 10^5 - j8.7 \times 10^5$
(d)	$-5e^{j120°}$	*ans.* $2.5 - j4.33$

D8-12

Find the polar form, $Ke^{j\theta}$ with $K \geq 0$ and $-360° < \theta < 360°$, of each of the following complex numbers. Use graphical aid where appropriate:

(a) $(6e^{j40°})(-2e^{j610°})$ *ans.* $12e^{j110°}$

(b) $\dfrac{5e^{j30°}(2-j2)}{-2e^{j60°}}$ *ans.* $7.08e^{j105°}$

(c) $\dfrac{7e^{-j50°}}{(-j4)(3-j4)}$ *ans.* $0.35e^{j93°}$

(d) $\dfrac{3e^{j70°}(-4+j3)}{-3e^{j150°}}$ *ans.* $5e^{j243°}$

Complex Algebra

A complex number in rectangular form

$$m = a + jb$$

consists of real part a and imaginary part b. The complex conjugate of a complex number m is denoted by m^* and is

$$m^* = a - jb$$

Complex algebra is identical to real number algebra, where $j = \sqrt{-1}$ is treated as a constant. Addition and subtraction of complex numbers are easiest to do in rectangular form and may be done graphically using the parallelogram rule.

A complex number may alternatively be expressed in polar form,

$$m = Ke^{j\theta}$$

where the magnitude is

$$K = |m| = \sqrt{a^2 + b^2}$$

and the angle of the complex number is

$$\theta = \angle m = \tan^{-1}\left(\frac{b}{a}\right)$$

Multiplication and division of complex numbers is easiest to perform in polar form.

Euler's relation

$$e^{+j\alpha} = \cos \alpha \pm j \sin \alpha$$

leads to the identities

$$\cos \alpha = \frac{e^{j\alpha} + e^{-j\alpha}}{2}$$

$$\sin \alpha = \frac{e^{j\alpha} - e^{-j\alpha}}{2j}$$

8.5   The Method of Components

8.5.1   Superposition of Euler Components

The Euler's relation identity

$$\cos \alpha = \tfrac{1}{2}e^{j\alpha} + \tfrac{1}{2}e^{-j\alpha}$$

allows expression of a sinusoidal source as the sum of two exponential components, which may be superimposed. Each component is exponential, so that impedance methods may be used for solution instead of involved trigonometric identities.

For example, in the network of Figure 8-9, the sinusoidal source has been expressed as the sum of two complex exponential components and the two terms, which involve different values of s, superimposed.

Figure 8-9   Superposition of complex exponential source components

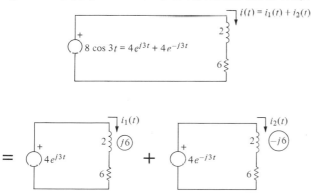

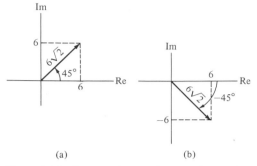

Figure 8-10 Rectangular to polar complex number conversions

For the first component problem, $s = j3$ and the current is

$$i_1(t) = \frac{4e^{j3t}}{6 + j6}$$

To perform the indicated division, convert the denominator to polar form, as indicated in Figure 8-10(a):

$$6 + j6 = 6\sqrt{2}e^{j45°}$$

Then

$$i_1(t) = \frac{4e^{j3t}}{6 + j6} = \frac{4e^{j3t}}{6\sqrt{2}e^{j45°}} = \frac{2}{3\sqrt{2}}e^{j(3t - 45°)}$$

Similarly, the second component problem has $s = -j3$ and the second current component is

$$i_2(t) = \frac{4e^{-j3t}}{6 - j6}$$

Converting to polar form, Figure 8-10(b),

$$6 - j6 = 6\sqrt{2}e^{-j45°}$$

and

$$i_2(t) = \frac{4e^{-j3t}}{6 - j6} = \frac{4e^{-j3t}}{6\sqrt{2}e^{-j45°}} = \frac{2}{3\sqrt{2}}e^{j(-3t + 45°)}$$

$$= \frac{2}{3\sqrt{2}}e^{-j(3t - 45°)}$$

The component currents, being complex functions of time, could not be measured in the laboratory; physical currents are real functions of time. Their sum, however, is the physical current of interest:

$$i(t) = i_1(t) + i_2(t)$$

$$= \frac{2}{3\sqrt{2}} e^{j(3t - 45°)} + \frac{2}{3\sqrt{2}} e^{-j(3t - 45)}$$

$$= \frac{2}{3\sqrt{2}} \left[e^{j(3t - 45°)} + e^{-j(3t - 45°)} \right]$$

$$= \frac{2}{3\sqrt{2}} [2 \cos (3t - 45°)]$$

$$- \frac{4}{3\sqrt{2}} \cos (3t - 45°)$$

8.5.2   Voltage-Source Example

In the more general problem, where the sinusoidal source function has a nonzero phase angle,

$$f(t) = A \cos (\omega t + \theta)$$

the Euler components are

$$f(t) = A \cos (\omega t + \theta) = \frac{A}{2} e^{j(\omega t + \theta)} + \frac{A}{2} e^{-J(\omega t + \theta)}$$

$$= \left(\frac{A}{2} e^{j\theta} \right) e^{j\omega t} + \left(\frac{A}{2} e^{-j\theta} \right) e^{-j\omega t}$$

The first component,

$$\left(\frac{A}{2} e^{j\theta} \right) e^{j\omega t}$$

is of the form Ke^{st} with

$$K = \frac{A}{2} e^{j\theta}$$

and $s = j\omega$. For the second component,

$$K = \frac{A}{2} e^{-j\theta}$$

and $s = -j\omega$.

A numerical network solution example is shown in Figure 8-11. The first current component is

$$i_1(t) = \frac{\frac{7}{2} e^{j(4t + 30°)}}{2 - j\frac{5}{4}}$$

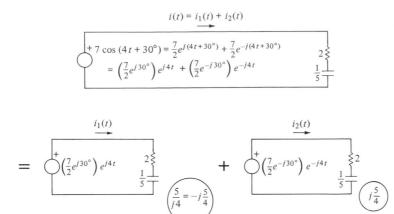

Figure 8-11 Component solution of a network containing a voltage source

To place $i_1(t)$ in polar form, the denominator is converted to the polar form

$$2 - j\tfrac{5}{4} = 2.36e^{-j32°},$$

as indicated graphically in Figure 8-12. Then

$$i_1(t) = \frac{3.5e^{j(4t+30°)}}{2.36e^{j(-32°)}} = 1.48e^{j(4t+62°)}$$

For the second component,

$$i_2(t) = \frac{\tfrac{7}{2}e^{-j(4t+30°)}}{2 + j\tfrac{5}{4}} = \frac{3.5e^{-j(4t+30°)}}{2.36e^{j32°}}$$

$$= 1.48e^{-j(4t+62°)}$$

which gives

$$i(t) = i_1(t) + i_2(t) = 1.48[e^{j(4t+62°)} + e^{-j(4t+62°)}]$$
$$= 1.48[2 \cos (4t + 62°)]$$
$$= 2.96 \cos (4t + 62°)$$

Figure 8-12 Rectangular to polar conversion for the voltage source problem

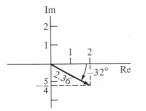

8.5.3 Current-Source Example

The network of Figure 8-13 involves a current source. In this example, the original problem is decomposed into the superposition of the two Euler component problems, to which impedance methods apply. Then equivalent circuits are used to solve each of the component problems by placing them in the form of current dividers.

The solution to the first component problem is

$$i_1(t) = \frac{(-j\frac{1}{6})(\frac{15}{2}e^{j(3t-100°)})}{-j\frac{1}{6} + 1 + j\frac{3}{4}}$$

$$= \frac{(\frac{1}{6}e^{-j90°})(\frac{15}{2}e^{j(3t-100°)})}{1 + j\frac{7}{12}}$$

The $-j\frac{1}{6}$ term in the numerator is easily converted to polar form. The denominator term is converted to polar form as indicated in Figure 8-14. Then

$$i_1(t) = \frac{(\frac{1}{6}e^{-j90°})(\frac{15}{2}e^{j(3t-100°)})}{1.34e^{j30.2°}}$$

$$= 0.93e^{j(3t-220.2°)}$$

Figure 8-13 Component solution of a network containing a current source

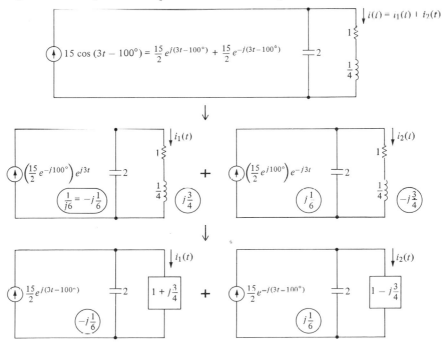

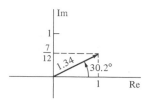

Figure 8-14 Rectangular to polar conversion for the current source problem

Similarly, using the current divider rule for the second component problem,

$$i_2(t) = \frac{(j\frac{1}{6})(\frac{15}{2}e^{-j(3t-100°)})}{j\frac{1}{6} + 1 - j\frac{3}{4}}$$

$$= \frac{(\frac{1}{6}e^{j90°})(\frac{15}{2}e^{-j(3t-100°)})}{1 - j\frac{7}{12}}$$

$$= \frac{(\frac{1}{6}e^{j90°})(\frac{15}{2}e^{-j(3t-100°)})}{1.34e^{-j30.2°}}$$

$$= 0.93e^{j(-3t+220.2°)}$$

$$= 0.93e^{-j(3t-220.2°)}$$

The current $i(t)$ is

$$i(t) = i_1(t) + i_2(t)$$
$$= 0.93[e^{j(3t-220.2°)} + e^{-j(3t-220.2°)}]$$
$$= 0.93[2\cos(3t - 220.2°)]$$
$$= 1.86\cos(3t - 220.2°)$$

Once a sinusoidally driven network problem is converted to equivalent component problems with impedances, any of the source-resistor network solution techniques are applicable.

D8-13

Expand the following sinusoidal functions into Euler components:

(a) $8\cos 7t$ ans. $4e^{j7t} + 4e^{-j7t}$

(b) $8\cos(6t + 58°)$ ans. $4e^{-j(6t+58°)} + 4e^{-j(6t+58°)}$

(c) $\left(\frac{1}{20}\right)\cos(100t - 80°)$ ans. $(1/40)(e^{-j80°})e^{j100t} +$
                                          $(1/40)(e^{j80°})(e^{-j100t})$

(d) $10\sin 5t$ ans. $(5e^{-j90°})e^{j5t} + (5e^{j90°})e^{-j5t}$

D8-14

Find the indicated signals using the method of components:

(a)

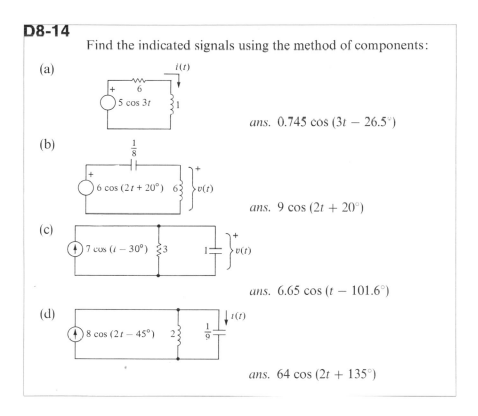

ans. $0.745 \cos (3t - 26.5°)$

(b)

ans. $9 \cos (2t + 20°)$

(c)

ans. $6.65 \cos (t - 101.6°)$

(d)

ans. $64 \cos (2t + 135°)$

The Method of Components

A sinusoidal source may be decomposed into the sum of two complex exponential sources via Euler's relation. The two complex exponential sources may be superimposed, and impedance methods may be used separately for each component.

8.6   The Method of Sinors

8.6.1   Redundancy of the Two Component Solutions

Considerable savings in the effort needed to find forced network response to a sinusoidal input may be made. The two Euler component responses, like the input components, are always complex conjugates of one another. That is, their real parts are identical and their imaginary parts are negatives

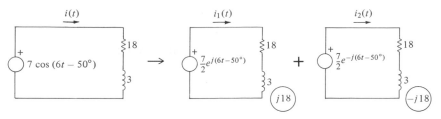

Figure 8-15   Component solution

of one another. Thus it is only necessary to *solve* for one of the two components; the second component may be easily found from the first.

As an example, the component solution of a network is outlined in Figure 8-15. The second component problem is related to the first by

$$i_2(t) = i_1{}^*(t).$$

The original problem solution is thus

$$i(t) = i_1(t) + i_2(t) = i_1(t) + i_1{}^*(t)$$
$$= 2 \operatorname{Re}\left[i_1(t)\right]$$

the sum of a complex number and its complex conjugate being twice the real part of either number.

8.6.2   The Sinor Solution

In the sinor solution method, the division by two to form the first component and the later multiplication by two (taking twice the real part) is eliminated. The response due to twice the first component is found, then one times the real part of that result is the desired signal.

The sinusoidal source,

(amplitude) cos (argument)

is replaced by the sinor source,

(amplitude) $e^{j(\text{argument})}$

converting the original problem into the equivalent sinor problem, as in Figure 8-16. Since inductor and capacitor impedances are imaginary

Figure 8-16   Sinor solution

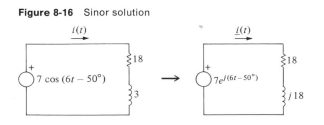

numbers, circling impedances to distinguish them from element values on the network diagram here is unnecessary.

Impedance methods apply to the sinor problem which, for the example, gives

$$\underline{i}(t) = \frac{7e^{j(6t-50°)}}{18+j18} = \frac{7e^{j(6t-50°)}}{18\sqrt{2}e^{j45°}}$$

$$= \frac{7}{18\sqrt{2}}e^{j(6t-95°)}$$

The sinor current is distinguished from the original current by the underbar. The original current is then

$$i(t) = \text{Re}\left[\underline{i}(t)\right]$$

$$= \text{Re}\left[\frac{7}{18\sqrt{2}}e^{j(6t-95°)}\right]$$

$$= \frac{7}{18\sqrt{2}}\cos(6t-95°)$$

Figure 8-17 shows another example of the sinor solution method. The original, sinusoidal, problem is replaced by the corresponding sinor network. Combining the individual impedances into a single equivalent impedance gives

$$\underline{v}(t) = \frac{[10e^{j((1/2)t+80°)}](j2)(3-j)}{(3+j)}$$

Next, $\underline{v}(t)$ is expressed in polar form:

$$\underline{v}(t) = \frac{[10e^{j((1/2)t+80°)}](2e^{j90°})(3.16e^{-j18.3°})}{(3.16e^{j18.3°})}$$

$$= 20e^{j((1/2)t+133.4°)}$$

Figure 8-17 Another network solution using the sinor method

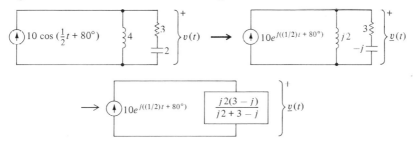

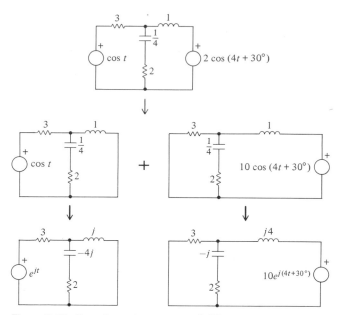

Figure 8-18 Superimposing sources of different frequency

The desired voltage $v(t)$ is then

$$v(t) = \operatorname{Re}\left[\underline{v}(t)\right] = 20 \cos\left(\tfrac{1}{2}t + 133.4°\right)$$

The real part of each voltage and each current sinor in the sinor network is the corresponding voltage or current of interest in the original, sinusoidal, network.

8.6.3 Multiple Sources

In a network with more than a single sinusoidal source, the sources *must* be superimposed if they are of different frequency. After all, the source sinors will have different values of s. An example of such a network is given in Figure 8-18. The sources are superimposed, and each of the two single-source problems are solved using sinors. For one source, $s = j1$ and for the other, $s = j4$.

If there are multiple sources of the same frequency, they need not be superimposed, because they involve the same value of s. The example of Figure 8-19 illustrates the solution of a multiple-source network, where the sources all have the same frequency.

Systematic simultaneous loop equations for the sinor network are as follows:

$$\begin{cases} (5 - j)\underline{i}_1(t) - (2 - j)\underline{i}_2(t) = e^{j4t} \\ -(2 - j)\underline{i}_1(t) + (2 + j3)\underline{i}_2(t) = 10e^{j(4t + 30°)} \end{cases}$$

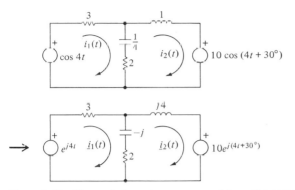

Figure 8-19   Sinor solution when sources all have the same frequency

These may be solved for $i_1(t)$ and $i_2(t)$. Other sinor voltages and currents are easily found from $i_1(t)$ and $i_2(t)$, and the real part of each sinor signal is the corresponding sinusoidal signal of interest.

8.6.4   Sources Expressed in Terms of the Sine Function

If a network source is expressed in terms of the sine function instead of cosine, it may be converted into a cosine expression by subtracting $90°$ from the argument:

$$A \sin (\omega t + \theta) = A \cos (\omega t + \theta - 90°)$$

Some savings in computational effort may be made by recognizing that if a source with phase angle θ produces a response with phase angle ϕ, then a source phase, instead, of $(\theta - 90°)$ will give response with phase angle $(\phi - 90°)$.

Figure 8-20 shows an example of a phase-shifted sinor technique. The source $7 \sin (4t - 20°)$ is treated as if it were $7 \cos (4t - 20°)$. Using the

Figure 8-20   Source and solution in terms of the sine function

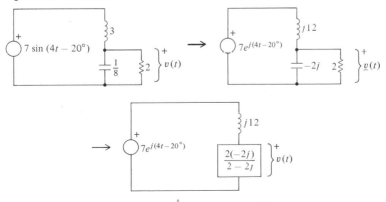

voltage divider rule,

$$\underline{v}(t) = \frac{\left(\dfrac{-4j}{2-2j}\right)7e^{j(4t-20°)}}{\dfrac{-4j}{2-2j}+12j} = \frac{(-4j)[7e^{j(4t-20°)}]}{24+20j}$$

$$= \frac{(4e^{-j90°})[7e^{j(4t-20°)}]}{31.2e^{j39.8°}} = 0.9e^{j(4t-149.8°)}$$

The solution for the desired voltage is then, replacing cosine by sine for the result,

$$v(t) = 0.9\sin(4t - 149.8°)$$

Alternatively, it may be interpreted that the *imaginary* parts of the sinor quantities are the actual voltages and currents of interest in this situation.

Of course, this technique will not work if the network has several sources, some expressed in terms of cosine and some in terms of sine. All of the sources must be placed in terms of one or the other, all cosine or all sine.

D8-15

Find the sinors which represent the following sinusoidal functions:

(a) $6\cos(3t + 47°)$ *ans.* $6e^{j(3t+47°)}$

(b) $8\cos\left(8t - \dfrac{3\pi}{4}\right)$ *ans.* $8e^{j[8t-(3\pi/4)]}$

(c) $10^5\cos 10^6 t$ *ans.* $10^5 e^{j10^6 t}$

(d) $2.6\cos(377t - 120°)$ *ans.* $2.6e^{j(377t-120°)}$

D8-16

Find the indicated signals using the method of sinors:

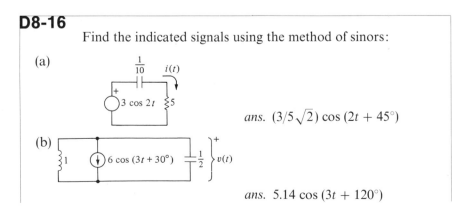

(a)

ans. $(3/5\sqrt{2})\cos(2t + 45°)$

(b)

ans. $5.14\cos(3t + 120°)$

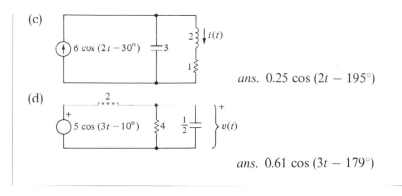

(c)

$6 \cos (2t - 30^\circ)$

ans. $0.25 \cos (2t - 195^\circ)$

(d)

$5 \cos (3t - 10^\circ)$

ans. $0.61 \cos (3t - 179^\circ)$

The Method of Sinors

The responses of the two Euler components of a sinusoidal source are always complex conjugates of one another. Thus it is only necessary to solve one of the component problems; the solution to the second component problem is the complex conjugate of the first.

In the sinor method, a sinusoidal source

$A \cos (\omega t + \theta)$

is replaced by the sinor source

$A e^{j(\omega t + \theta)}$

The real part of every voltage and current in the sinor network is the solution for the corresponding signal in the sinusoidal network.

The sinor may alternatively be interpreted as the addition of the imaginary part $jA \sin (\omega t + \theta)$ to the actual sinusoidal source, $A \cos (\omega t + \theta)$. The real part of every sinor network voltage and current is the solution due to the real part of the source.

8.7   The Method of Phasors

8.7.1   "Snapshots" of Sinors

When plotted on the complex plane, the sinor

$\underline{y}(t) = A e^{j(\omega t + \theta)}$

circles counterclockwise about the origin at the angular rate of ω rad/s as indicated in the drawing of Figure 8-21(a). The projection of the sinor on the real axis, its real part, is the desired sinusoidal function, Figure 8-21(b).

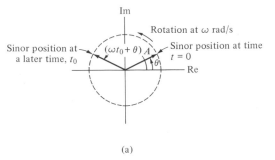

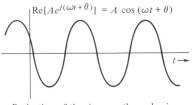

Projection of the sinor on the real axis, as a function of time

(a)

(b)

Figure 8-21 Sinor rotating on the complex plane

All of the sinors for the various voltages and currents in a sinusoidally driven network involve the same radian frequency ω and so rotate together on the complex plane at the same rate of ω rad/s. These sinors maintain the same angular relationships to one another as they rotate.

The complex value of a sinor at time $t = 0$ is called the *phasor* for the corresponding network signal:

$$\mathbf{Y} = \underline{y}(0) = Ae^{j\theta}$$

Phasor quantities are identified by boldface type in printed text. Hand written characters representing phasors are underlined with a wavy line (like a cycle of a sinusoid) to distinguish them. The phasor magnitude is the amplitude of the sinusoidal signal it represents,

$$|\mathbf{Y}| = A,$$

and its angle is the sinusoidal phase angle:

$$\angle \mathbf{Y} = \theta$$

8.7.2 The Phasor Solution

The phasor solution of a sinusoidally driven network is the solution of the sinor network at time $t = 0$. The sinors at time $t = 0$ convey all of the needed information about the solutions for the various sinusoidal network signals, their amplitudes and their phase angles. Their frequencies are all the same as that of the source or sources.

In the phasor solution method, the original problem is replaced by the phasor problem, which is the sinor network at time $t = 0$, as in the example of Figure 8-22. The net effect is to drop the factors of $e^{j\omega t}$ from each of the sinors to form the phasors.

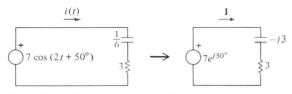

Figure 8-22 Phasor network solution

The sinusoidal source, $7 \cos (2t + 50°)$, is replaced by the phasor source, $7e^{j50°}$, with magnitude equal to the source amplitude and angle the source's phase angle. The impedances are evaluated at $s = j\omega = j2$, as with sinors, resulting in a complex number source-resistor network.

The phasor $\mathbf{I}$ is the value of the sinor $\underline{i}(t)$ at time $t = 0$. It is a complex number with magnitude equal to the amplitude of the corresponding current of interest $i(t)$, and angle equal to the phase angle of $i(t)$. The single-loop equation for $\mathbf{I}$ is

$$(3 - j3)\mathbf{I} = 7e^{j50°}$$

$$\mathbf{I} = \frac{7e^{j50°}}{3 - j3} = \frac{7e^{j50°}}{3\sqrt{2}e^{-j45°}} = \frac{7}{3\sqrt{2}} e^{j95°}$$

The current $i(t)$ thus has amplitude $7/3\sqrt{2}$ and phase angle $95°$. Its radian frequency is the same as the network source, $\omega = 2$:

$$i(t) = \frac{7}{3\sqrt{2}} \cos (2t + 95°)$$

Another example is shown in Figure 8-23. The original network, in which it is desired to find $i(t)$ and $v(t)$, is replaced by the phasor network. Equivalent circuits will be used to solve for $\mathbf{I}$ and $\mathbf{V}$ in the phasor network.

Figure 8-23 Phasor solution using equivalent circuits

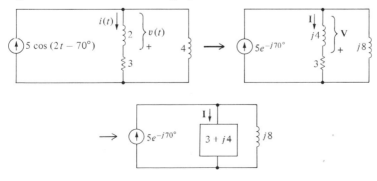

Combining the impedances 3 and $j4$ in series and using the current divider rule gives

$$I = \frac{j8(5e^{-j70°})}{3 + j4 + j8} \quad \frac{(8e^{j90°})(5e^{-j70°})}{(12.37e^{j56.3°})}$$

$$= 3.23e^{-j36.3°}$$

In the original network, then

$$i(t) = 3.23 \cos (2t - 36.3°)$$

The phasor V is now found:

$$V = -(j4)I$$

The minus sign in the relation between V and I is due to the source relationship between the references for V and I.

$$V = -j4I = (4e^{-j90°})(3.23e^{-j36.3°})$$
$$= 12.92e^{-j126.3°}$$

So

$$v(t) = 12.92 \cos (2t - 126.3°)$$

8.7.3   Simultaneous Equations

Mesh and nodal equations are written for phasor networks exactly as for source-resistor networks. The phasor sources and impedances are generally complex numbers, so that complex algebra is involved, but the equations and their solutions proceed just as they did for source-resistor networks.

Suppose it is desired to find the current $i_1(t)$ in the network of Figure 8-24. The systematic, simultaneous mesh equations for the phasor network are as follows:

$$\begin{cases} (9 + j9)I_1 - (4 + j6)I_2 = 0 \\ -(4 + j6)I_1 + (4 + j5)I_2 = -7e^{j40°} \end{cases}$$

Figure 8-24   Phasor solution using mesh equations

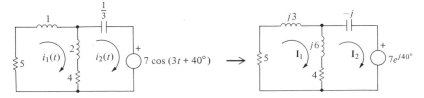

The phasor $\mathbf{I}_1$ may be found via Cramer's rule. The algebra involved is rather lengthy, but straightforward:

$$\mathbf{I}_1 = \frac{\begin{vmatrix} 0 & -(4 + j6) \\ -7e^{j40°} & (4 + j5) \end{vmatrix}}{\begin{vmatrix} (9 + j9) & -(4 + j6) \\ -(4 + j6) & (4 + j5) \end{vmatrix}} = \frac{(7e^{j40°})(-4 - j6)}{(9 + j9)(4 + j5) - (4 + j6)(4 + j6)}$$

$$= \frac{(7e^{j40°})(-4 - j6)}{36 - 45 + j36 + j45 - 16 + 36 - j24 - j24} = \frac{(7e^{j40°})(-4 - j6)}{11 + j33}$$

Converting to polar form,

$$\mathbf{I}_1 = \frac{(7e^{j40°})(7.75e^{j236.5°})}{(34.8e^{j71.7°})} = 1.56e^{j204.8°}$$

Then

$$i_1(t) = 1.56 \cos (3t + 204.8°)$$

A more complicated example involving coupled inductors is given in Figure 8-25. The coupled inductors are first replaced by their controlled

Figure 8-25 Phasor solution involving inductive coupling

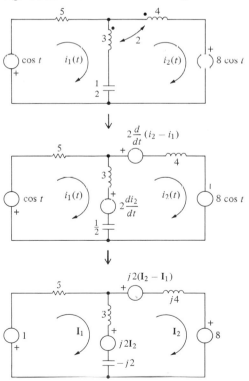

source equivalent, then the phasor network is drawn. In the language of phasors, $d/dt(\ )$ becomes $j\omega(\ )$, so the controlled sources' controlling signals may be expressed in terms of the phasors.

Systematic simultaneous loop equations are as follows:

$$\begin{cases} (5+j)\mathbf{I}_1 - \quad j\mathbf{I}_2 = -1 - j2\mathbf{I}_2 \\ \quad -j\mathbf{I}_1 + (j5)\mathbf{I}_2 = -8 + j2\mathbf{I}_2 - j2(\mathbf{I}_2 - \mathbf{I}_1) \end{cases}$$

Rearranging,

$$\begin{cases} (5+j)\mathbf{I}_1 + \quad j\mathbf{I}_2 = -1 \\ \quad -j3\mathbf{I}_1 + j5\mathbf{I}_2 = -8 \end{cases}$$

If it is desired to find the current $i_2(t)$,

$$\mathbf{I}_2 = \frac{\begin{vmatrix} (5+j) & -1 \\ -j3 & -8 \end{vmatrix}}{\begin{vmatrix} (5+j) & j \\ -j3 & j5 \end{vmatrix}} = \frac{-40 - j8 - j3}{-5 + j25 - 3}$$

$$= \frac{-40 - j11}{-8 + j25} = \frac{41.5e^{j195.4°}}{26.25e^{j107.8°}}$$

$$= 1.58e^{j87.6°}$$

Then

$$i_2(t) = 1.58 \cos(t + 87.6°)$$

A nodal equation example is given in Figure 8-26. The original network is replaced by the phasor network, and systematic simultaneous nodal equations are as follows:

$$\begin{cases} (3+j4)\mathbf{V}_1 - \quad j8\mathbf{V}_2 = 1 \\ \quad -j8\mathbf{V}_1 + (5+j8)\mathbf{V}_2 = -3e^{j45°} \end{cases}$$

If it is desired to find the current $i(t)$, the node voltage phasor $\mathbf{V}_1$ is first found:

$$\mathbf{V}_1 = \frac{\begin{vmatrix} 1 & -j8 \\ -3e^{j45°} & (5+j8) \end{vmatrix}}{\begin{vmatrix} (3+j4) & -j8 \\ -j8 & (5+j8) \end{vmatrix}} = \frac{(5+j8) + (-j8)(3e^{j45°})}{15 + j24 + j20 - 32 + 64}$$

$$= \frac{5 + j8 + (-j8)(2.12 + j2.12)}{47 + j44} = \frac{22 - j9}{47 + j44}$$

$$= \frac{23.8e^{-j22.2°}}{64.38e^{j43.1°}} = 0.37e^{-j65.3°}$$

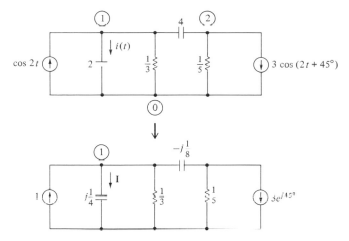

Figure 8-26 Phasor solution using nodal equations

The phasor $\mathbf{I}$ is then

$$\mathbf{I} = \frac{\mathbf{V}_1}{-j\frac{1}{4}} = j4\mathbf{V}_1 = (4e^{j90°})(0.37e^{-j65.3°})$$

$$- 1.48e^{j24.7°}$$

So

$$i(t) = 1.48 \cos(2t + 24.7°)$$

8.7.4 Sources Expressed in Terms of the Sine Function

For networks in which all sources are expressed in terms of the sine function, the same phasors may be used, as if the sines were cosines. Then phasor magnitudes represent the sinusoidal amplitudes and phasor angles represent the phase angles within the sine function rather than within the cosine.

Of course, if there is a mixture of cosine- and sine-function sources, one or the other function will have to be used exclusively as the basis for all of the phase angles.

Consider the network of Figure 8-27. The source $6 \sin(3t + 110°)$ is represented by the phasor $6e^{j110°}$. To solve the phasor network, a Thévenin-Norton transformation has been made to convert it to an equivalent two-node problem.

The two-node equation for $\mathbf{V}$ is

$$\left(\frac{1}{2} - \frac{1}{j3} - \frac{1}{j4}\right)\mathbf{V} = 2e^{j200°}$$

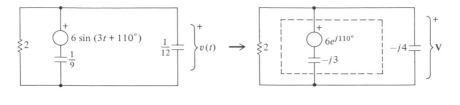

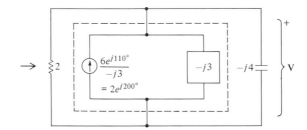

Figure 8-27 Phasor solution in terms of the sine function

Multiplying each side of this equation by $j12$ gives

$$(j6 - 7)\mathbf{V} = (j12)2e^{j200°}$$

$$\mathbf{V} = \frac{(j12)(2e^{j200°})}{7 - j6} = \frac{(12e^{j90°})(2e^{j200°})}{9.22e^{j139.4°}}$$

$$= 2.6e^{j150.6°}$$

Since the original source was described in terms of the sine function, the solution involves sine:

$$v(t) = 2.6 \sin(3t + 150.6°)$$

8.7.5 Perspective on the Three Solution Methods

In the method of components, a sinusoidal source is simply decomposed into its two complex exponential components, which are then superimposed. The components being exponential, impedance methods apply, and the corresponding exponential responses may be found by source-resistor network methods. The added complexity of complex numbers is generally far less complicated than the alternative of dealing directly with the sinusoidal functions and trigonometric identities.

In more advanced work, where more general source functions are expanded into an infinite series of sinusoids of different frequencies, components are particularly attractive for solution because the sum of components is the function itself; no conversions from sinor or phasor representations are required.

The method of sinors is perhaps the most widely used solution method for sinusoidally driven linear systems of all kinds. The method is particularly popular in such fields as quantum mechanics, communications, control systems, and electromagnetics.

The phasor method requires slightly less writing than sinors, $e^{j\omega t}$ factors being eliminated. Network analysists, particularly those concerned with electrical power, deal almost exclusively in terms of phasors. Phasors have a substantial advantage in computer-aided network analysis in that only complex constants, not functions of time, occur as equation driving functions; and phasors are easily plotted and manipulated graphically on the complex plane.

D8-17

Find the phasors that represent the following sinusoidal functions:

(a) $7\cos(10t - 80°)$ *ans.* $7e^{-j80°}$

(b) $10^4\cos(t + \pi/3)$ *ans.* $10^4 e^{j\pi/3}$

(c) $1.2\cos(0.1t - 60°)$ *ans.* $1.2e^{\;j60°}$

(d) $50\cos 10^4 t$ *ans.* 50

D8-18

Find the indicated signals using phasor methods:

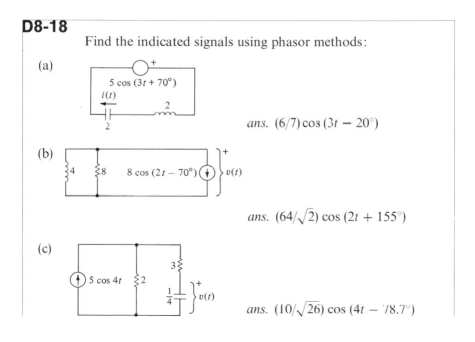

(a)   *ans.* $(6/7)\cos(3t - 20°)$

(b)   *ans.* $(64/\sqrt{2})\cos(2t + 155°)$

(c)   *ans.* $(10/\sqrt{26})\cos(4t - 78.7°)$

(d)

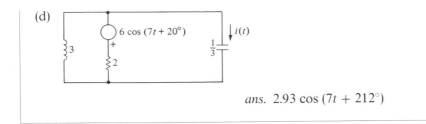

ans. $2.93 \cos (7t + 212°)$

D8-19

Find the two indicated signals in each network using phasor methods with equivalent circuits:

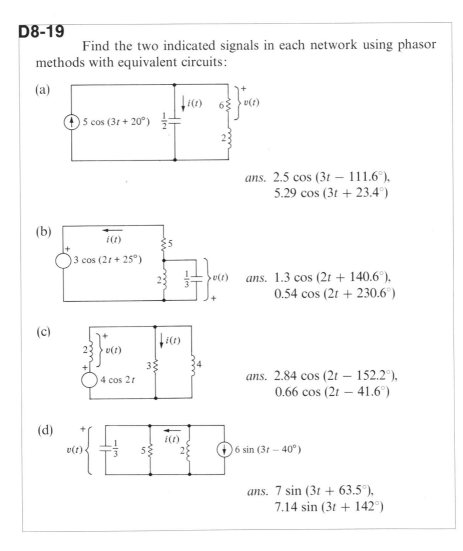

(a)

ans. $2.5 \cos (3t - 111.6°)$,
$5.29 \cos (3t + 23.4°)$

(b)

ans. $1.3 \cos (2t + 140.6°)$,
$0.54 \cos (2t + 230.6°)$

(c)

ans. $2.84 \cos (2t - 152.2°)$,
$0.66 \cos (2t - 41.6°)$

(d)

ans. $7 \sin (3t + 63.5°)$,
$7.14 \sin (3t + 142°)$

D8-20

Find the indicated signal using phasor methods and mesh equations:

(a)

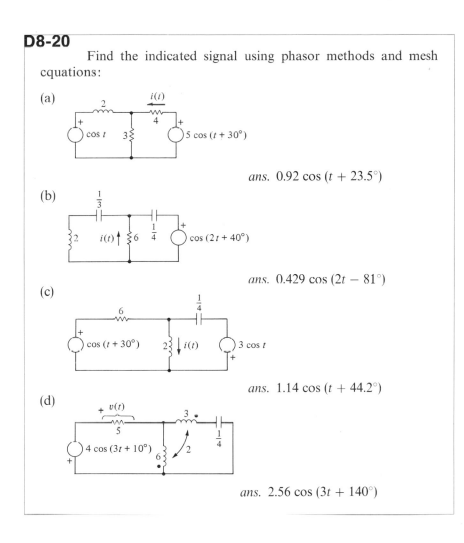

ans. $0.92 \cos (t + 23.5°)$

(b)

ans. $0.429 \cos (2t - 81°)$

(c)

ans. $1.14 \cos (t + 44.2°)$

(d)

ans. $2.56 \cos (3t + 140°)$

D8-21

Find the indicated signal using phasor methods and nodal equations:

(a)

ans. $0.65 \cos (2t + 49.4°)$

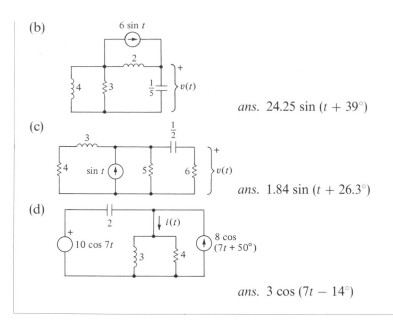

(b)

6 sin t

$v(t)$

ans. 24.25 sin $(t + 39°)$

(c)

$sin\ t$

$v(t)$

ans. 1.84 sin $(t + 26.3°)$

(d)

2

$i(t)$

10 cos $7t$

8 cos
$(7t + 50°)$

ans. 3 cos $(7t - 14°)$

The Method of Phasors

In the phasor method, a sinusoidal source $A \cos(\omega t + \theta)$ is replaced by the phasor source $Ae^{j\theta}$ which is the source sinor at time $t = 0$. Network impedances are evaluated at $s = j\omega$, as with sinors.

The magnitude of each phasor signal is the amplitude of the corresponding sinusoidal signal in the original network, and each phasor angle is the corresponding signal's phase angle.

8.8 Graphical Phasor Methods

8.8.1 Graphical Addition of Phasors

Phasors offer an easy graphical method of calculating sums of same-frequency sinusoidal signals with various amplitudes and phase angles. For example, the phasors $\mathbf{Y}_1$ and $\mathbf{Y}_2$ which represent the sinusoidal signals

$$y_1(t) = 8 \cos(100t - 40°)$$

and

$$y_2(t) = 5 \cos(100t + 120°)$$

are drawn on the complex plane in Figure 8-28.

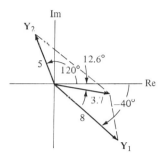

Figure 8-28 Graphical addition of the phasors representing two sinusoidal signals

The sum of the two phasors, which may be found by graphical addition, is

$$\mathbf{Y}_1 + \mathbf{Y}_2 = 3.7e^{-j12.6°}$$

so

$$y_1(t) + y_2(t) = 3.7 \cos (100t - 12.6°)$$

Consider the sum

$$y_3(t) + y_4(t) = 10 \sin (10^4 t) - 10 \cos (10^4 t)$$

The sine function is related to the cosine by

$$A \sin (\omega t + \phi) = A \cos (\omega t + \phi - 90°)$$

so

$$y_3(t) = 10 \sin (10^4 t)$$

has the phasor along the negative imaginary axis, as in Figure 8-29. From that diagram,

$$y_3(t) + y_4(t) = 10 \sin (10^4 t) + 10 \cos (10^4 t + 180°)$$
$$= 10\sqrt{2} \cos (10^4 t + 225°)$$

Figure 8-29 Sine and cosine sum

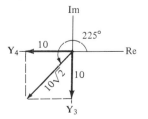

8.8.2   Graphical Relations for R, L, and C

The sink reference voltage and current phasors for resistors, inductors, and capacitors have the simple relations shown in Figure 8-30.

For the resistor, the voltage and current phasors have the same angle (the voltage and current are said to be *in phase*), and the length of the voltage phasor is R times the length (or magnitude) of the current phasor.

For the inductor,

$$\frac{\mathbf{V}_L}{\mathbf{I}_L} = j\omega L$$

or

$$\mathbf{V}_L = j\omega L \mathbf{I}_L = (\omega L)e^{j90°}\mathbf{I}_L$$

The length of the phasor $\mathbf{V}_L$ is (ωL) times the length of $\mathbf{I}_L$, and the angle of $\mathbf{V}_L$ is 90° larger than that of $\mathbf{I}_L$. The current is thus said to *lag* the voltage by 90°.

For the capacitor,

$$\frac{\mathbf{V}_C}{\mathbf{I}_C} = \frac{1}{j\omega C}$$

$$\mathbf{V}_C = \left(\frac{1}{\omega C}\right)e^{-j90°}\mathbf{I}_C$$

The capacitor voltage phasor has length $(1/\omega C)$ times that of the capacitor current phasor. The capacitor current leads the capacitor voltage by 90°.

Figure 8-30   Phasor relationships for the basic elements

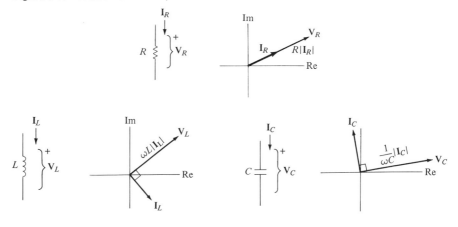

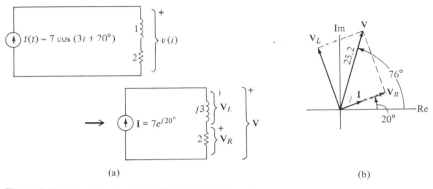

(a)   (b)

Figure 8-31   Graphical solution of a series network

8.8.3   Solution of Trivial Networks

The R, L, and C phasor relationships may be used to solve simple sinusoidal network problems graphically. For example, for the network in Figure 8-31(a), the source phasor is plotted on the phasor diagram, Figure 8-31(b). I is the current through the inductor, so the inductor voltage magnitude is ωL times the magnitude of the current. The angle of V_L is 90° greater than that of the current. The resistor voltage phasor V_R has the same angle as I and twice its magnitude.

The voltage V is

$$V = V_L + V_R$$

which is found by graphically summing V_L and V_R with a parallelogram. Using a ruler and a protractor, it is found to be approximately

$$V = 25.2e^{j76°}$$

so

$$v(t) = 25.2 \cos (3t + 76°)$$

For the network of Figure 8-32(a) the phasor V is plotted on the complex plane in Figure 8-32(b). I_R is in the same direction as V and has one-third of its magnitude. The phasor I_L is of magnitude

$$|I_L| = \left(\frac{1}{\omega L}\right)|V| - |V|$$

and lags V by 90°. The phasor I_C has magnitude

$$|I_C| = \omega C|V| = \tfrac{1}{2}|V|$$

and angle 90° greater than that of V.

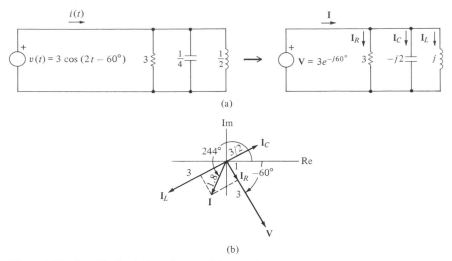

(a)

(b)

Figure 8-32 Graphical solution of a parallel network

The net current

$$I = I_R + I_L + I_C$$

is constructed graphically on the phasor diagram and is found to be approximately

$$I = 1.8e^{j244°}$$

So

$$i(t) = 1.8 \cos (2t + 244°)$$

8.8.4 The Trial Source Method

It is only in rather trivial network problems that the phasors of interest are directly related to the source function as in the previous examples. To solve more involved networks graphically, use is made of the fact that for a single-source network the solution for any phasor signal is proportional to the source phasor. Thus if the magnitude of the source is changed by some factor, the magnitudes of all network phasors are changed by the same factor. Doubling the source magnitude, for instance, doubles all the other phasor magnitudes; and if the source phase angle is changed, all phasor angles are changed by the same amount.

For the network of Figure 8-33, the voltages V_R and V_L are not simply related to the source V. If the current I were known, finding V would be easy, but it is V that is known. Although the impedances 2 and $j3$ could be combined and dealt with as a single element, to do so would be compli-

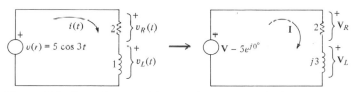

Figure 8-33 Example network for the trial source method

cated in comparison to using the simple in-phase and right-angle phasor relationships of the basic elements.

Instead, the graphical method is to try a convenient current **I**, perhaps a guess at what it might be. Using the simple phasor relations, the corresponding **V** may be found. The **V** found, which is unlikely to be the correct value, is the source voltage which would produce the assumed current. Then the correct current and the correct values of the other phasors are found by considering how the magnitude and angle of the trial **I** must be changed to give the correct **V**.

For the example, a convenient trial current

$$I = 1e^{j0°}$$

is first chosen. The resistor and inductor phasor voltages are, with this current,

$$V_R = 2e^{j0°}$$
$$V_L = 3e^{j90°}$$

as plotted on the trial phasor diagram of Figure 8-34.

The sum of the resistor voltage and the inductor voltage is the source voltage **V**, which if the current were

$$I = 1e^{j0°}$$

would have to be

$$V = 3.6e^{j56.3°}$$

Figure 8-34 Trial phasor diagram

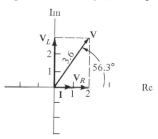

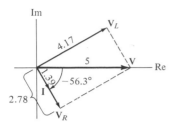

Figure 8-35 Revised phasor diagram

Of course, the source voltage is not the calculated value based on $\mathbf{I}$; it would be a great coincidence if it were.

It is now easy to determine the correct value for $\mathbf{I}$ however. If $\mathbf{I}$ were at the angle $-56.3°$, all other phasors would be rotated clockwise on the phasor diagram by $56.3°$, and $\mathbf{V}$ would be at $0°$, as it should be.

If $\mathbf{I}$ were $(5/3.6)$ units long instead of 1 unit long, all the other phasors on the diagram would be $(5/3.6)$ times as long and, in particular, the magnitude of $\mathbf{V}$ would be the correct 5 units instead of 3.6 units.

A corrected phasor diagram for the network is given in Figure 8-35.

Another example of the trial solution method involves the network of Figure 8-36(a). Trying

$$\mathbf{I} = 1e^{j0°}$$

gives

$$\mathbf{V}_R = 3\mathbf{I}_1 = 3e^{j0°}$$
$$\mathbf{V}_L = j4\mathbf{I}_1 = 4e^{j90°}$$

and

$$\mathbf{V} = \mathbf{V}_R + \mathbf{V}_L$$

as shown on the trial phasor diagram, Figure 8-36(b).

Figure 8-36 Network and trial phasor diagram

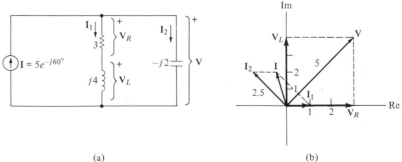

(a)   (b)

From the capacitor voltage $\mathbf{V}$, the capacitor current is

$$\mathbf{I}_2 = \frac{\mathbf{V}}{-j2} = \left(\frac{1}{2}e^{j90°}\right)\mathbf{V}$$

which is half the magnitude of $\mathbf{V}$ and leads it by 90°. The source current is then

$$\mathbf{I} = \mathbf{I}_1 + \mathbf{I}_2$$

as shown, which with the assumed value for $\mathbf{I}_1$ is found graphically to be approximately

$$\mathbf{I} = 1.8e^{j164°}$$

To achieve

$$\mathbf{I} = 5e^{-j60°}$$

all phasors on the trial phasor diagram must be rotated clockwise 184° and expanded in length by the factor (5/1.8), as shown in Figure 8-37. The correct current $\mathbf{I}_1$ is thus approximately

$$\mathbf{I}_1 = 2.78e^{-j224°}$$

The trial source method will generally not work for networks with more than a single source because it depends on the proportionality of phasor network signals to the source. For the graphical solution of multiple-source networks, the sources should be superimposed.

The choice of starting signal, $\mathbf{I}_1$ in the previous example, is made by thinking through the problem: From $\mathbf{I}_1$, $\mathbf{V}_R$ and $\mathbf{V}_L$ may be constructed; $\mathbf{V}$ may be constructed from $\mathbf{V}_R$ and $\mathbf{V}_L$; $\mathbf{I}_2$ may be found from $\mathbf{V}$; and $\mathbf{I}$ is the sum $\mathbf{I}_1 + \mathbf{I}_2$. Alternatively, the solution could have started with a trial $\mathbf{V}_R$ or a trial $\mathbf{V}_L$, but to start with $\mathbf{I}_2$ or $\mathbf{V}$ would not allow an easy construction for $\mathbf{I}_1$.

Figure 8-37 Actual phasor diagram

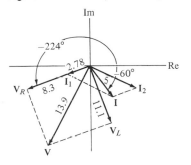

Occasionally, in more complicated networks, an impasse is reached, where no initial trial phasor leads simply back to the source. In this event, some numerical calculation is necessary, or more than a single rotation and stretching or compression of phasors on the diagram is expedient.

D8-22

Find the sums of the following sinusoidal signals using graphical addition of phasors on the complex plane. Express in the form $A \cos(\omega t + \theta)$, with $A \geq 0$, $\omega \geq 0$, and $-360° < \theta < 360°$:

(a) $3 \cos 4t + 4 \cos(4t + 30°)$ *ans.* $6.8 \cos(4t + 17°)$

(b) $\cos(10^5 t + 60°) + \cos(10^5 t - 60°)$ *ans.* $\cos 10^5 t$

(c) $3 \cos(377t + 10°) - \sin(377t + 45°)$ *ans.* $2.6 \cos(377t + 28°)$

(d) $2 \sin(7t - 30°) + 3 \sin 7t$ *ans.* $4.8 \cos(7t - 102°)$

D8-23

Using entirely *graphical* methods, construct phasor diagrams for each of the following networks, showing every voltage and every current phasor:

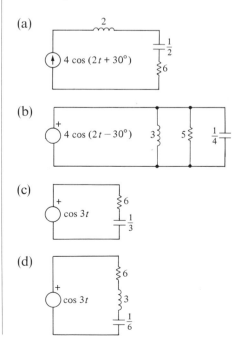

(a)

(b)

(c)

(d)

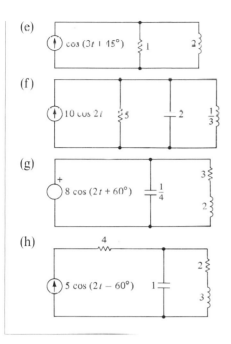

(e) $\cos (3t + 15°)$, 1, 2

(f) $10 \cos 2t$, 5, 2, $\frac{1}{3}$

(g) $8 \cos (2t + 60°)$, $\frac{1}{4}$, 3, 2

(h) 4, $5 \cos (2t - 60°)$, 1, 2, 3

Graphical Phasor Methods

The sum of two sinusoidal functions of the same frequency may be found by adding their phasors to obtain the phasor for the sum.

The phasor relationships for the sink reference relation voltage and current for R, L, and C elements are as follows:

Resistor: Current is in phase with the voltage, with magnitude $(1/R)$ times the voltage magnitude.

Inductor: Current lags the voltage by 90°, with magnitude $(1/\omega L)$ times the voltage magnitude.

Capacitor: Current leads the voltage by 90°, with magnitude (ωC) times the voltage magnitude.

The graphical solution of a single-source phasor network involves drawing a phasor diagram upon which are plotted the phasors for each network voltage and current. In trivial cases the other phasors are found by starting with the source phasor and proceeding through the network, using the R, L, and C phasor relationships.

In general, beginning with a trial voltage or current elsewhere in the network, the source phasor corresponding to the trial signal is computed. Then the resulting phasor diagram is rotated and scaled to give the source phasor and the true solution.

Chapter Eight Problems

Basic Problems

Sinusoidal Functions

1. Sketch plots of the following sinusoidal functions. Find their amplitude, radian frequency, phase angle (in terms of the cosine function), and period:
(a) $-10 \cos (10^6 t + 1000)$
(b) $9 \sin (60° - 7t)$
(c) $60 \cos (-377t - 140°)$

(d) $-100 \sin \left(8t - \dfrac{16\pi}{5} \right)$

2. Convert to the form $A \cos (\omega t + \theta)$ with $A \geq 0$, $\omega > 0$, and $-360° < \theta < 360°$:
(a) $f(t) = 3 \sin (80t - 300°)$
(b) $f(t) = 20 \cos 10t + 5 \sin 10t$
(c) $f(t) = -8 \cos 377t - 9 \sin 377t$
(d) $f(t) = 100 \cos (3t + 90°) - 50 \sin 3t$

Differential Equations with Sinusoidal Driving Functions

3. Find the general solutions of each of the following differential equations:

(a) $\dfrac{dy}{dt} = 10 \cos 20t$

(b) $2\dfrac{dy}{dt} + y = -6 \sin 3t$

(c) $\dfrac{dy}{dt} + 4y = 5 \cos (2t + 45°)$

(d) $\dfrac{d^2 y}{dt^2} + 5\dfrac{dy}{dt} + 6y = 7 \sin 2t$

Complex Algebra

4. Find the indicated complex numbers. Express in both rectangular form and in polar form, and sketch the location of each number on the complex plane:
(a) $(3 - j4)(-5 + j6)$

(b) $\dfrac{(4 + j5)}{(2 - j3)}$

(c) $6e^{j30°} + j7$

(d) $3e^{j100°} + 5e^{-j40°}$

(e) $\dfrac{(6 - j7)(-3 - j8)}{(-50 + j40)}$

Method of Components

5. Use the method of components to find the indicated signals. There are two signals to find in each network.

(a)

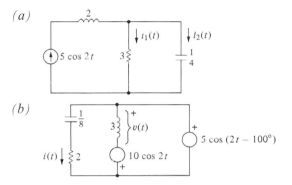

(b)

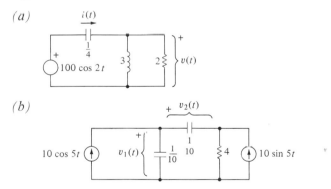

Method of Sinors

6. Use the method of sinors to find the indicated signals. There are two signals to find in each network.

(a)

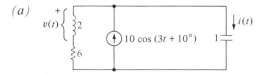

(b)

Phasor Solutions

7. Use the method of phasors and equivalent circuits to find the indicated signals. There are two signals to find in each network.

(a)

(b)

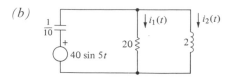

8. Use phasor methods and mesh equations to find the indicated signals:

(a)

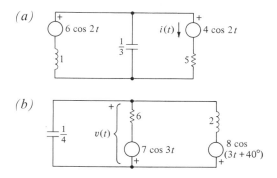

(b)

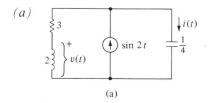

9. Use phasor methods and nodal equations to find the indicated signals:

(a)

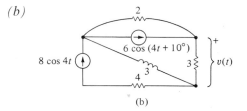

(a)

(b)

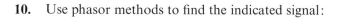

(b)

Networks with Coupled Inductors

10. Use phasor methods to find the indicated signal:

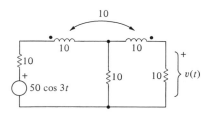

Networks with Controlled Sources

11. Use phasor methods to find the indicated signal:

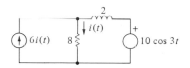

Sources with Different Frequencies

12. Use phasor methods to find the indicated signal:

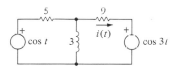

Graphical Phasor Methods

13. Using graphical methods with phasors, find the constants A, B, C, and θ:

(a) $9 \cos (10t + 130°) = A \cos 10t + B \sin 10t$
(b) $6 \cos (4t - 40°) - 7 \cos (4t + 80°) = A \cos 4t + B \sin 4t = C \cos (4t + \theta)$
(c) $5 \sin (10^4 t - 70°) - 3 \sin (10^4 t) = A \cos 10^4 t + B \sin 10^4 t$

14. Using entirely graphical methods, find the indicated signals:

(a)

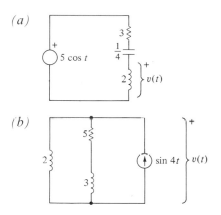

(b)

Practical Problems

AC Ammeters and Voltmeters

When a sinusoidal signal is applied to a dc meter such as a permanent magnet ammeter or voltmeter, the instrument will respond only if the frequency is very low. At higher frequencies, the inertia of the mechanical

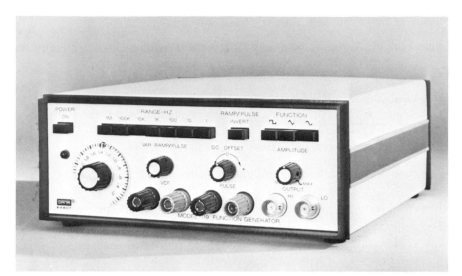

Closeup view of a function generator. Square-wave, triangular-wave, and sinusoidal signals are selectable with the pushbuttons at the top right. The frequency is controlled by other push-button switches and the dial at the left, which turns a potentiometer. The knob at the right controls the amplitude of the output signal. This model will produce signals with frequencies from less than 1 Hz to over 1 million Hz. (*Photo courtesy of Exact Electronics, Inc.*)

movement results in a deflection proportional to the average force which, for a sinusoid, is zero.

Meters suitable for indicating the amplitude of sinusoidal currents and voltages average the square or the magnitude of the sinusoidal quantity and produce a reading of the corresponding sinusoidal amplitude.

A *two-coil* meter uses the attraction between two current-carrying coils to produce a force proportional to the average of the square of the current. An *iron vane* instrument uses the attraction between a current-carrying coil and the induced magnetization of a piece of soft iron to produce a similar force. The scales of these meters tend to be compressed at the low end and

Figure 8-38 Practical Problem 2

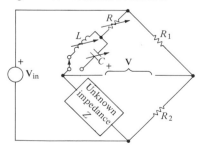

spread out at high values of current or voltage because of the nonlinear force relation.

Most electronic ac meters respond to the average of the magnitude (or absolute value) of the measured voltage or current.

1. An ac ammeter responds with a deflection proportional to the average of the square of the current. If a constant applied current of 10 A produces a certain deflection, what must be the amplitude of sinusoidal current to produce the same deflection?

Impedance Bridge

2. The adjustable network of Figure 8-38 is one type of impedance bridge. The resistor R and inductance or capacitance, L or C, are adjusted until $V = 0$. The bridge is then said to be balanced.

Show that, at balance, the impedance $(R + j\omega L)$ or $[R - (j/\omega C)]$ is proportional to the unknown impedance Z at the frequency of V_{in}.

An electronic frequency counter. Precise measurement of the frequency of a signal is made by counting the number of cycles in a measured interval of time. (*Photo courtesy of Hewlett-Packard, Inc.*)

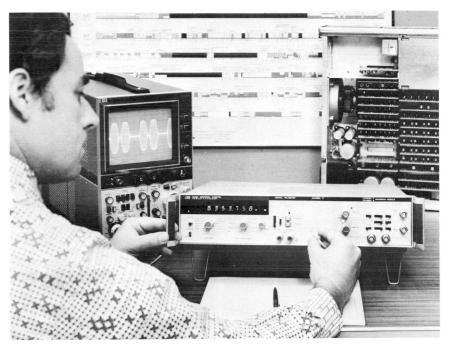

Phase-Difference Metering

3. The phase difference between two sinusoidal signals may be measured by adjusting the amplitudes of each to a fixed value A, adding them, and measuring the amplitude of their sum. The sum has amplitude $2A$ if the signals are in phase and zero amplitude if the signals differ in phase by $180°$.

What is the relationship between the phase difference of the two signals and the amplitude of their sum?

4. Another method of phase-difference measurement is to pass one of the sinusoidal signals through an additional, adjustable, calibrated phase shift and adjust this phase shift until the resulting signals are exactly in phase. The phase difference is then the amount of added phase shift of one signal necessary to make the phases equal.

Computer-aided engineering design. The electronic pen on the drafting table to the rear is used to enter graphical data into the computer. Results are displayed on the video screen and may be printed in ink on the plotter in the foreground. (*Photo courtesy of Tektronix, Inc.*)

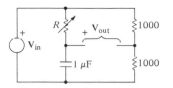

Figure 8-39 Practical Problem 4

For the phase-shifting network of Figure 8-39, show that

$$|\mathbf{V}_{out}| = \tfrac{1}{2}|\mathbf{V}_{in}|$$

at every frequency, and find the amount of phase shift,

$$\angle\mathbf{V}_{out} - \angle\mathbf{V}_{in}$$

as a function of R.

Phasor Diagrams

5. Using entirely graphical methods, construct phasor diagrams for each of the networks of Figure 8-40, showing every voltage phasor and every current phasor.

Advanced Problems

Switched Sinusoidal Networks

1. Find $i(t)$ in the network of Figure 8-41 before and after time $t - 0$ and sketch it.

Figure 8-40 Practical Problem 5

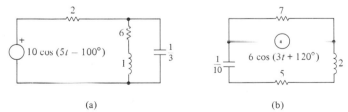

(a)   (b)

Figure 8-41 Advanced Problem 1

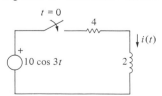

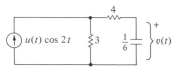

Figure 8-42 Advanced Problem 2

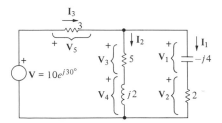

Figure 8-43 Advanced Problem 5

2. Find $v(t)$ in the network of Figure 8-42 before and after time $t = 0$ and sketch it.

Phasors

3. The equation

$$f(t) = Re[\underline{f}(t)]$$

relates a sinor $\underline{f}(t)$ to the time function $f(t)$ that it represents. Find the equation that relates a phasor to the time function it represents.

4. Show that if the rectangular components of a phasor are

$$\mathbf{V} = a + jb$$

the quadrature components of the corresponding sinusoid are

$$v(t) = a \cos \omega t + (-b) \sin \omega t$$

Graphical Solutions

5. Use entirely *graphical* methods to construct a phasor diagram for the network of Figure 8-43.

CHAPTER NINE
Resonance and Frequency Response

9.1 Introduction

Having learned the fundamentals of sinusoidal response calculation, it is now feasible to apply those techniques to the analysis and understanding of resonance in networks. The phenomenon of resonance is basic to such applications as radio, radar, and telephone transmission, and it is also exceedingly useful in a variety of other fields.

The chapter begins by considering resonance in series and parallel *RLC* networks. Then general resonance concepts are investigated, and single-frequency equivalences are examined.

When you complete this chapter, you should know

1. The characteristics of resonance in series and in parallel *RLC* networks.
2. What frequency response plots are, and how to construct them.
3. The general definitions and meaning of resonance and bandwidth.
4. About the quality factor *Q* of a resonant network, and its relationship to bandwidth.
5. How to represent an impedance at a single frequency as a series combination and as a parallel combination of resistance and reactance.
6. How to connect a series or parallel inductor or capacitor to a network to make it resonant at a given frequency.

9.2 The Series RLC Network

9.2.1 Series Resonance

In two-terminal networks containing both inductors and capacitors, there are one or more sinusoidal frequencies for which the network impedance, $Z(s = j\omega)$, is entirely real, as it is for a resistor. This phenomenon is termed *resonance*.

In the vicinity of such a resonant frequency, the impedance at the network terminals generally exhibits a peak or a dip in magnitude. Networks may thus be designed to emphasize frequencies near their resonance. In a radio receiver, for example, the resonant frequency of a network may be adjusted by varying an element value (often the value of a capacitor) to "tune in" the range of frequencies sent by a transmitting station, greatly attenuating other frequencies sent by other stations.

Perhaps the most simple resonant network is the series RLC one, diagramed in Figure 9-1. For sinusoidal signals with radian frequency ω, the impedance as viewed from the network terminals is

$$Z(s = j\omega) = R + j\omega L + \frac{1}{j\omega C}$$

$$= R + j\left(\omega L - \frac{1}{\omega C}\right)$$

The frequency $\omega = \omega_r$, for which

$$\omega_r L = \frac{1}{\omega_r C}$$

$$\omega_r^2 LC = 1$$

$$\omega_r = \frac{1}{\sqrt{LC}}$$

is called the network's *resonant frequency*. It is the frequency at which the capacitor and inductor impedances add to zero. That is, at the resonant frequency, the series combination of inductor and capacitor behave as a short circuit.

Figure 9-1 Series RLC network

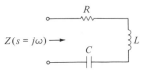

The resonant frequency in hertz (cycles per second) for the series *RLC* network is

$$f_r = \frac{\omega_r}{2\pi} = \frac{1}{2\pi\sqrt{LC}}$$

9.2.2 Behavior Above and Below Resonance

At frequencies lower than the resonant frequency,

$$\frac{1}{\omega C} > \omega L$$

so that

$$Z(s = j\omega) = R + j\left(\omega L - \frac{1}{\omega C}\right)$$

$$= R + j(\text{negative number})$$

At such frequencies, the sink reference network current leads the network voltage, as in an *RC* network.

At resonance,

$$Z(s = j\omega) = R + j(0) = R$$

and the network voltage and current are in phase with one another.

At frequencies higher than the resonant frequency,

$$\omega L > \frac{1}{\omega C}$$

and

$$Z(s = j\omega) = R + j\left(\omega L - \frac{1}{\omega C}\right)$$

$$= R + j(\text{positive number})$$

The sink reference network current lags the voltage, as in an *RL* network.

The magnitude of the impedance,

$$Z(s = j\omega) = \sqrt{R^2 + \left(\omega L - \frac{1}{\omega C}\right)^2}$$

is minimum at resonance, which is to say that for constant applied voltage amplitude, the resulting current is of greatest amplitude at the resonant frequency.

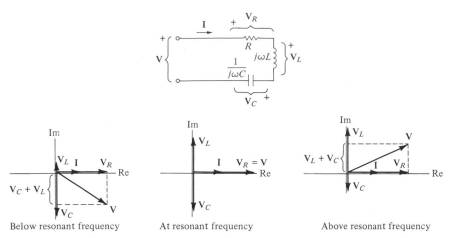

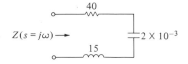

Below resonant frequency   At resonant frequency   Above resonant frequency

Figure 9-2   Phasor relations in the series *RLC* network

9.2.3   Phasor Relations

Phasor diagrams for the series *RLC* network, in terms of the current **I**, are shown in Figure 9-2. Relations for a frequency below resonance, at resonant frequency, and at a frequency above resonance are shown.

9.2.4   A Numerical Example

The series *RLC* network of Figure 9-3 has impedance, for sinusoidal signals with radian frequency ω,

$$Z(s = j\omega) = R + j\left(\omega L - \frac{1}{\omega C}\right)$$

$$= 40 + j\left(15\omega - \frac{1000}{2\omega}\right)$$

It is resonant at the frequency

$$\omega_r = \frac{1}{\sqrt{LC}} = \frac{1}{\sqrt{30 \times 10^{-3}}} = 5.77 \text{ rad/s}$$

Figure 9-3   Series resonance example

or

$$f_r = \frac{1}{2\pi\sqrt{LC}} = 0.92 \text{ Hz}$$

At the resonant frequency,

$$Z(s = j5.77) = R + j(0) = 40 \ \Omega$$

At a frequency below resonance, say one-half the resonant frequency,

$$Z\left(s = j\frac{5.77}{2}\right) = 40 + j[43.3 - 173.3]$$

$$= 40 - j130$$

The inductor impedance cancels part, but not all, of the capacitor's impedance.

At frequencies above resonance, the reverse occurs. At twice the resonant frequency, for example,

$$Z(s = j11.54) = 40 + j(173.3 - 43.3)$$

$$= 40 + j130$$

D9-1

   Find the (hertz) resonant frequency of each of the following series *RLC* networks. Find the impedance of each network at resonance, at half the resonant frequency, and at twice the resonant frequency:

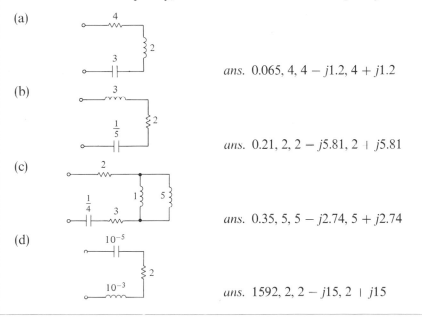

(a)

    ans. 0.065, 4, 4 − j1.2, 4 + j1.2

(b)

    ans. 0.21, 2, 2 − j5.81, 2 + j5.81

(c)

    ans. 0.35, 5, 5 − j2.74, 5 + j2.74

(d)

    ans. 1592, 2, 2 − j15, 2 + j15

Series *RLC* Networks

The impedance for sinusoidal signals of a series *RLC* network is

$$Z = R + j\omega L + \frac{1}{j\omega C} = R + j\left(\omega L - \frac{1}{\omega C}\right)$$

where ω is the source radian frequency. The network is said to be resonant at the frequency

$$\omega_r = \frac{1}{\sqrt{LC}}, \qquad f_r = \frac{1}{2\pi\sqrt{LC}}$$

for which the series combination of inductor and capacitor is equivalent to a short circuit, and

$$Z = R$$

As a function of frequency, the sink reference network voltage is in phase with the current at resonance. The current phase leads the voltage (as in an *RC* network) below the resonant frequency and lags it (as in an *RL* network) above resonance.

For constant network voltage amplitude, the sink reference network current is maximum and in phase with the voltage at resonance.

9.3   The Parallel *RLC* Network

9.3.1   Parallel Resonance

Another simple resonant network is the parallel *RLC* network, Figure 9-4. It has impedance, for sinusoidal signals,

$$Z(s = j\omega) = \frac{1}{\dfrac{1}{R} + j\left(\omega C - \dfrac{1}{\omega L}\right)}$$

The *admittance* of this network (admittance is the inverse of impedance),

$$Y(s = j\omega) = \frac{1}{Z(s = j\omega)} = \frac{1}{R} + j\left(\omega C - \frac{1}{\omega L}\right)$$

Figure 9-4   Parallel *RLC* network

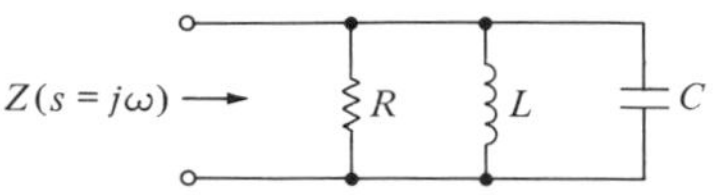

is of the same form as the impedance of the series *RLC* network, with the roles of *L* and *C* interchanged, and *R* replaced by $1/R$.

The frequency $\omega = \omega_r$, for which

$$\omega_r C = \frac{1}{\omega_r L}$$

$$\omega_r = \frac{1}{\sqrt{LC}} \text{ rad/s}$$

is the network's resonant frequency, where

$$Z(s = j\omega_r) = \frac{1}{\dfrac{1}{R} + j(0)} = R$$

The parallel inductor and capacitor impedances combine to give an *open* circuit at the resonant frequency.

9.3.2   Behavior Above and Below Resonance

Multiplying the numerator and denominator of *Z* by the complex conjugate of the denominator places *Z* in rectangular form, where its real and imaginary parts are evident:

$$Z(s = j\omega) = \frac{1}{\dfrac{1}{R} + j\left(\omega C - \dfrac{1}{\omega L}\right)}$$

$$= \frac{\dfrac{1}{R} - j\left(\omega C - \dfrac{1}{\omega L}\right)}{\left[\dfrac{1}{R} + j\left(\omega C - \dfrac{1}{\omega L}\right)\right]\left[\dfrac{1}{R} - j\left(\omega C - \dfrac{1}{\omega L}\right)\right]}$$

$$= \frac{\dfrac{1}{R} + j\left(\dfrac{1}{\omega L} - \omega C\right)}{\left(\dfrac{1}{R}\right)^2 + \left(\omega C - \dfrac{1}{\omega L}\right)^2}$$

At frequencies below the resonant frequency,

$$\omega C < \frac{1}{\omega L}$$

so that

$$Z(s = j\omega) = \frac{\dfrac{1}{R} + j(\text{positive number})}{\text{positive number}}$$

At such frequencies, the sink reference network current lags the network voltage, as in an *RL* network.

At frequencies above resonance,

$$\omega C > \frac{1}{\omega L}$$

and

$$Z(s = j\omega) = \frac{\frac{1}{R} + j(\text{negative number})}{\text{positive number}}$$

The sink reference current leads the voltage, as in an *RC* network.

The magnitude of the impedance is maximum at resonance, since the magnitude of

$$Y(s = j\omega) = \frac{1}{R} + j\left(\omega C - \frac{1}{\omega L}\right) = \frac{1}{Z(s = j\omega)}$$

is minimum at resonance. For constant applied voltage amplitude the resulting current has the smallest amplitude at the resonant frequency.

9.3.3 Phasor Relations

Phasor diagrams for the parallel *RLC* network are shown in Figure 9-5 in terms of the voltage **V**.

Figure 9-5 Phasor relations in the parallel *RLC* network

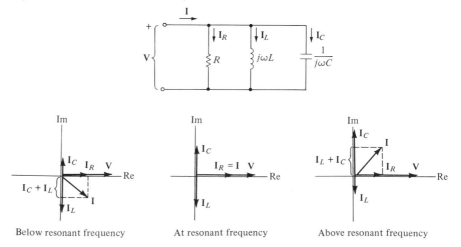

Below resonant frequency   At resonant frequency   Above resonant frequency

9.3.4 A Numerical Example

The parallel *RLC* network in Figure 9-6 has impedance for sinusoidal signals

$$Z(s = j\omega) = \cfrac{1}{\cfrac{1}{R} + j\omega C + \cfrac{1}{j\omega L}}$$

$$= \cfrac{1}{\cfrac{1}{50} + j\left(2 \times 10^{-3}\omega - \cfrac{1}{4\omega}\right)}$$

It is resonant at the frequency

$$\omega_r = \frac{1}{\sqrt{LC}} = \frac{1}{\sqrt{8 \times 10^{-3}}} = 11.2 \text{ rad/s},$$

or

$$f_r = \frac{1}{2\pi\sqrt{LC}} = 1.78 \text{ Hz}$$

At the resonant frequency,

$$Z(s = j11.2) = \cfrac{1}{\cfrac{1}{R} + j(0)} = R = 50 \ \Omega$$

At a frequency below resonance, say one-half the resonant frequency,

$$Z(s = j5.6) = \cfrac{1}{0.02 + j(0.0112 - 0.0446)}$$

$$= \cfrac{1}{0.02 - j0.0335} = 13.1 + j22.0$$

The sink reference network current lags the voltage, as in an *RL* network. At frequencies above resonance, current leads voltage, as in an *RC* network. At twice the resonant frequency,

$$Z(s = j22.4) = \cfrac{1}{0.02 + j(0.0448 - 0.01116)}$$

$$= \cfrac{1}{0.02 + j0.0335} = 13.1 - j22.0$$

Figure 9-6 Parallel resonance example

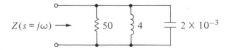

$Z(s = j\omega) \longrightarrow \quad 50 \quad 4 \quad 2 \times 10^{-3}$

D9-2

Find the (hertz) resonant frequencies of each of the following parallel *RLC* networks. Find the impedance of each network at resonance.

(a)

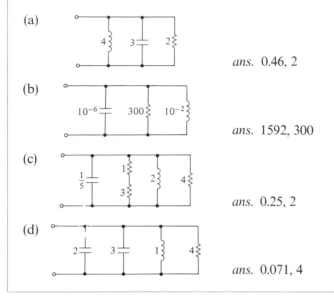

ans. 0.46, 2

(b)

ans. 1592, 300

(c)

ans. 0.25, 2

(d)

ans. 0.071, 4

D9-3

For the parallel *RLC* network below, find the impedance at resonance, at one-tenth the resonant frequency, and at ten times the resonant frequency:

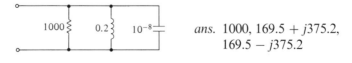

ans. 1000, $169.5 + j375.2$,
$169.5 - j375.2$

Parallel *RLC* Networks

The impedance for sinusoidal signals of a parallel *RLC* network is

$$Z = \cfrac{1}{\cfrac{1}{R} + \cfrac{1}{j\omega L} + j\omega C} = \cfrac{1}{\cfrac{1}{R} + j\left(\omega C - \cfrac{1}{\omega L}\right)}$$

where ω is the source radian frequency. The network is resonant at the frequency

$$\omega_r = \frac{1}{\sqrt{LC}}, \qquad f_r = \frac{1}{2\pi\sqrt{LC}}$$

where the parallel LC combination is equivalent to an open circuit, and

$$Z = R$$

At frequencies below resonance, the effect of the inductor dominates that of the capacitor, and the sink reference voltage phase leads the current phase. Above resonance, the voltage lags the current.

For constant voltage amplitude, the sink reference network current has minimum amplitude and is in phase with the voltage at resonance.

9.4 Frequency Response

9.4.1 Amplitude Ratio and Phase Shift

For sinusoidal signals, the impedance of a two-terminal network, Figure 9-7, has magnitude that is the ratio of voltage amplitude to current amplitude:

$$|Z(s = j\omega)| = \frac{|\mathbf{V}|}{|\mathbf{I}|} = \frac{\text{amplitude of voltage}}{\text{amplitude of current}}$$

The angle of the impedance is the difference in phase, or *phase shift*, between the sink reference voltage and current:

$$\angle Z(s = j\omega) = \angle \mathbf{V} - \angle \mathbf{I}$$
$$= (\text{voltage phase angle}) - (\text{current phase angle})$$

Plots of $|Z(s = j\omega)|$ and $\angle Z(s = j\omega)$, as functions of ω, are called *frequency response plots* of the impedance.

9.4.2 Frequency Response of the Series RLC Network

For the series RLC network,

$$Z(s = j\omega) = R + j\left(\omega L - \frac{1}{\omega C}\right)$$

Figure 9-7 Impedance for sinusoidal signals

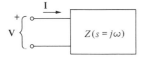

The ratio of amplitudes of the network voltage to the network current is

$$\frac{|\mathbf{V}|}{|\mathbf{I}|} = |Z(s=j\omega)| = \sqrt{R^2 + \left(\omega L - \frac{1}{\omega C}\right)^2}$$

a sketch of which is shown in Figure 9-8(a).

The difference in phase angle between the series RLC network's sink reference voltage and current is

$$\angle \mathbf{V} - \angle \mathbf{I} = \angle Z(s=j\omega) = \tan^{-1}\left(\frac{\omega L - \dfrac{1}{\omega C}}{R}\right)$$

which is sketched in Figure 9-8(b). For small values of ω, the effect of the series capacitor dominates that of the inductor, and the impedance angle approaches $-90°$. At resonance, it is $0°$; and for large ω, where the series inductor dominates, the impedance angle approaches $+90°$.

9.4.3 Frequency Response of Other Signals

In general, frequency response may express the amplitude ratio and phase shift of any two sinusoidal signals in a network, as a function of frequency. One could plot the frequency response of an admittance, or the relation between two voltages or between two network currents.

Figure 9-8   Frequency response plots of the impedance of a series RLC network

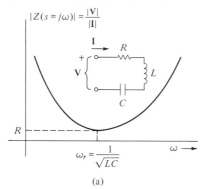

(a)

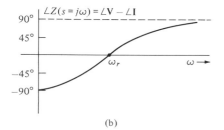

(b)

For example, the ratio of the inductor voltage to the total voltage in the series RLC network in Figure 9-9(a) is

$$\frac{\mathbf{V}_L}{\mathbf{V}} = \frac{j\omega L}{R + j\omega L + \dfrac{1}{j\omega C}} = \frac{-\omega^2 LC}{-\omega^2 LC + j\omega RC + 1}$$

as given by the voltage-divider rule. The amplitude ratio of these two voltages is

$$\frac{|\mathbf{V}_L|}{|\mathbf{V}|} = \frac{\omega^2 LC}{\sqrt{(1 - \omega^2 LC)^2 + \omega^2 RC}}$$

$$= \frac{1}{\sqrt{L^2 C^2 \omega^4 + (R^2 C^2 - 2LC)\omega^2 + 1}}$$

and the phase shift is

$$\angle \mathbf{V}_L - \angle \mathbf{V} = 180° - \tan^{-1}\left(\frac{\omega RC}{1 - \omega^2 LC}\right)$$

The shapes of this frequency response and that of

$$\frac{\mathbf{V}_C}{\mathbf{V}} = \frac{\dfrac{1}{j\omega C}}{R + j\omega L + \dfrac{1}{j\omega C}} = \frac{1}{-\omega^2 LC + j\omega RC + 1}$$

for the series network are sketched in Figure 9-9(b).

Figure 9-9 Frequency response of other quantities in a series RLC network

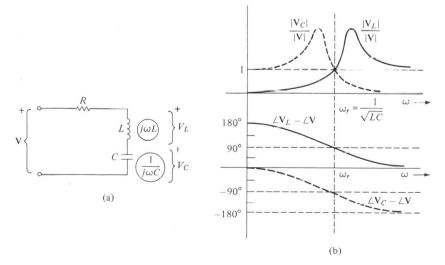

9.4.4   Parallel RLC Networks

The impedance of the parallel RLC network is

$$Z(s = j\omega) = \cfrac{1}{\cfrac{1}{R} + j\left(\omega C - \cfrac{1}{\omega L}\right)}$$

The amplitude ratio of voltage to current at the terminals is

$$|Z(s = j\omega)| = \cfrac{1}{\sqrt{\left(\cfrac{1}{R}\right)^2 + \left(\omega C - \cfrac{1}{\omega L}\right)^2}}$$

and the phase shift between the two is

$$\angle Z(s = j\omega) = -\tan^{-1} \cfrac{\omega C - \left(\cfrac{1}{\omega L}\right)}{\left(\cfrac{1}{R}\right)}$$

The frequency response curves are sketched in Figure 9-10. The amplitude curve is the inverse of a series RLC network amplitude curve since the admittance of the parallel network has the same form as the impedance of a series network.

Figure 9-10   Frequency response of the impedance of a parallel RLC network

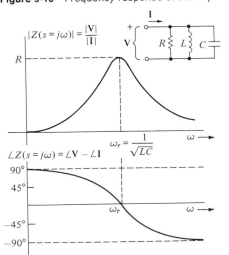

D9-4

Sketch frequency response plots (both magnitude and phase shift) to scale for the impedances of the following RLC networks:

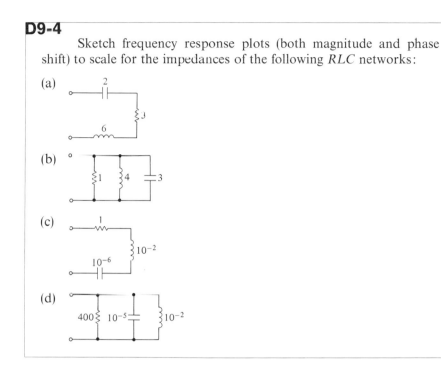

(a)

(b)

(c)

(d)

Frequency Response

For sinusoidal signals, the magnitude of an impedance is the ratio of voltage amplitude to current amplitude. The impedance angle is the phase angle of the voltage minus the phase angle of the current, the phase shift between voltage and current.

Frequency response plots consist of a plot of the ratio of the amplitudes of two sinusoidal signals in a network, as a function of frequency, and a plot of the difference in phase between the two signals, as a function of frequency.

9.5 Bandwidth

9.5.1 Series Network Half-Power Frequencies and Bandwidth

In the series RLC network the frequency $\omega = \omega_1$ for which

$$\frac{1}{\omega_1 C} - \omega_1 L = R$$

gives

$$Z(s = j\omega_1) = R + j\left(\omega_1 L - \frac{1}{\omega_1 C}\right) = R - jR$$

At this frequency,

$$\angle Z(s = j\omega_1) = -45°$$

and

$$|Z(s = j\omega_1)| = \sqrt{R^2 + (-R)^2} = R\sqrt{2}$$

At the frequency $\omega = \omega_2$, for which

$$\omega_2 L - \frac{1}{\omega_2 C} = R$$

$$Z(s = j\omega_2) = R + jR$$

and

$$\angle Z(s = j\omega_2) = 45°$$

$$|Z(s = j\omega_2)| = \sqrt{R^2 + R^2} = R\sqrt{2}$$

The frequencies ω_1 and ω_2 (or their hertz counterparts $f_1 = \omega_1/2\pi$ and $f_2 = \omega_2/2\pi$) are termed the *half-power frequencies* of the series *RLC* network. Since electrical power is related to the square of current, increasing the impedance by a factor of $\sqrt{2}$, at constant voltage, halves power. The half-power frequencies of the series *RLC* network are indicated in Figure 9-11.

Figure 9-11   Half-power frequencies and bandwidth of a series *RLC* network

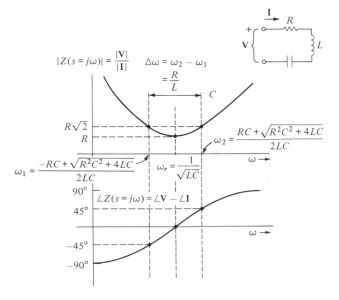

The interval of frequency between the half-power frequencies is called the *bandwidth* of a resonant network. For the series RLC network, where

$$\frac{1}{\omega_1 C} - \omega_1 L = R$$

$$\omega_1^2 LC + \omega_1 RC - 1 = 0$$

$$\omega_1 = \frac{-RC \pm \sqrt{R^2 C^2 + 4LC}}{2LC}$$

The solution for positive ω_1 is

$$\omega_1 = \frac{-RC + \sqrt{R^2 C^2 + 4LC}}{2LC}$$

Similarly, for ω_2,

$$\omega_2 L - \frac{1}{\omega_2 C} = R$$

$$\omega_2^2 LC - \omega_2 RC - 1 = 0$$

$$\omega_2 = \frac{RC \pm \sqrt{R^2 C^2 + 4LC}}{2LC}$$

and the solution for positive ω_2 is

$$\omega_2 = \frac{RC + \sqrt{R^2 C^2 + 4LC}}{2LC}$$

The difference between ω_2 and ω_1,

$$\Delta\omega = \omega_2 - \omega_1 = \frac{R}{L}$$

or

$$\Delta f = \frac{\Delta\omega}{2\pi} = \frac{R}{2\pi L}$$

as indicated in Figure 9-11.

9.5.2 Bandwidth of the Parallel Network

In the parallel RLC network, for which

$$Z(s = j\omega) = \frac{1}{\dfrac{1}{R} + j\left(\omega C - \dfrac{1}{\omega L}\right)}$$

the amplitude ratio $|Z(s = j\omega)|$ exhibits a maximum at resonance. The half-power points are then defined to be the frequencies for which the magnitude of Z is reduced by a factor of $\sqrt{2}$ from its value at resonance. This occurs for ω_1, for which

$$\frac{1}{\omega_1 L} - \omega_1 C = \frac{1}{R}$$

and for ω_2, for which

$$\omega_2 C - \frac{1}{\omega_2 L} = \frac{1}{R}$$

The solutions for ω_1 and ω_2 are

$$\omega_1 = \frac{-\dfrac{L}{R} + \sqrt{\left(\dfrac{L}{R}\right)^2 + 4LC}}{2LC}$$

$$\omega_2 = \frac{\dfrac{L}{R} + \sqrt{\left(\dfrac{L}{R}\right)^2 + 4LC}}{2LC}$$

and the bandwidth of this network is

$$\Delta\omega = \omega_2 - \omega_1 = \frac{1}{RC} \text{ rad/s}$$

or

$$\Delta f = \frac{\Delta\omega}{2\pi} = \frac{1}{2\pi RC} \text{ Hz}$$

Figure 9-12   Half-power frequencies and bandwidth of a parallel RLC network

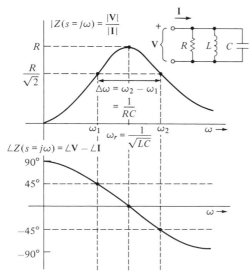

The phase shift between voltage and current is $\pm 45°$ at the half-power frequencies, as illustrated in the sketch of Figure 9-12.

D9-5

Find the bandwidth in hertz of each of the resonant networks of problem D9-4.

ans. 0.08; 0.053; 15.9; 39.81

Bandwidth

The bandwidth of a resonant minimum in frequency response amplitude is the frequency interval, extending from one side of the minimum to the other, for which the amplitude ratio is $\sqrt{2}$ as large as it is at the minimum. The frequencies at which the amplitude ratio equals $\sqrt{2}$ times the minimum value are called *half-power frequencies*.

The series RLC network has bandwidth

$$\Delta\omega = \frac{R}{L}, \qquad \Delta f = \frac{R}{2\pi L}$$

For a resonant maximum in frequency response amplitude, the half-power frequencies are those for which the amplitude ratio equals $(1/\sqrt{2})$ times the maximum value. The bandwidth is the difference between the half-power frequencies.

The parallel RLC network has bandwidth

$$\Delta\omega = \frac{1}{RC}, \qquad \Delta f = \frac{1}{2\pi RC}$$

For the series and the parallel RLC networks, the phase shifts between network voltage and current are $\pm 45°$ at the half-power frequencies.

9.6 Resonance in General

9.6.1 Definitions of Resonant Frequency

More complicated RLC networks than the series and parallel ones also exhibit resonance. In fact, if there are several L's and C's, there may be two or more resonant frequencies.

In the more involved networks, frequencies for which

$$\angle Z(s = j\omega) = 0$$

are not precisely the same frequencies for which $|Z(s = j\omega)|$ has maxima or minima. In practice, both the zero-angle and the magnitude extrema are used as definitions of resonance, the choice depending on the application at hand.

9.6.2   Example of Zero-Angle Resonance

The RLC network of Figure 9-13 commonly occurs when a physical inductor and capacitor are connected in parallel. The winding resistance of the inductor is not negligible and is represented by the resistance R. This network is resonant, but it is neither a series nor a parallel RLC network.

For sinusoidal signals, the impedance of this network at the terminals is

$$Z(s = j\omega) = \frac{(R + j\omega L)\left(\dfrac{1}{j\omega C}\right)}{R + j\omega L + \dfrac{1}{j\omega C}} = \frac{R + j\omega L}{1 - \omega^2 LC + j\omega RC}$$

The real and imaginary parts of Z are as follows:

$$Z(s = j\omega) = \frac{(R + j\omega L)(1 - \omega^2 LC - j\omega RC)}{(1 - \omega^2 LC + j\omega RC)(1 - \omega^2 LC - j\omega RC)}$$

$$= \frac{R - \omega^2 RLC + \omega^2 LRC}{(1 - \omega^2 LC)^2 + \omega^2 R^2 C^2} - \frac{\omega^3 L^2 C + \omega R^2 C - \omega L}{(1 - \omega^2 LC)^2 + \omega^2 R^2 C^2}$$

The angle of Z is zero, and thus the network is zero-angle resonant for ω for which the imaginary part of Z is zero:

$$\omega^3 L^2 C^2 + \omega(R^2 C - L) = 0$$

$$\omega = 0, \quad \pm\sqrt{\frac{1 - \left(\dfrac{R^2 C}{L}\right)}{LC}}$$

Ordinarily the obvious $\omega = 0$ solution for a network, corresponding to constant voltages and currents, is not considered to be a resonant frequency;

Figure 9-13   A series-parallel RLC network

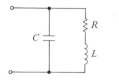

and only positive values of ω are sought, so

$$\omega_r = \sqrt{\frac{1 - \left(\dfrac{R^2C}{L}\right)}{LC} - \sqrt{1 - \left(\frac{R^2C}{L}\right)}\left(\frac{1}{\sqrt{LC}}\right)}$$

for this network.

For $R = 0$, the network is parallel LC, and

$$\omega_r = \frac{1}{\sqrt{LC}}$$

as expected. For significant R, the resonant frequency is lowered by the factor $\sqrt{1 - (R^2C/L)}$.

9.6.3 Resonance and Natural Oscillation Frequency

Almost everyone encounters resonance at an early age. Children learn to pump a swing at a frequency for which the oscillations increase rapidly; and they learn to slosh the water back and forth in a bathtub at just the rate that puts most of it on the floor in minimum time.

In these and many other cases, the system's resonant frequency is very nearly the rate at which it tends to oscillate naturally. For example, one pumps a swing at just about the rate it swings when the rider is aboard but not pumping.

The situation is similar for resonant electrical networks: Resonant frequencies tend to be very nearly the frequencies of a network's oscillatory natural response.

D9-6

     Find the zero-angle resonant frequencies (in radians per second) for each of the following networks:

(a)

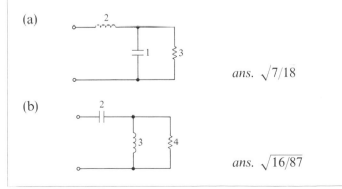

        ans. $\sqrt{7/18}$

(b)

        ans. $\sqrt{16/87}$

Resonance

There are two common definitions of resonance. In the first, a two-terminal network with impedance $Z(s = j\omega)$ is said to be resonant at any frequency for which

$$\angle Z(s = j\omega) = 0°$$

that is, any frequency at which Z is entirely real. At a resonant frequency, the sink reference network voltage and current are in phase, and the network is equivalent to a resistor.

A network is resonant according to the second definition at any frequency for which $|Z(s = j\omega)|$ exhibits an extrema (a maximum or a minimum).

For the series and the parallel RLC networks, both definitions yield the same resonant frequency (which is the same $\omega_r = 1/\sqrt{LC}$ for each type of network). For more involved networks, the two definitions generally give differing results.

9.7 Quality Factor

The quality factor Q of a resonant network is a measure of the narrowness of its bandwidth, that is, the "sharpness" of the resonant maximum or minimum of $|Z(s = j\omega)|$. It is defined to be

$$Q = 2\pi \frac{\begin{array}{c}\text{energy stored in the network} \\ \text{at resonant frequency}\end{array}}{\begin{array}{c}\text{energy dissipated in the network} \\ \text{per sinusoidal cycle at resonance}\end{array}}$$

For the series RLC network, energy is stored in the inductor in amount

$$W_L(t) = \tfrac{1}{2} L i_L{}^2(t)$$

and in the capacitor,

$$W_C(t) = \tfrac{1}{2} C v_C{}^2(t)$$

Suppose the resonant-frequency sinusoidal current in the network is

$$i(t) = A \cos(\omega_r t)$$

as in the drawing of Figure 9-14. The current $i(t)$ is the inductor current, so

$$W_L = \frac{1}{2} L i^2(t) = \frac{LA^2}{2} \cos^2 \omega_r t$$

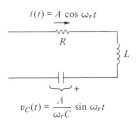

Figure 9-14   Series *RLC* network at resonance

The capacitor voltage is then

$$V_C(t) = \frac{A}{\omega_r C} \sin \omega_r t$$

giving

$$W_C = \frac{1}{2} C v_C^2 = \frac{A^2}{2\omega_r^2 C} \sin^2 \omega_r t$$

Subsituting $\omega_r^2 = 1/LC$,

$$W_C = \frac{LA^2}{2} \sin^2 \omega_r t$$

The total stored energy is

$$W = W_L + W_C = \frac{LA^2}{2} (\sin^2 \omega_r t + \cos^2 \omega_r t)$$

$$= \frac{LA^2}{2}$$

which is constant. Energy is alternately stored more in the inductor, then more in the capacitor, in such a way that the total amount of stored energy is constant.

At resonance, energy is dissipated in the resistor in the series network according to

$$p_{\text{into}}(t) = R i^2(t) = R A^2 \cos^2 \omega_r t$$

In one sinusoidal cycle, the net energy dissipated is

$$P = \int_0^{2\pi/\omega_r} R A^2 \cos^2 \omega_r t \, dt = R A^2 \int_0^{2\pi/\omega_r} \left(\frac{1}{2} + \frac{1}{2} \cos 2\omega_r t \right) dt$$

$$= \frac{R A^2}{2} \left[1 + \frac{\sin 2\omega_r t}{2\omega_r} \right]_0^{2\pi/\omega_r} = \frac{R A^2}{2} \left[\frac{2\pi}{\omega_r} + 0 \right]$$

$$= \frac{\pi R A^2}{\omega_r}$$

The Q of a series RLC network is thus

$$Q = \frac{2\pi L A^2 / 2}{\pi R A^2 / \omega_r} = \frac{\omega_r L}{R} = \sqrt{\frac{L}{R^2 C}}$$

A similar calculation of Q for RLC in parallel gives

$$Q = \frac{R}{\omega_r L} = \sqrt{\frac{R^2 C}{L}}$$

for that network.

For both the series and the parallel RLC networks, bandwidth and Q are related by

$$\frac{\Delta\omega}{\omega_r} = \frac{\Delta f}{f_r} = \frac{1}{Q}$$

and this relationship applies to other resonant networks that have sufficiently high Q to be approximated well in their behavior near the resonant frequency by a series or parallel network.

D9-7

Find the Q of each of the resonant networks of problem D9-4.

ans. $\sqrt{1/3}$; $\sqrt{3/4}$; 100, $\sqrt{160}$

Quality Factor

The quality factor Q of a resonant network is defined to be

$$Q = 2\pi \frac{\text{energy stored in the network}}{\text{energy dissipated in the network per sinusoidal cycle}}$$

at the resonant frequency.

For the series RLC network,

$$Q = \frac{\omega_r L}{R} = \sqrt{\frac{L}{R^2 C}}$$

whereas for the parallel RLC network,

$$Q = \frac{R}{\omega_r L} = \sqrt{\frac{R^2 C}{L}}$$

In the series and the parallel RLC networks,

$$\frac{\Delta\omega}{\omega_r} = \frac{\Delta f}{f_r} = \frac{1}{Q}$$

This relationship is closely approximated in other resonant networks with sufficiently high Q

9.8 Resistive and Reactive Components of an Impedance

9.8.1 Series Resistance-Reactance Equivalent

At any fixed frequency, ω_0, the impedance of a two-terminal network is a complex number

$$Z(s = j\omega_0) = \mathscr{R} + j\mathscr{X}$$

At frequency ω_0, the same impedance could be produced by a resistor of resistance $\mathscr{R}$, in series with an inductor if $\mathscr{X}$ is positive, or a capacitor if $\mathscr{X}$ is negative.

As an example consider the network of Figure 9-15(a). At the frequency $\omega = 10$ rad/s, the impedance at the terminals is

$$Z(s = j10) = j5 + \frac{(10 + j20)(\ j10)}{10 + j20 - j10}$$

$$= j5 + \frac{(200 - j100)(10 - j10)}{(10 + j10)(10 - j10)}$$

$$= j5 + \frac{1000 - j4000}{200}$$

$$= 5 - j15$$

At that frequency, $\omega = 10$, a 5-Ω resistor in series with a capacitor for which

$$-\frac{j}{10C} = -j15$$

$$C = \frac{1}{150}$$

will produce the same impedance. The original network and the RC network are thus equivalent for 10 rad/s sinusoidal signals, as indicated in Figure 9-15(b).

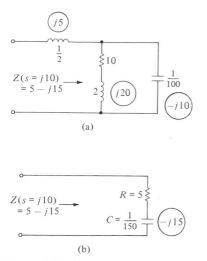

(a)

(b)

Figure 9-15 Network and series equivalent at a single frequency

At another frequency, of course, the network would have an impedance with different real and imaginary parts, and the series equivalent would generally be different. For example, for $\omega = 2$,

$$Z(s = j2) = j1 + \frac{(10 + j4)(-j50)}{10 + j4 - j50}$$

$$= j1 + \frac{(200 - j500)(10 + j46)}{(10 - j46)(10 + j46)}$$

$$= j1 + \frac{25{,}000 + j4200}{2216}$$

$$= 11.28 + j2.9$$

At this frequency, the impedance is produced by an 11.28-Ω resistor in series with an inductor for which

$$j2L = j2.9$$

$$L = 1.45$$

as in Figure 9-16.

9.8.2 Parallel Resistance-Reactance Equivalent

The term *reactance* is used to specify the entirely imaginary impedance for sinusoidal signals produced by inductors and capacitors. The reactance X of an inductor is

$$X = \omega L$$

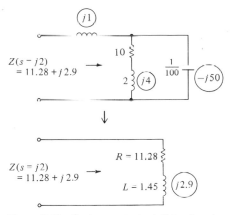

Figure 9-16 Series equivalent at another frequency

and of a capacitor is

$$X = -\frac{1}{\omega C}$$

At a given frequency, an arbitrary impedance may also be obtained from a resistance in *parallel* with a reactance, Figure 9-17. Denoting the reactance of an inductor or capacitor by X,

$$Z = \frac{R(jX)}{R + jX} = \frac{jRX(R - jX)}{(R + jX)(R - jX)}$$

$$= \frac{RX^2}{R^2 + X^2} + j\frac{R^2X}{R^2 + X^2}$$

for the parallel connection.

For Z to have given real part $\mathscr{R}$ and given imaginary part $\mathscr{X}$,

$$\begin{cases} \dfrac{RX^2}{R^2 + X^2} = \mathscr{R} \\[2ex] \dfrac{R^2X}{R^2 + X^2} = \mathscr{X} \end{cases}$$

Figure 9-17 Parallel resistance and reactance

Solving for R and X, the parallel resistance and reactance in terms of the real and imaginary parts of Z is

$$\begin{cases} R = \dfrac{\mathscr{R}^2 + \mathscr{X}^2}{\mathscr{R}} \\[2ex] X = \dfrac{\mathscr{R}^2 + \mathscr{X}^2}{\mathscr{X}} \end{cases}$$

An impedance

$$Z = \mathscr{R} + j\mathscr{X}$$

may thus be obtained with the parallel connection of a resistance R and a reactance jX, where the values of R and X are given by these relations.

Consider the network in Figure 9-18 which, at the frequency $\omega = 4$, has impedance

$$\begin{aligned} Z(s = j4) &= \frac{(7 + j8)(1 - j20)}{7 + j8 + 1 - j20} \\[2ex] &= \frac{(195 - j132)(8 + j12)}{(8 - j12)(8 + j12)} \\[2ex] &= \frac{3144 + j1284}{64 + 144} \\[2ex] &= 15.1 + j6.17 \end{aligned}$$

The parallel connection of a resistor and an inductor of appropriate values will also have this impedance for $\omega = 4$.

Using

$$\mathscr{R} = 15.1, \qquad \mathscr{X} = 6.17$$

Figure 9-18 Network and parallel equivalent at a single frequency

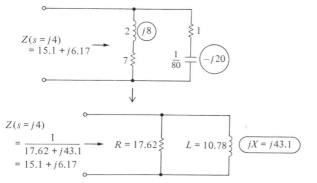

the equivalent parallel resistance and reactance are

$$R = \frac{\mathcal{R}^2 + \mathcal{X}^2}{\mathcal{R}} = \frac{228 + 38.07}{15.1} = 17.62$$

$$X = \frac{\mathcal{R}^2 + \mathcal{X}^2}{\mathcal{X}} = \frac{266.1}{6.17} = 43.1$$

The inductor with the required reactance at $\omega = 4$ has value

$$j4I. = j43.1$$
$$L = 10.78$$

as shown in Figure 9-18.

At a different frequency, $\omega = 10$ for example, the original network has impedance

$$Z(s = j10) = \frac{(7 + j20)(1 - j8)}{7 + j20 + 1 - j8}$$

$$= \frac{(167 - j36)(8 - j12)}{(8 + j12)(8 - j12)}$$

$$= \frac{904 - j2292}{64 + 144}$$

$$= 4.35 - j11$$

At this frequency, the impedance has a negative imaginary part that will involve a capacitor instead of an inductor in the parallel equivalent.

The equivalent parallel resistance and reactance are

$$R = \frac{\mathcal{R}^2 + \mathcal{X}^2}{\mathcal{R}} = \frac{(4.35)^2 + (-11)^2}{4.35} = 32.2$$

$$X = \frac{\mathcal{R}^2 + \mathcal{X}^2}{\mathcal{X}} = \frac{(4.35)^2 + (-11)^2}{-11} = -12.7$$

Figure 9-19 Parallel equivalent at another frequency

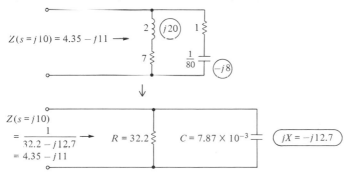

$Z(s = j10) = 4.35 - j11 \longrightarrow$

$Z(s = j10)$
$= \dfrac{1}{32.2 - j12.7} \longrightarrow$
$= 4.35 - j11$

The capacitor with the required reactance at $\omega = 10$ has value

$$\frac{-j}{10C} = -j12.7$$

$$C = 7.87 \times 10^{-3}$$

and is shown in Figure 9-19.

9.8.3   Obtaining Series Resonance

At a single frequency, any network, with impedance $\mathscr{R} + j\mathscr{X}$ at that frequency, may be made zero-angle resonant by the addition of a series reactive impedance $-j\mathscr{X}$.

For example, consider the behavior of the network in Figure 9-20(a) for sinusoidal signals of radian frequency $\omega = 10$. The impedance of this network is

$$Z(s = j10) = \frac{(40 + j20)(-j30)}{40 + j20 - j30} = \frac{600 - j1200}{40 - j10}$$

$$= \frac{(600 - j1200)(40 + j10)}{(40 - j10)(40 + j10)}$$

$$= \frac{(24000 + 12000) - j(48000 - 6000)}{1600 + 100}$$

$$= 21.2 - j24.7$$

At this single frequency, $\omega = 10$, the same impedance would be produced by a resistor and capacitor in series, as shown in Figure 9-20(b).

Figure 9-20   Series resonating a network

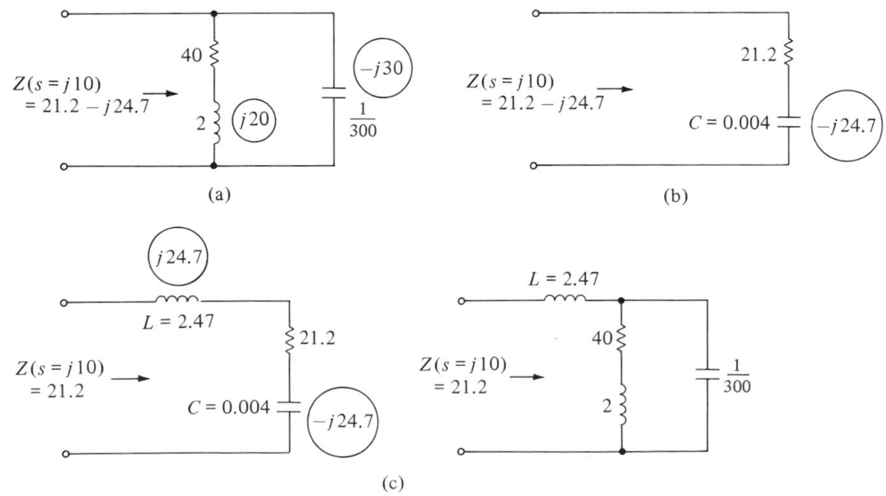

(a)

(b)

(c)

The addition of an inductor with impedance $j24.7$, Figure 9-20(c), will make the series equivalent network, and thus the original network, resonant at frequency $\omega = 10$. The required inductor value is given by

$$j\omega L = j24.7$$

$$L = \frac{24.7}{10} = 2.47$$

9.8.4   Obtaining Parallel Resonance

Using the parallel single-frequency equivalent, any network may also be made zero-angle resonant at that frequency by the addition of an appropriate parallel reactance.

For the example of the previous section, at the given frequency $\omega = 10$, a parallel connection of a resistor and a reactance also has the same impedance as the given network. Using the series-parallel relations,

$$R = \frac{\mathscr{R}^2 + \mathscr{X}^2}{\mathscr{R}} = \frac{(21.2)^2 + (-24.7)^2}{21.2} = 50$$

$$X = \frac{\mathscr{R}^2 + \mathscr{X}^2}{\mathscr{X}} = \frac{(21.2)^2 + (-24.7)^2}{-24.7} = -42.9$$

The parallel equivalent, and thus the original network, will be resonant at the frequency $\omega = 10$ if a parallel inductor with reactance $+42.9$ is placed across the terminals. The value of the added parallel inductor that causes resonance is given by

$$j\omega L = j42.9$$

$$L = \frac{42.9}{10} = 4.29$$

as shown in Figure 9-21.

Figure 9-21   Parallel resonating a network

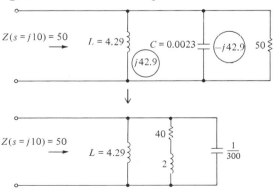

D9-8

Find the series resistive and reactive components of each of the following networks, at the given frequency:

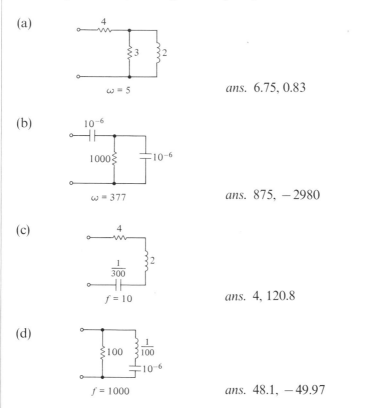

(a)

$\omega = 5$ *ans.* 6.75, 0.83

(b)

$\omega = 377$ *ans.* 875, -2980

(c)

$f = 10$ *ans.* 4, 120.8

(d)

$f = 1000$ *ans.* 48.1, -49.97

D9-9

Find the parallel resistive and reactive components of each of the networks of Problem D9-8.

ans. 6.85, 55.7; 11,024, -3237; 3654, 121; 100, -96.3

D9-10

Find the values of additional series inductance or capacitance that will make each of the networks of Problem D9-8 resonant (with zero phase angle) at the given frequencies.

ans. $C = 0.24$; $L = 7.9$; $C = 1.3 \times 10^{-4}$; $L = 7.96 \times 10^{-3}$

D9-11

Find the values of additional parallel inductance or capacitance that will make each of the networks of Problem D9-8 resonant (with zero phase angle) at the given frequencies.

ans. $C = 3.6 \times 10^{-3}; L = 8.58; C = 1.3 \times 10^{-4}; L = 0.015$

Resistive and Reactive Components of an Impedance

With sinusoidal signals at a single frequency, an impedance is a complex number

$$Z = \mathcal{R} + j\mathcal{X}$$

that could be produced by a resistor in series with an inductor or a capacitor, depending on the algebraic sign of $\mathcal{X}$.

The real part $\mathcal{R}$ of the impedance is called its *resistive* component, and the imaginary part $\mathcal{X}$ is called the *reactive* component of Z. $\mathcal{X}$ is said to be an inductive reactance if positive and a capacitive reactance if negative.

Alternatively, at a single frequency, any impedance may be produced by an appropriate resistor R in parallel with an inductor or a capacitor of reactance X, according to

$$\begin{cases} \mathcal{R} = \dfrac{RX^2}{R^2 + X^2} \\[3mm] \mathcal{X} = \dfrac{R^2 X}{R^2 + X^2} \end{cases}$$

or

$$\begin{cases} R = \dfrac{\mathcal{R}^2 + \mathcal{X}^2}{\mathcal{R}} \\[3mm] X = \dfrac{\mathcal{R}^2 + \mathcal{X}^2}{\mathcal{X}} \end{cases}$$

At a single frequency a reactance $-\mathcal{X}$ in series with an impedance

$$Z = \mathcal{R} + j\mathcal{X}$$

will make the combination resonant (with zero phase angle) at that frequency. Similarly the parallel combination of Z and the reactance

$$-X = -\dfrac{\mathcal{R}^2 + \mathcal{X}^2}{\mathcal{X}}$$

is resonant (with zero phase angle) at that frequency.

Chapter Nine Problems

Basic Problems

Series and Parallel *RLC* Networks

1. Find the resonant frequency of each of the following networks, their impedance, bandwidth, and Q at resonance:

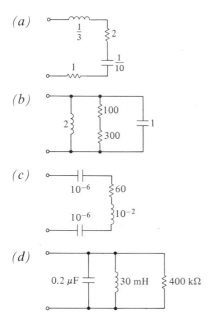

(a)

(b)

(c)

(d)

2. For a series *RLC* network with $R = 2\ \Omega$, $L = 100$ mH, and $C = 0.02\ \mu$F, find the impedance at resonance, one-fifth, one-half, twice, and five times the resonant frequency.

3. For a parallel *RLC* network with $R = 1000\ \Omega$, $L = 200$ mH, and $C = 0.3\ \mu$F, find the impedance at resonance, one-fifth, one-half, twice, and five times the resonant frequency.

4. For the network below, draw phasor diagrams showing the voltage and each current
(a) At resonance.
(b) At one-fourth the resonant frequency.
(c) At four times the resonant frequency.

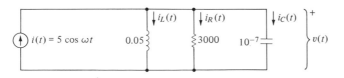

Frequency Response

5. Sketch frequency response plots for the impedance

$$Z(s) = \frac{s + 2}{s^2 + s + 10}$$

General Resonance

6. Find the zero-angle resonant frequencies of each of the following networks:

(a)

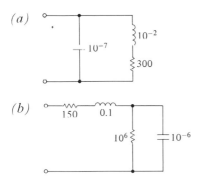

(b)

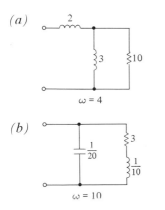

Resistive and Reactive Impedance Components

7. Find the series resistive and reactive components and the parallel resistive and reactive components of each of the following networks at the given frequency:

(a)

(b)

8. Find the values of additional series inductance or capacitance and the values of additional parallel inductance or capacitance that will make each of the networks resonant (with zero phase angle) at the given frequencies:

(a)

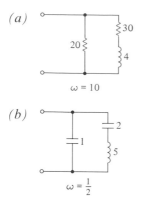

$\omega = 10$

(b)

$\omega = \frac{1}{2}$

Practical Problems

Adjustable Tuning

1. It is desired to construct a parallel resonant network with a 15-mH inductor and an adjustable capacitor that may be adjusted to be resonant at any frequency within the standard broadcast band of 550–1600 kHz. Over what range of capacitance must the capacitor be adjustable? The

A sweep frequency function generator that can produce a sinusoidal signal with slowly increasing frequency for frequency-response measurements. *(Photo courtesy of Wavetek.)*

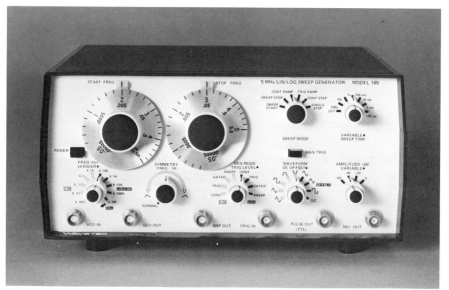

largest capacitance, C_{max}, will give resonance at 550 kHz and the smallest capacitance, C_{min}, will give resonance at 1600 kHz. Find the resonant frequency for the midvalue of capacitance,

$$\frac{C_{max} - C_{min}}{2}$$

The tuning scales of many radio receivers are "spread out" at the lower frequencies because of the nonlinear relation between capacitance (or inductance) and resonant frequency.

Resonance Measurements

2. A physical inductor with an inductance of 10 mH and 30-Ω winding resistance is connected in series with a 0.01-μF capacitor and a voltage source of constant amplitude but adjustable frequency.

(a) Will the inductor voltage be maximum at a frequency above or below the resonant frequency?

Spectrum analyzer shows the frequency response of an electrical network. The oscilloscope screen at the lower left displays amplitude versus frequency. (*Photo courtesy Systron-Donner Corp.*)

(b)   If the maximum network current amplitude is 15 mA, what is the source voltage amplitude?

(c)   At what frequency will the network current and the source voltage be in phase with one another?

(d)   Over how wide a range of frequency will the network current and source voltage phases be within $\pm 45°$ of one another?

3.   A physical inductor with an inductance of 20 mH and 500-Ω winding resistance is connected in parallel with a 300-pF capacitor with a 10-MΩ leakage resistance. Find the precise (zero-angle) resonant frequency of this series-parallel network.

Resonance of a Coil

A coil of wire, in addition to having inductance and resistance, has capacitances between each wire segment, as indicated schematically in the sketch of Figure 9-22(a). At low frequencies, these stray capacitances, which depend on the construction of the coil, are negligible. At sufficiently high frequencies, the stray capacitances become important and may be modeled by a single capacitor in parallel with the coil inductance and resistance, Figure 9-22(b). A coil of wire thus acts as a resonant circuit at very high frequencies, being inductive below resonance and *capacitive* above its resonance.

At still higher frequencies, a more complicated model, involving several inductors and capacitors, is needed.

4.   A certain coil has an inductance of 10 mH and a resistance of 50 Ω, when measured at frequencies that are small compared to the coil's resonant frequency. What is the effective capacitance if the coil is self-resonant at 3 MHz?

5.   A coil with negligible winding resistance has inductance 200 μH and is self-resonant at 25 MHz. What is the coil impedance at 20 MHz? At 30 MHz?

Figure 9-22   Practical Problem 5

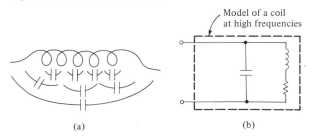

Model of a coil
at high frequencies

(a)   (b)

Domestic communications satellite receiving station. Communications satellites use a wide band of frequencies in order to accommodate a large number of voice, data, and video channels. *(Photo courtesy of A. T. & T. Long Lines.)*

Advanced Problems

Series and Parallel *RLC* Networks

1. Find and sketch the frequency response magnitudes and phase shifts
(a) $\mathbf{I/V}$
(b) $\mathbf{V}_L/\mathbf{V}$
(c) $\mathbf{V}_C/\mathbf{V}$
(d) $\mathbf{V}_{RL}/\mathbf{V}$

for the network of Figure 9-23.

Figure 9-23   Advanced Problem 1

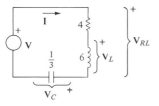

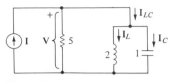

Figure 9-24 Advanced Problem 2

2. Find and sketch the frequency response magnitudes and phase shifts
(*a*) $\mathbf{V}/\mathbf{I}$
(*b*) $\mathbf{I}_L/\mathbf{I}$
(*c*) $\mathbf{I}_C/\mathbf{I}$
(*d*) $\mathbf{I}_{LC}/\mathbf{I}$

for the network of Figure 9-24.

Quality Factor

3. Show that the total stored energy in a parallel *RLC* network is constant at the resonant frequency.

4. Show that, for a parallel *RLC* network,

$$Q = \sqrt{\frac{R^2 C}{L}}$$

Resistive and Reactive Components

5. Starting with the relations

$$\begin{cases} \dfrac{RX^2}{R^2 + X^2} = \mathscr{R} \\[4mm] \dfrac{R^2 X}{R^2 + X^2} = \mathscr{X} \end{cases}$$

algebraically derive the relations

$$\begin{cases} R = \dfrac{\mathscr{R}^2 + \mathscr{X}^2}{\mathscr{R}} \\[4mm] X = \dfrac{\mathscr{R}^2 + \mathscr{X}^2}{\mathscr{X}} \end{cases}$$

CHAPTER TEN
Power in Sinusoidally Driven Networks

10.1 Introduction

This final chapter concerns electrical power flow in networks with sinusoidal voltages and currents. The relations developed are fundamental to the areas of power, electronic design, and communications, and have applications in other areas as well.

Networks designed for the efficient transmission of electrical power are then considered, and special attention is given to straightforward solution methods for three-phase networks.

When you complete this chapter, you should know

1. How power flow in networks with sinusoidal signals may be expressed as an average part plus a fluctuating part.
2. Relations for average power flow in the basic elements and in general.
3. What reactive power is, and reactive power relations for the basic elements and in general.
4. The concept of complex power and complex power relations.
5. How rms values are used in power calculations.
6. The configuration of a single-phase power transmission system and how to solve related problems.

7. Why power factor correction is important in electrical power transmission and how to use resonance methods to solve such problems.
8. What three-phase power transmission is, its advantages, and various source and load configurations.
9. How to use source and load delta-wye transformations to convert balanced three-phase power networks to equivalent single-phase problems.

10.2   Sinusoidal Power

10.2.1   Sinusoidal Power in General

The electrical power flow into an element is the product of the element's sink reference voltage and current. For sinusoidal voltage.

$$v(t) = A \cos (\omega t + \theta)$$

and a sinusoidal element current,

$$i(t) = B \cos (\omega t + \phi)$$

the power flow is

$$p_{\text{into}}(t) = v(t)i(t) = AB \cos (\omega t + \theta) \cos (\omega t + \phi)$$

Using the trigonometric identity

$$\cos x \cos y = \tfrac{1}{2} \cos (x - y) + \tfrac{1}{2} \cos (x + y)$$

$$p_{\text{into}}(t) = \frac{AB}{2} \cos (\theta - \phi) + \frac{AB}{2} \cos (2\omega t + \theta + \phi)$$

The power consists of a constant component that depends on the difference in phase angles between voltage and current, $(AB/2) \cos (\theta - \phi)$, plus a sinusoidal, fluctuating component with frequency twice that of the voltage or current, $(AB/2) \cos (2\omega t + \theta + \phi)$.

10.2.2   Sinusoidal Power in a Resistor

In a resistor, the sink reference voltage and current are in phase. If

$$v_R(t) = A \cos (\omega t + \theta)$$

then

$$i_R(t) = \frac{v_R}{R} = \frac{A}{R} \cos (\omega t + \theta)$$

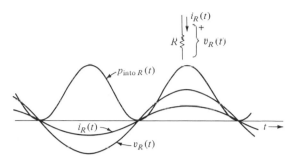

Figure 10-1   Voltage, current, and power for a resistor

and

$$p_{\text{into } R}(t) = \frac{A^2}{R} \cos^2 (\omega t + \theta)$$

$$= \frac{A^2}{2R} + \frac{A^2}{2R} \cos (2\omega t + 2\theta)$$

As shown in the sketch of typical voltage, current, and power in Figure 10-1, the power flow consists of equal-amplitude constant and fluctuating components so that $p_{\text{into } R}(t)$ is never negative.

10.2.3   Sinusoidal Power in an Inductor

In an inductor, sink reference voltage and current are 90° out of phase. If

$$v_L(t) = A \cos (\omega t + \theta)$$

the forced sinusoidal inductor current is

$$i_L(t) = \frac{A}{\omega L} \cos (\omega t + \theta - 90°)$$

and

$$p_{\text{into } L}(t) = \frac{A^2}{\omega L} \cos (\omega t + \theta) \cos (\omega t + \theta - 90°)$$

$$= \frac{A^2}{2\omega L} \cos 90° + \frac{A^2}{2\omega L} \cos (2\omega t + 2\theta - 90°)$$

$$= \frac{A^2}{2\omega L} \cos (2\omega t + 2\theta - 90°)$$

There is no constant component to the power flow, as indicated in the sketch of Figure 10-2. Power flows into and out of the inductor, back and forth, twice each cycle of the voltage or current.

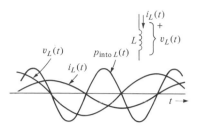

Figure 10-2 Voltage, current, and power for an inductor

10.2.4 Sinusoidal Power in a Capacitor

The sink reference voltage lags the current by 90° in a capacitor. If

$$v_C(t) = A \cos (\omega t + \theta)$$

the forced sinusoidal capacitor current is

$$i_C(t) = A\omega C \cos (\omega t + \theta + 90°)$$

$$p_{\text{into}}(t) = A^2 \omega C \cos (\omega t + \theta) \cos (\omega t + \theta + 90°)$$

$$= \frac{A^2 \omega C}{2} \cos (-90°) + \frac{A^2 \omega C}{2} \cos (2\omega t + \theta + 90°)$$

$$= \frac{A^2 \omega C}{2} \cos (2\omega t + \theta + 90°)$$

As with the inductor, the capacitor power flow contains no constant component, as illustrated in Figure 10-3. Power is alternately stored in and released from the capacitor, twice during each cycle of the voltage or current.

10.2.5 Conservation of Energy

In a network the net power flowing out of sources equals the net power flow into the other elements at every instant of time.

For example, consider the network of Figure 10-4, in which all voltages and currents have been found and are indicated on the network diagram. The

Figure 10-3 Voltage, current, and power for a capacitor

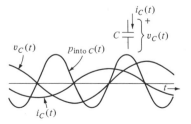

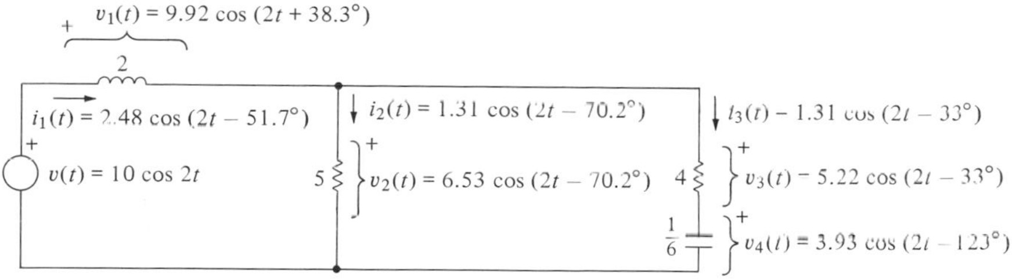

$v_1(t) = 9.92 \cos (2t + 38.3°)$

$i_1(t) = 2.48 \cos (2t - 51.7°)$ $i_2(t) = 1.31 \cos (2t - 70.2°)$ $i_3(t) - 1.31 \cos (2t - 33°)$

$v(t) = 10 \cos 2t$ $5 \quad v_2(t) = 6.53 \cos (2t - 70.2°)$ $4 \quad v_3(t) - 5.22 \cos (2t - 33°)$

$\dfrac{1}{6} \quad v_4(t) = 3.93 \cos (2t - 123°)$

Figure 10-4 Network with sinusoidal signals indicated

electrical power flow into the inductor is

$$P_{\text{into } L}(t) = v_1(t)i_1(t)$$
$$= [9.92 \cos (2t + 38.3°)][2.48 \cos (2t - 51.7°)]$$
$$= 12.39 \cos (4t - 13.4°)$$

The electrical power flow into the capacitor is

$$P_{\text{into } C}(t) = v_4(t)i_3(t)$$
$$= [3.93 \cos (2t - 123.3°)][1.31 \cos (2t - 33.3°)]$$
$$= 2.57 \cos (4t - 156.6°)$$

The electrical power flow into the 5-Ω resistor is

$$P_{\text{into } 5}(t) = v_2(t)i_2(t)$$
$$= [6.53 \cos (2t - 70.2°)][1.31 \cos (2t - 70.2°)]$$
$$= 4.28 + 4.28 \cos (4t - 140.4°)$$

and the electrical power flow into the 4-Ω resistor is

$$P_{\text{into } 4}(t) = v_3(t)i_3(t)$$
$$= [5.22 \cos (2t - 33.3°)][1.31 \cos (2t - 33.3°)]$$
$$= 3.42 + 3.42 \cos (4t - 66.6°)$$

The electrical power flow out of the source,

$$P_{\text{out}}(t) = v(t)i_1(t)$$
$$= [10 \cos 2t][2.48 \cos (2t - 51.7°)]$$
$$= 12.4 \cos (51.7°) + 12.4 \cos (4t - 51.7°)$$
$$= 7.7 + 12.4 \cos (4t - 51.7°)$$

is equal to the sum of the power flow into the other four elements:

$$P_{\text{out}}(t) = P_{\text{into } L}(t) + P_{\text{into } C}(c) + P_{\text{into } 5}(t) + P_{\text{into } 4}(t)$$
$$= 12.39 \cos (4t - 13.4°) + 2.57 \cos (4t - 156.6°)$$
$$+ 4.28 + 4.28 \cos (4t - 140.4°) + 3.42$$
$$+ 3.42 \cos (4t - 66.6°)$$

That the instantaneous power flow out of the source equals the net power flow into the other elements may be verified as follows: The constant parts add to equal the constant part of $p_{out}(t)$, 7.7. The fluctuating part of this sum consists of the sum of four sinusoidal components (with frequencies twice that of the voltages and currents) with various amplitudes and phase angles. Decomposing each into quadrature components (Section 8.2.3), there results

$$12.39 \cos (4t - 13.4°) = 12.05 \cos 4t + 2.86 \sin 4t$$
$$2.57 \cos (4t - 156.6°) = -2.35 \cos 4t + 1.02 \sin 4t$$
$$4.28 \cos (4t - 140.4°) = -3.3 \cos 4t + 2.74 \sin 4t$$
$$3.42 \cos (4t - 66.6°) = 1.35 \cos 4t + 3.15 \sin 4t$$

the sum of which is

$$7.73 \cos 4t + 9.77 \sin 4t = 12.4 \cos (4t - 51.7°)$$

which equals the corresponding component in the source power flow.

It should be noted that there are times when the power flow out of the source is negative; that is, there are intervals of time when energy is flowing from the rest of the network back into the source. This situation is typical of sinusoidally driven networks. Energy flows from the source to the other network elements. Energy that flows into resistors is completely dissipated. Energy that flows into reactive elements, L's and C's, is stored, then released, allowing a return of some energy to the source.

If a network has more than one source, energy may be exchanged by the sources, and it sometimes happens that one source mainly supplies energy while another source mainly absorbs it.

D10-1
Element voltages and currents for the networks below have been found and appear on the network diagrams. For each network, find the electrical power flow into each element as a function of time and write the equation relating power flow out of sources to power flow into the other network elements.

(a)

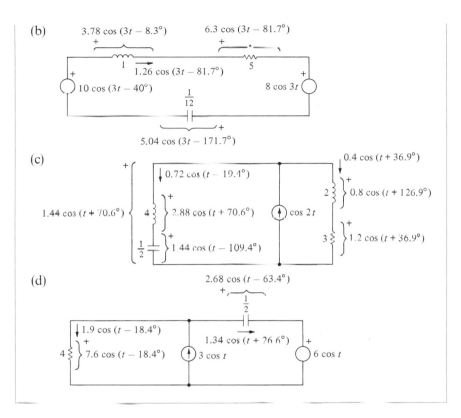

Sinusoidal Power

The electrical power flow into an element is the product of its sink reference voltage and current.

The product of a sinusoidal voltage,

$$v(t) = A \cos (\omega t + \theta)$$

and a sinusoidal current,

$$i(t) = B \cos (\omega t + \phi)$$

is

$$p(t) = v(t)i(t) = \frac{AB}{2} \cos (\theta - \phi) + \frac{AB}{2} \cos (2\omega t + \theta + \phi)$$

The sum of electrical power flows out of sources equals the sum of electrical power flows into the other elements in a network at every instant of time.

10.3   Average Power Flow

10.3.1   Average Power in General

For sinusoidal, sink reference voltage and current,

$$v(t) = A \cos (\omega t + \theta)$$
$$i(t) = B \cos (\omega t + \phi)$$

an element's power flow is

$$p_{into}(t) = v(t)i(t)$$

$$= \frac{AB}{2} \cos (\theta - \phi) + \frac{AB}{2} \cos (2\omega t + \theta + \phi)$$

The average power flow is the constant term

$$P_{into} = \frac{AB}{2} \cos (\theta - \phi)$$

Average power is denoted by a capital letter P, to distinguish it from instantaneous power $p(t)$.

The phasors representing $v(t)$ and $i(t)$ are

$$\mathbf{V} = Ae^{j\theta}$$
$$\mathbf{I} = Be^{j\phi}$$

and in terms of these, the average power is

$$P_{into} = \tfrac{1}{2}|\mathbf{V}||\mathbf{I}| \cos (\angle \mathbf{V} - \angle \mathbf{I})$$

10.3.2   Average Power in Resistors, Inductors, and Capacitors

For a resistor, voltage and current are in phase, so

$$P_{into\,R} = \frac{AB}{2} \cos (0°) = \frac{AB}{2}$$

one-half the product of the voltage amplitude with the current amplitude.
In terms of the phasors describing the resistor voltage and current,

$$P_{into\,R} = \tfrac{1}{2}|\mathbf{V}_R||\mathbf{I}_R|$$

For an inductor, the current lags the voltage by 90°, giving

$$P_{into\,L} = \frac{AB}{2} \cos (90°) = 0$$

For a capacitor, where current leads voltage by 90°,

$$P_{into\,C} = \frac{AB}{2} \cos (-90°) = 0$$

Energy flows back and forth, into and out of inductors and capacitors, but the average flows are zero.

10.3.3   Conservation of Average Energy

In a network the power flow out of sources equals the power flow into the other elements at each instant of time. The *average* power flow from sources thus equals the *average* power flow into the other elements. In many practical situations it is the average power rather than the detailed, instantaneous power that is of interest so far as element ratings, capabilities, and efficiency are concerned.

As a numerical example of average power flow relations, consider again the example network of Figure 10-4. The average electrical power flow into the 5-Ω resistor is

$$P_{\text{into } 5} = \frac{(6.53)(1.31)}{2} = 4.28$$

The 4-Ω resistor, similarly, has an average power flow of

$$P_{\text{into } 4} = \frac{(5.22)(1.31)}{2} = 3.42$$

The average power flows into the inductor and capacitor are each zero:

$$P_{\text{into } L} = 0$$
$$P_{\text{into } C} = 0$$

The average power flow out of the source is

$$P_{\text{out}} = \frac{(10)(2.48)}{2} \cos 51.7° = 7.7$$

which equals the sum of the average power flows into the other network elements:

$$P_{\text{out}} = P_{\text{into } 5} + P_{\text{into } 4} + P_{\text{into } L} + P_{\text{into } C}$$

D10-2

For the networks of Problem D10-1, find the average electrical power flow into each element and verify that the net average power flow out of sources equals the net average power flow into the other elements.

ans. 10, 0, 0, -10; 0, 3.97, 0, -4.7,
0.73; 0, 0, 0, 0.24, -0.24; 7.22,
0, -10.82, 3.6

Average Power

In general, the average power flow into an element with sink reference sinusoidal voltage and current

$$v(t) = A \cos(\omega t + \theta), \qquad \mathbf{V} = A e^{j\theta}$$
$$i(t) = B \cos(\omega t + \phi), \qquad \mathbf{I} = B e^{j\phi}$$

is

$$P_{\text{into}} = \frac{AB}{2} \cos(\theta - \phi)$$

$$= \tfrac{1}{2} |\mathbf{V}| \, |\mathbf{I}| \cos(\angle \mathbf{V} - \angle \mathbf{I})$$

Average power relations for the basic elements are as follows:

Resistor

$$P_{\text{into}} = \frac{AB}{2} = \frac{1}{2} |\mathbf{V}_R| \, |\mathbf{I}_R|$$

Inductor

$$P_{\text{into}} = 0$$

Capacitor

$$P_{\text{into}} = 0$$

The sum of average electrical power flows out of sources equals the sum of average electrical power flows into the other elements of a network.

10.4 Maximum Power Transfer

Consider the network of Figure 10-5, for which it is desired to choose the impedance

$$Z_L = R_L + jX_L$$

for maximum average electrical power flow into Z_L. The impedance Z_L is commonly termed the "load impedance" in such a situation.

The average power flow into Z_L is

$$P = \frac{|\mathbf{I}|^2 R_L}{2}$$

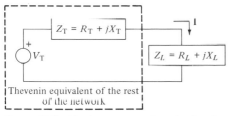

Figure 10-5 Maximum power transfer for sinusoidal signals

and the current phasor is

$$I = \frac{V_T}{Z_T + Z_L} = \frac{V_1}{(R_T + R_L) + j(X_T + X_L)}$$

giving

$$P = \frac{|V_T|^2 R_L}{2[(R_T + R_L)^2 + (X_T + X_L)^2]}$$

Equating the partial derivatives of P with respect to R_L and X_L to zero, to find the minimum, there results

$$\begin{cases} \dfrac{\partial P}{\partial R_L} = \dfrac{2[(R_T + R_L)^2 + (X_T + X_L)^2]|V_T|^2 - |V_T|^2 R_L \cdot 4(R_T + R_L)}{2[(R_1 + R_L)^2 + (X_T + X_L)^2]^2} \\[2ex] \qquad = \dfrac{|V_T|^2[R_T^2 - R_L^2 + (X_1 + X_L)^2]}{2[(R_T + R_L)^2 + (X_T + X_L)^2]^2} = 0 \\[2ex] \dfrac{\partial P}{\partial X_L} = \dfrac{-|V_T|^2 R_L \cdot 4(X_T + X_L)}{2[(R_T + R_L)^2 + (X_T + X_L)^2]^2} = 0 \end{cases}$$

The above conditions are satisfied if and only if

$$X_L = -X_T \qquad \text{and} \qquad R_L = R_T$$

This is to say that to adjust a load impedance for maximum average power flow into it, choose the real part of the load impedance equal to the real part of the Thévenin impedance of the rest of the network. And choose the imaginary part of the load impedance to be the negative of the Thévenin impedance's imaginary part:

$$Z_L = Z_T{}^*$$

The series circuit is made resonant by canceling the Thévenin reactance with the load reactance. Then the load resistance is made equal to the Thévenin resistance, as in the result for resistive networks.

As a numerical example, consider the network in Figure 10-6(a), in which it is desired to chose the impedance Z for maximum average power flow into Z. The Thévenin impedance of the portion of the network connected

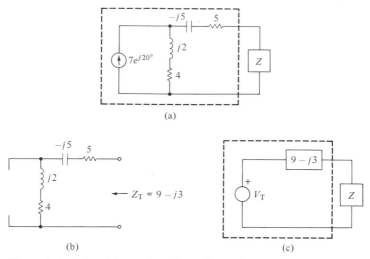

(a)

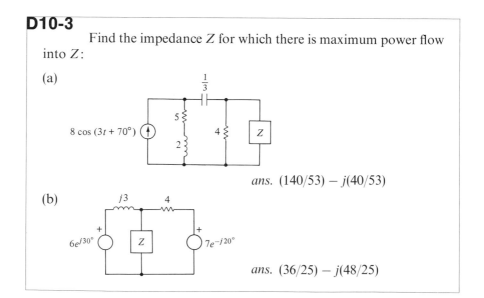

(b)   (c)

Figure 10-6 Determining an impedance for maximum power transfer

to Z is computed by finding the terminal impedance, with the source set to zero, in Figure 10-6(b). The Thévenin voltage $\mathbf{V}_T$ could also be found, but it is not needed for the calculation at hand. In Figure 10-6(c) the network connected to Z is replaced by the Thévenin equivalent, which is equivalent so far as Z is concerned.

Maximum average power will flow into Z if it is chosen to be

$$Z = 9 + j3$$

D10-3

Find the impedance Z for which there is maximum power flow into Z:

(a)

ans. $(140/53) - j(40/53)$

(b)

ans. $(36/25) - j(48/25)$

Maximum Power Transfer

For sinusoidal signals and an adjustable load impedance Z_L, maximum average power is transferred from a two-terminal network with Thévenin impedance Z_T when

$$Z_L = Z_T{}^*$$

10.5   Reactive Power and Network Elements

10.5.1   Resistive and Reactive Components of Instantaneous Power

For sinusoidal voltage and current,

$$v(t) = A \cos (\omega t + 0)$$
$$i(t) = B \cos (\omega t + \phi)$$

a general power flow is

$$p(t) = v(t)i(t)$$

$$-\frac{AB}{2} \cos (\theta - \phi) + \frac{AB}{2} \cos (2\omega t + \theta + \phi)$$

Using the trigonometric identity

$$\cos (x + y) = \cos x \cos y - \sin x \sin y$$

$$\begin{aligned}
\cos (2\omega t + \theta + \phi) &= \cos (\theta - \phi) \cos (2\omega t + 2\phi) \\
&\quad - \sin (\theta - \phi) \sin (2\omega t + 2\phi) \\
&= \cos (\theta - \phi) \cos (2\omega t + 2\phi) \\
&\quad + \sin (\theta - \phi) \cos (2\omega t + 2\phi + 90°)
\end{aligned}$$

giving

$$p(t) = \left[\frac{AB}{2} \cos (\theta - \phi) \right] [1 + \cos (2\omega t + 2\phi)]$$

$$+ \left[\frac{AB}{2} \sin (\theta - \phi) \right] \cos (2\omega t + 2\phi + 90°)$$

Power flow is thus expressible as a sum,

$$p(t) = p_{\mathscr{R}}(t) + p_{\mathscr{X}}(t)$$

of a resistor power flow

$$p_{\mathscr{R}}(t) = \left[\frac{AB}{2} \cos{(\theta - \phi)} \right] [1 + \cos{(2\omega t + 2\phi)}]$$

plus a reactive (inductor or capacitor) power flow

$$p_{\mathscr{X}}(t) = \left[\frac{AB}{2} \sin{(\theta - \phi)} \right] \cos{(2\omega t + 2\phi + 90°)}$$

The resistive component of power flow has average value

$$P = \frac{AB}{2} \cos{(\theta - \phi)}$$

as indicated in the sketch of Figure 10-7, whereas the reactive component of the instantaneous power has zero average value, and amplitude

$$Q = \frac{AB}{2} \sin{(\theta - \phi)}$$

as shown.

10.5.2 General Reactive Power

The reactive power flow Q into an element is defined as

$$Q_{\text{into}} = \tfrac{1}{2} AB \sin{(\theta - \phi)}$$

where the element's sink reference voltage and current are

$$v(t) = A \cos{(\omega t + \theta)}$$
$$i(t) = B \cos{(\omega t + \phi)}$$

Figure 10-7 Resistive and reactive components of instantaneous power flow

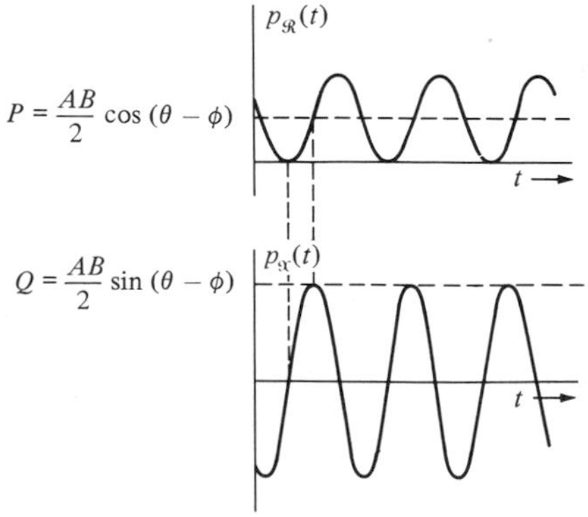

Reactive power is the amplitude of the reactive component of the instantaneous power flow, which is alternately stored and returned to the rest of the network every half cycle of the voltage and current.

The unit of reactive power is generally taken to be the VAR (which stands for volt-amperes reactive). Reactive power Q should not be confused with the Q of a resonant circuit; the two are different quantities with the same symbol.

10.5.3   Reactive Power in Resistors, Inductors, and Capacitors

For a resistor the voltage and current are in phase and

$$Q_{\text{into } R} = \frac{AB}{2} \sin (0^\circ) = 0$$

All of the energy that flows into a resistor is dissipated; none is alternately stored then released.

For an inductor, the current lags the voltage by 90°, so

$$Q_{\text{into } L} = \frac{AB}{2} \sin (90^\circ) = \frac{AB}{2}$$

The reactive power is one-half the product of voltage amplitude with current amplitude. In terms of the phasors describing the inductor voltage and current,

$$Q_{\text{into } L} = \tfrac{1}{2}|\mathbf{V}_L||\mathbf{I}_L|$$

For a capacitor, where current leads voltage by 90°,

$$Q_{\text{into } C} = \frac{AB}{2} \sin (-90^\circ) = -\frac{AB}{2}$$

In terms of the voltage and current phasors,

$$Q_{\text{into } C} = -\tfrac{1}{2}|\mathbf{V}_C||\mathbf{I}_C|$$

The negative nature of reactive power for a capacitor indicates a negative amplitude for the reactive component of instantaneous power. This is to say that a capacitor releases energy during the time interval an inductor stores it, and vice versa.

10.5.4   Conservation of Reactive Energy

The reactive power oscillations are 90° out of phase with the power oscillations associated with energy dissipation, and represent energy that is interchanged between sources and energy-storing elements every half cycle of the voltage or current.

The sum of the reactive powers out of sources in a network equals the sum of reactive powers into the other elements.

As a numerical example, consider the network of Figure 10-4 once again. The reactive power flow into the inductor is

$$Q_{\text{into } L} = \frac{(9.92)(2.48)}{2} = 12.3$$

The capacitor has reactive power flow

$$Q_{\text{into } C} = -\frac{(3.93)(1.31)}{2} = -2.57$$

The resistors each have zero reactive power flow:

$$Q_{\text{into } 5} = 0$$
$$Q_{\text{into } 4} = 0$$

The reactive power flow out of the source is

$$Q_{\text{out}} = \frac{(10)(2.48)}{2} \sin 51.7° = 9.73$$

which equals the sum of reactive power flows into the other network elements:

$$Q_{\text{into } 5} + Q_{\text{into } 4} + Q_{\text{into } L} + Q_{\text{into } C} = Q_{\text{out}}$$

D10-4

For the network of Problem D10-1, find the reactive power flow into each element and verify that the net reactive power flow out of sources equals the net reactive power flow into the other elements.

ans. 0, −5, 8.35, −3.35; 2.38, 0,
−3.17, −4.19, 4.98; 1.04, −0.52,
0.16, 0, −0.68; 0, −1.8, 3.6, −1.8

Reactive Power

Reactive power Q (not to be confused with the Q of a resonant circuit) is a measure of the energy that flows into an element, is stored, and then flows out of the element each half cycle.

In general, the reactive power flow into an element with sink reference sinusoidal voltage and current

$$v(t) = A \cos (\omega t + \theta), \qquad \mathbf{V} = A e^{j\theta}$$
$$i(t) = B \cos (\omega t + \phi), \qquad \mathbf{I} = B e^{j\phi}$$

is

$$Q_{into} - \frac{AB}{2} \sin (\theta - \phi)$$

$$= \tfrac{1}{2}|\mathbf{V}||\mathbf{I}| \sin (\angle \mathbf{V} - \angle \mathbf{I})$$

Reactive power relations for the basic elements are as follows:

Resistor

$$Q_{into} = 0$$

Inductor

$$Q_{into} = \frac{AB}{2} = \frac{1}{2}|\mathbf{V}_L||\mathbf{I}_L|$$

Capacitor

$$Q_{into} = -\frac{AB}{2} = -\frac{1}{2}|\mathbf{V}_C||\mathbf{I}_C|$$

The sum of reactive power flows out of sources equals the sum of reactive power flows into the other elements of a network.

10.6 Complex Power

In performing power calculations it is often helpful to combine the average and reactive power for an element into a complex quantity which is termed *complex power*:

$$\mathscr{P} = P + jQ$$

If the element's sink reference voltage and current are

$$v(t) = A \cos (\omega t + \theta)$$
$$i(t) = B \cos (\omega t + \phi)$$

then

$$\mathscr{P} = \frac{AB}{2} \cos (\theta - \phi) + j \frac{AB}{2} \sin (\theta - \phi)$$

$$= \frac{AB}{2} e^{j(\theta - \phi)}$$

The real part of $\mathscr{P}$ is the average power, and the imaginary part of $\mathscr{P}$ is the reactive power.

Average power is sometimes called "real power" or "active power" and reactive power is sometimes called "imaginary power," to give a symmetry to these terms comparable to their mathematical symmetry.

In terms of the sink reference voltage and current phasors for an element, $\mathbf{V}$ and $\mathbf{I}$, its complex power is

$$\mathscr{P} = \tfrac{1}{2}|\mathbf{V}||\mathbf{I}|e^{j(\angle \mathbf{V} - \angle \mathbf{I})} = \tfrac{1}{2}\mathbf{V}\mathbf{I}^*$$

where $\mathbf{I}^*$ denotes the complex conjugate of $\mathbf{I}$.

Conservation of average energy and conservation of reactive energy in a network are simultaneously expressed by stating that the net complex power flow out of sources equals the net complex power flow into the other network elements.

D10-5

For the networks of Problem D10-1, find the complex power flow into each element and verify that the net complex power flow out of sources equals the net complex power into the other elements.

> *ans.* $10, -j5, j8.35, -10 - j3.35;$
> $j2.38, 3.97, -j3.17, -4.7, -j4.19,$
> $0.73 + j4.98; j1.04, -j0.52, j0.16,$
> $0.24, -0.24 - j0.68; 7.22, -j1.8,$
> $-10.82 + j3.6, 3.6 - j1.8$

Complex Power

With sink reference sinusoidal voltage and current

$$v(t) = A \cos(\omega t + \theta), \qquad \mathbf{V} = Ae^{j\theta}$$
$$i(t) = B \cos(\omega t + \phi), \qquad \mathbf{I} = Be^{j\phi}$$

the complex power flow into an element is

$$\mathscr{P}_{\text{into}} = P_{\text{into}} + jQ_{\text{into}}$$
$$= \tfrac{1}{2}AB \cos(\theta - \phi) + j\tfrac{1}{2}AB \sin(\theta - \phi)$$
$$= \tfrac{1}{2}ABe^{j(\theta - \phi)}$$
$$= \tfrac{1}{2}|\mathbf{V}||\mathbf{I}|e^{j(\angle \mathbf{V} - \angle \mathbf{I})}$$
$$= \tfrac{1}{2}\mathbf{V}\mathbf{I}^*$$

Complex power is used to combine concisely average (or *active* or *real*) and reactive (or *imaginary*) power into a single mathematical quantity.

The sum of complex power flows out of sources equals the sum of complex power flows into the other elements of a network.

10.7 Power Calculations Using RMS Values

10.7.1 RMS Values

Over the years, considerable effort has been devoted to expressing the power relations as simply as possible. One concern has been the factors of one-half in the expressions for P and Q. For *constant* voltages and currents (termed dc or "direct current" conditions), power is the voltage-current product; but the product involves this factor of one-half in the sinusoidal (ac or "alternating current") case.

The root-mean-square (rms) of a sinusoid is related to its amplitude by

$$(\text{rms value}) = \frac{1}{\sqrt{2}} (\text{amplitude})$$

so in terms of voltage and current

$$v(t) = A \cos(\omega t + \theta)$$
$$i(t) = B \cos(\omega t + \phi)$$

the rms values are

$$V_{\text{rms}} = \frac{A}{\sqrt{2}}$$

$$I_{\text{rms}} = \frac{B}{\sqrt{2}}$$

giving

$$P = \tfrac{1}{2}AB \cos(\theta - \phi) = V_{\text{rms}}I_{\text{rms}} \cos(\theta - \phi)$$
$$Q = \tfrac{1}{2}AB \sin(\theta - \phi) = V_{\text{rms}}I_{\text{rms}} \sin(\theta - \phi)$$

Although defining away a factor of one-half may seen uncalled for with today's mathematical sophistication, it was expedient to do so at the turn of the century, when most technical people viewed alternating current (ac) in direct current (dc) terms.

Root-mean-square (rms) means the square root of the average (or mean) of the square of a function. For a sinusoidal function such as

$$v(t) = A \cos(\omega t + 0)$$

the average of the square is

$$\frac{1}{\text{period of } v(t)} \int_{\substack{\text{integral over} \\ \text{one period of } v(t)}} v^2(t)\, dt = \frac{\omega}{2\pi} \int_0^{2\pi/\omega} A^2 \cos^2(\omega t + \theta)\, dt$$

$$= \frac{\omega}{2\pi} \int_0^{2\pi/\omega} \left[\frac{A^2}{2} + \frac{A^2}{2} \cos(2\omega t + 2\theta) \right] dt$$

$$= \frac{A^2}{2}$$

The root-mean-square (rms) of the sinusoid is the square root of the above, which is

$$V_{rms} = \frac{A}{\sqrt{2}}$$

10.7.2 RMS Phasors

In situations in which ac power calculations are used extensively, it is helpful to redefine phasors so that phasor magnitudes are all the rms values of the sinusoids they represent, instead of their amplitudes. The sinusoidal signal

$$v(t) = A \cos(\omega t + \theta)$$

is represented by

$$\mathbf{V}_{rms} = \frac{A}{\sqrt{2}} e^{j\theta}$$

instead of by

$$\mathbf{V} = A e^{j\theta}$$

Scaling all phasors by the factor $1/\sqrt{2}$ does not change the relations between them in a network; only the translation between phasor and time function is involved. An impedance relating two ordinary phasors also relates the rms phasors:

$$Z = \frac{\mathbf{V}}{\mathbf{I}} = \frac{\mathbf{V}/\sqrt{2}}{\mathbf{I}/\sqrt{2}} = \frac{\mathbf{V}_{rms}}{\mathbf{I}_{rms}}$$

In practice, both ordinary and rms phasors are used in sinusoidal response calculation. The electric power industry uses rms phasors almost exclusively, whereas in communications, control systems, and electronics, ordinary (or peak) phasors seem to have the edge. It certainly is a bother to be continually dealing with factors of $\sqrt{2}$ if power calculations are not to be made.

10.7.3 Volt-Amperes and Power Factor

By defining volt-amperes (VA) and power factor (PF) as

$$VA = V_{rms} I_{rms}$$

and

$$PF = \cos(\angle \mathbf{V} - \angle \mathbf{I})$$

average power may be expressed as

$$P = V_{rms}I_{rms} \cos (\angle V - \angle I) = (VA)(PF)$$

The volt-amperes are simply the product of an element's rms voltage and current, which may be read with ac meters, the scales of which read rms values for sinusoids. Volt-amperes is sometimes called "apparent power" because for constant (dc) voltages and currents, the power would simply be voltage times current.

The power factor (PF) is a number between zero and unity, which converts the "apparent power" to actual power.

Suppose a 220-V rms voltage is applied to an impedance

$$Z = 3 + j2$$

The rms of the current is given by

$$I_{rms} = \frac{V_{rms}}{|Z|} = \frac{220}{\sqrt{3^2 + 2^2}} = \frac{220}{\sqrt{13}}$$

Hence the volt-amperes are

$$VA = V_{rms}I_{rms} = \frac{(220)^2}{\sqrt{13}}$$

The power factor is

$$PF = \cos (\angle V - \angle I) = \cos (\angle Z)$$

$$= \cos \left(\tan^{-1} \frac{2}{3} \right) = \frac{3}{\sqrt{13}}$$

The average power flow into the impedance is

$$P = (VA)(PF) = \frac{(220)^2 \cdot 3}{13} = 11{,}169$$

Although the phase difference between voltage and current in an element may have any value in a 360° range, negative power factors are not used. Instead, one deals with power flows out of elements that supply average power and with power flows into elements that dissipate average power.

Even within a voltage-current phase difference range of $\pm 90°$, the algebraic sign of the phase difference cannot be determined from the power factor since

$$\cos (\theta - \phi) = \cos (\phi - \theta)$$

To convey the sense of the phase difference, a power factor is said to be *inductive* or *lagging* (current lags voltage) if $\angle V - \angle I$ is positive, and *capacitive* or *leading* (current leads voltage) if $\angle V - \angle I$ is negative.

Let the power into an impedance with magnitude 10 Ω and power factor 0.5 leading be 600 W. The angle of the impedance is

$$\angle Z = -\cos^{-1} 0.5 = -60°$$

the minus sign applying because current leads voltage. Thus

$$Z = 10e^{-j60°} = 5 - j8.66$$

In terms of rms values, the average power is

$$P = V_{\text{rms}} I_{\text{rms}}(\text{PF})$$
$$600 = V_{\text{rms}} I_{\text{rms}}(0.5)$$

so that

$$V_{\text{rms}} I_{\text{rms}} = (\text{VA}) = 1200$$

The ratio of rms voltage to current is the magnitude of the impedance,

$$\frac{V_{\text{rms}}}{I_{\text{rms}}} = 10$$

and substituting into the power relation gives

$$V_{\text{rms}} = 109.5$$
$$I_{\text{rms}} = 10.95$$

The reactive power is

$$Q = (\text{VA}) \sin (\angle Z) = 1200 \sin (-60°)$$
$$= -1039.2$$

D10-6

The following are representative of a class of problems found on many engineer-in-training (EIT) examinations. Here, rms values, sink reference relations for impedances (loads), and source references for sources (generators) are implied unless otherwise stated. "Impedance" often means the magnitude of the impedance.

(a) A certain load has impedance 10 Ω and power factor 0.9, lagging. If the load voltage is 110, what are the current, the volt-amperes, the power, and the reactive power?

ans. 11, 1210, 1089, 528.5

(b) A generator supplies 3000 VA to a load at 220 V with a capacitive power factor of 0.85. Find the current, the real power, and the reactive power.

ans. 13.64, 2550, −1576

(c)   A load of $10 + j4$ is connected through wires of total resistance $2\,\Omega$ to a 440-V generator. What are the power factor, the volt-amperes, and the power of the load and of the generator? What power is lost in the wires?

ans. 0.928, 13032, 12100, 0.949, 15305, 14520, 2420

(d)   A 10.6-Ω impedance with capacitive power factor 0.7 is connected in parallel with 4.8 Ω and 0.9 power factor, lagging. What is the impedance and what is the power factor of the combination?

ans. 3.93, 0.9959 inductive

Power Calculations with RMS Values

The rms value of a sinusoidal signal is $(1/\sqrt{2}) \cong 0.707$ times its amplitude. In terms of rms values,

$$P_{into} = V_{rms} I_{rms} \cos\left(\angle \mathbf{V} - \angle \mathbf{I}\right)$$

$$Q_{into} = V_{rms} I_{rms} \sin\left(\angle \mathbf{V} - \angle \mathbf{I}\right)$$

$$|Z| = \frac{|\mathbf{V}|}{|\mathbf{I}|} = \frac{V_{rms}}{I_{rms}}$$

When dealing extensively with rms values, as in the electric power industry, it is convenient to modify the definitions of phasors so that phasor magnitudes are rms values of the sinusoidal signals rather than their amplitudes (peak values).

Average power flow is commonly expressed in the form

$$P_{into} = (\text{volt-amperes})(\text{power factor}) = (\text{VA})(\text{PF})$$

The volt-amperage of an element is the product of its rms voltage and current:

$$\text{VA} = V_{rms} I_{rms}$$

The power factor of an element is

$$\text{PF} = \cos\left(\angle \mathbf{V} - \angle \mathbf{I}\right) = \cos\left(\angle \mathbf{I} - \angle \mathbf{V}\right)$$

A power factor is said to be *inductive* or *lagging* if $\angle \mathbf{V} - \angle \mathbf{I}$ is between $0°$ and $90°$. It is said to be *capacitive* or *leading* if $\angle \mathbf{V} - \angle \mathbf{I}$ is between $0°$ and $-90°$.

10.8 Single-Phase Power Transmission

10.8.1 A Power System Model

A simple ac power transmission system consists of a generator, intercon-
necting wires or "lines," and the device or devices using the power, called
the *load*. As illustrated in the drawing of Figure 10-8, the generator may be
modeled by a Thévenin equivalent, the lines by an impedance, often resistive,
and the load by another impedance.

Most power systems in North America operate at a frequency of 60 Hz;
50 Hz is commonly used in other parts of the world.

The *efficient* transmission of power is very different than the transfer of
maximum power discussed in Section 10.4. Maximum power transfer is
desirable in applications such as communications processing, where the
signals involve a small amount of power and are competing with undesirable
noise signals. Efficient transfer is important in power systems where large,
expensive amounts of energy are involved. Energy losses in the generator
and the transmission lines are wasted energy not delivered to the load.

10.8.2 Power Factor Correction

For a given amount of average power flow into the load,

$$P_{\text{into load}} = \tfrac{1}{2}|\mathbf{V}_L||\mathbf{I}|\cos(\angle Z_L)$$

and a fixed-voltage amplitude, $|\mathbf{V}_L|$, the load-current amplitude is smallest
when the power factor of the load is unity so that the load voltage and current
are in phase.

This load current flows through the lines and generator. The average
power losses in the lines,

$$P_{\text{into lines}} = \tfrac{1}{2}|\mathbf{I}|^2 R_T$$

Figure 10-8 Single-phase power transmission model

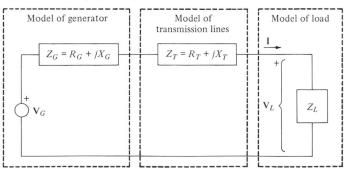

and in the generator,

$$P_{\text{into generator}} = \tfrac{1}{2}|\mathbf{I}|^2 R_G$$

are proportional to its square. Thus in the interest of reducing power loss in the generator and the lines, a purely resistive, unity power factor, load impedance is desirable.

In practice, many loads involving large amounts of power as in an industrial installation do not have a unity power factor. The usual situation is that of an inductive load, typical of induction motors. In these cases it is advantageous to modify the load impedance by adding a parallel reactance so that the composite load has unity power factor.

The problem of finding the desired parallel reactance is that of obtaining parallel resonance of the load at the power system frequency, as discussed in Section 9.8.4.

Consider the original inductive load impedance Z_L in Figure 10-9. A parallel capacitive reactance is to be added so that the parallel combination is purely resistive. A direct solution for the desired reactance X is as follows:

$$Z_0 = \frac{(-jX)(10 + j3)}{10 + j3 - jX} = \frac{3X - j10X}{10 + j(3 - X)}$$

$$= \frac{(3X - j10X)[10 - j(3 - X)]}{100 + (3 - X)^2}$$

$$= \frac{10X^2 + j(3X^2 - 109X)}{100 + (3 - X)^2}$$

Setting

$$3X^2 - 109X = 0$$

the nonzero capacitive reactance of

$$X = \frac{109}{3}$$

will give unity power factor for the overall load.

Figure 10-9   Load power factor correction

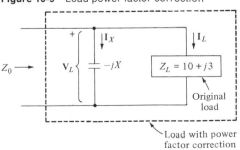

Load with power
factor correction

At a power line frequency of 60 Hz, the value of capacitance needed is given by

$$\frac{1}{\omega C} = X$$

$$C = \frac{1}{2\pi f X} = \frac{3}{(6.28)(60)(109)} = 7.3 \times 10^{-5}$$

$$= 73 \ \mu F$$

The parallel reactance for correction of the load power factor to unity may also be easily calculated by considering reactive powers. The reactive power flow into Z_L is

$$Q_{\text{into } Z_L} = \tfrac{1}{2}|\mathbf{I}_L|^2 \cdot 3$$

$$= \frac{1}{2}\left(\frac{|\mathbf{V}_L|}{|10 + j3|}\right)^2 \cdot 3$$

$$= \frac{3|\mathbf{V}_L|^2}{2(109)}$$

The reactive power flow into the parallel capacitive reactance is

$$Q_{\text{into } X} = -\tfrac{1}{2}|\mathbf{I}_X|^2 \cdot X$$

$$= -\frac{1}{2}\left(\frac{|\mathbf{V}_L|}{X}\right)^2 \cdot X$$

$$= -\frac{|\mathbf{V}_L|^2}{2X}$$

When the overall load impedance Z_0 is purely resistive, the net reactive power is zero, giving

$$\frac{3|\mathbf{V}_L|^2}{2(109)} - \frac{|\mathbf{V}_L|^2}{2X} = 0$$

$$X = \frac{109}{3}$$

Power factor correction is thus seen to be the connection of additional reactance to the load to bring its net reactive power to zero. A nonzero load reactive power means that energy is being stored in the load then returned to the source, back and forth, twice each cycle, giving rise to additional power losses. The added reactance confines this interchange of reactive power to the load itself.

D10-7

The following are representative of a class of problems found on many EIT examinations:

(a) In a 440-V ac circuit, an inductive load draws 60 A at a power factor of 0.8. Find the volt-amperes, the power, the reactive volt-amperes, and the load impedance.

ans. 26400, 21120, 15840, 5.87 + j4.4

(b) A circuit consisting of a 3-Ω resistance in series with a 7-Ω reactance is connected to a 110-V, 60-Hz source. Find the current, the real volt-amperes, the reactive volt-amperes, and the total volt-amperes of the load.

ans. 20.08, 8701, 20302, 22088

(c) A 10-kW, 60-Hz, 120-V load has a power factor of 0.9, leading. Find the current, the volt-amperes, and the reactive volt-amperes.

ans. 83.33, 11111, −4836

(d) Two loads are connected in parallel across a 120-V, 50-Hz ac line. The first load is 6000 VA at a power factor of 0.9, leading. The second load is 9000 VA with a 0.8 power factor, lagging. Find the total volt amperes, the overall power factor, the line current, and the net real and reactive powers.

ans. 12904, 0.976, 107.5, 12600, 2785

(e) It is desired to correct the power factor of a 240-V, 5-kW, 6-kVA inductive load to unity. What reactance should be connected in parallel?

ans. −17.37

(f) A 12-kW, 60-Hz, 220-V load has a power factor of 0.75, lagging. It is desired to correct the power factor to 0.9, lagging. What value of capacitor should be connected in parallel with the load?

ans. 2.62×10^{-4}

Single-Phase Power Transmission

The major elements of a single-phase power transmission system consist of the generator, modelled by a Thévenin equivalent, the lines, modeled by an impedance (usually a pure resistance), and the load, modeled by another impedance.

Correcting the power factor of a load impedance reduces power losses in the generator and lines.

The most common power factor correction problem involves finding a capacitance to place in parallel with an inductive load such that their combination is entirely real.

10.9 Power Transformers

10.9.1 A Transformer Model

The term *power transformer* generally means a set of coupled inductors with a relatively large degree of magnetic flux coupling between the coils, intended for efficient transmission of electrical power.

One coil is termed the *primary* and the other is called the *secondary*, although which is called which is somewhat arbitrary. Usually the label "primary" is attached to the coil that is to be connected closest to the source of power, so that average electrical power flows from the primary to the secondary.

For power applications, a more tractable transformer model than the coupled inductors, Figure 10-10(a), or their controlled source equivalent, Figure 10-10(b), may be derived.

Figure 10-10 Coupled inductor models

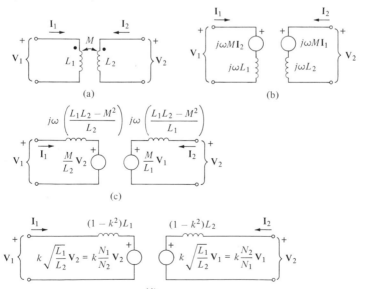

(a)

(b)

(c)

(d)

For coupled inductors, Figure 10-10(a) or (b), the equations for V_1 and V_2, in terms of I_1 and I_2, are

$$\begin{cases} V_1 = j\omega L_1 I_1 + j\omega M I_2 \\ V_2 = j\omega M I_1 + j\omega L_2 I_2 \end{cases}$$

Eliminating I_2 between these two equations gives

$$V_1 = \frac{M}{L_2} V_2 + j\omega \left(\frac{L_1 L_2 - M^2}{L_2} \right) I_1$$

Similarly, eliminating I_1 between the same equations,

$$V_2 = \frac{M}{L_1} V_1 + j\omega \left(\frac{L_1 L_2 - M^2}{L_1} \right) I_2$$

These two new equations are equivalent to the original two, and give the alternative representation of Figure 10-10(c), where it is seen that each winding involves some inductance in series with a voltage proportional to the voltage applied to the opposite winding.

In terms of the coupling coefficient of the transformer,

$$k = \frac{M}{\sqrt{L_1 L_2}}$$

these equations are

$$\begin{cases} V_1 = k \sqrt{\frac{L_1}{L_2}} V_2 + j\omega(1 - k^2) L_1 I_1 \\ V_2 = k \sqrt{\frac{L_2}{L_1}} V_1 + j\omega(1 - k^2) L_2 I_2 \end{cases}$$

The quantity

$$\sqrt{\frac{L_2}{L_1}} = \frac{N_2}{N_1}$$

is called the *turns ratio* of the transformer; N_2/N_1 is approximately the ratio of the number of turns of the L_2 coil to the number of turns of the L_1 coil.

The model is relabeled in terms of k and turns ratios in Figure 10-10(d). The equivalent inductances, which become zero for perfect coupling of $k = 1$, are termed *leakage inductances*, referring to imperfect coupling caused by leakage of magnetic flux.

10.9.2   Ideal Transformers

An idealized transformer model, for which the magnetic coupling coefficient k is maximum,

$$k = 1, \qquad M = \sqrt{L_1 L_2}$$

is a useful approximation for many practical transformers. It is called the *ideal transformer* and is symbolized by two inductor symbols and a turns ratio, as in the example of Figure 10-11. If the senses of the windings are of importance to the application, they should also be indicated.

For the ideal transformer, $k = 1$, the leakage inductances are zero and the winding voltages are related by the turns ratio:

$$\frac{V_2}{V_1} = \sqrt{\frac{L_2}{L_1}} = \frac{N_2}{N_1}$$

In general, the ratio of currents in the two windings is, from the equations of the previous section,

$$\frac{I_2}{I_1} = \frac{L_1}{L_2} \cdot \frac{V_2 - k\sqrt{\dfrac{L_2}{L_1}}\,V_1}{V_1 - k\sqrt{\dfrac{L_1}{L_2}}\,V_2}$$

$$= -\sqrt{\frac{L_1}{L_2}} \left[\frac{V_2 - k\sqrt{\dfrac{L_2}{L_1}}\,V_1}{kV_2 - \sqrt{\dfrac{L_2}{L_1}}\,V_1} \right]$$

For k near unity, the factor in brackets approaches unity, giving

$$\frac{I_2}{I_1} = -\sqrt{\frac{L_1}{L_2}} = -\frac{N_1}{N_2}$$

for perfect coupling. The currents I_1 and I_2 are related by the negative inverse of the turns ratio. The minus sign indicates that I_1 and I_2 are in opposite senses.

In an ideal transformer when the voltage is stepped up or stepped down by the turns ratio,

$$V_2 = \frac{N_2}{N_1} V_1$$

Figure 10-11   Voltage relations for an ideal transformer

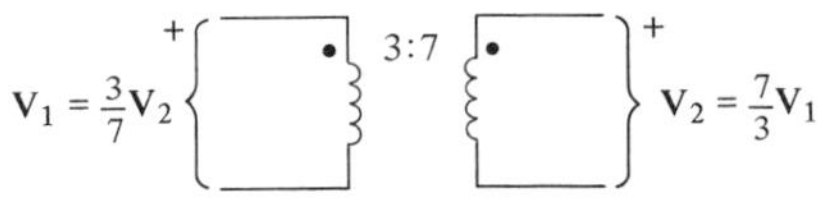

$V_1 = \frac{3}{7}V_2$ 3:7 $V_2 = \frac{7}{3}V_1$

the current is stepped in inverse proportion,

$$\mathbf{I}_2 = -\frac{N_1}{N_2}\mathbf{I}_1$$

so that the voltage-current product is unchanged. All of the electrical power flowing into one winding flows out of the other winding.

As an example of these relations, consider the network model of Figure 10-12, which contains an ideal transformer with turns ratio 5:2. The secondary voltage is two-fifths that of the primary, so the secondary current (with reference direction reversed from that used previously) is

$$\mathbf{I}_2 = \frac{\mathbf{V}_2}{4 + j2} = \frac{\tfrac{2}{5}\mathbf{V}_1}{4 + j2} = \frac{40e^{j50°}}{4.47e^{j26.6°}}$$

$$= 8.95e^{j23.4°}$$

The primary current is two-fifths that of the secondary:

$$\mathbf{I}_1 - \tfrac{2}{5}\mathbf{I}_2 - 3.58e^{j23.4°}$$

Transformers are commonly used to increase or decrease generated voltages in power systems. Power is generated by rotating machines at voltages (typically 5 to 50 kV) that are convenient for the generator construction. Transformers are used to step up the generator voltage, while stepping down the current, for long-distance transmission.

Since the power dissipated in a transmission line is the line resistance times the square of the rms line current, the smaller the current, the smaller the losses. Long-distance power transmission is done at the highest practical voltage, typically 200 to 400 kV.

As the power transmission lines branch to localized distribution areas, transformers successively step down the line voltages to values suitable for the area, eventually reaching the low voltages common in residential distribution.

Power lines servicing residential areas typically involve line voltages of 2 to 10 kV which are stepped down to a center tapped 220 V, allowing the home user to connect to either 110 V for low-power appliances or to 220 V for greater efficiency in supplying devices having larger power requirements.

Figure 10-12 Network involving an ideal transformer

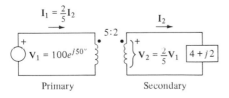

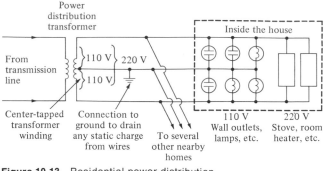

Figure 10-13 Residential power distribution

The drawing of Figure 10-13 illustrates home power distribution and utilization.

10.9.3 Impedance Reflection

When the secondary of an ideal transformer is connected to an impedance, as in the diagram of Figure 10-14, the ratio of secondary phasor voltage to current is that impedance. The primary voltage and current are related to the secondary quantities by the turns ratio as shown, so the impedance at the primary winding terminals is

$$Z_1 = \frac{\mathbf{V}_1}{\mathbf{I}_1} = \frac{\dfrac{N_1}{N_2}\mathbf{V}_2}{\dfrac{N_2}{N_1}\mathbf{I}_2} = \left(\frac{N_1}{N_2}\right)^2 \frac{\mathbf{V}_2}{\mathbf{I}_2}$$

$$= \left(\frac{N_1}{N_2}\right)^2 Z_2$$

Viewed through the transformer, the impedance connected to the secondary is magnified by the turns ratio squared. This result is independent of the transformer winding senses.

Impedance reflection is a useful technique in equivalent circuit solution of networks involving transformers. For example, the impedance as viewed

Figure 10-14 Impedance reflection by the ideal transformer

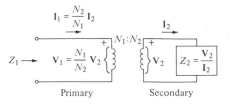

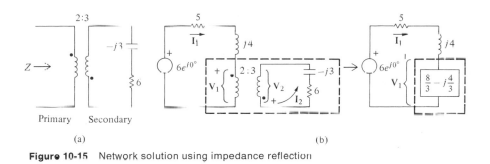

Figure 10-15   Network solution using impedance reflection

from the primary terminals of the transformer in Figure 10-15(a) is

$$Z = (\tfrac{2}{3})^2(6 - j3) = \tfrac{8}{3} - j\tfrac{4}{3}$$

If the primary of this transformer is connected as shown in Figure 10-15(b), so far as the rest of the network is concerned, the primary terminals present the impedance Z as in the equivalent network. So

$$I_1 = \frac{6e^{j0^\circ}}{\dfrac{23}{3} + j\dfrac{8}{3}} = \frac{6e^{j0^\circ}}{8.1e^{j19.2^\circ}} = 0.74e^{-j19.2^\circ}$$

$$V_1 = 6e^{j0^\circ}\,\frac{\tfrac{8}{3} - j\tfrac{4}{3}}{\tfrac{23}{3} + j\tfrac{8}{3}} = \frac{6e^{j0^\circ}(2.98e^{-j26.6^\circ})}{8.1e^{j19.2^\circ}}$$

$$= 2.2e^{-j45.8^\circ}$$

Using the turns ratio,

$$V_2 = \tfrac{3}{2}V_1 = \tfrac{3}{2}(2.2e^{-j45.8^\circ})$$

$$= 3.3e^{-j45.8^\circ}$$

and

$$I_2 = \tfrac{2}{3}I_1 = \tfrac{2}{3}(0.74e^{-j19.2^\circ})$$

$$= 0.49e^{-j19.2^\circ}$$

D10-8

Find the indicated impedances. The transformers are ideal:

(a)

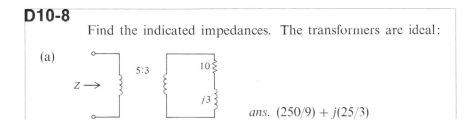

ans. $(250/9) + j(25/3)$

(b)

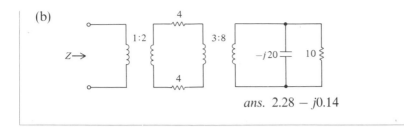

$$ans.\ 2.28 - j0.14$$

D10-9

For the following networks involving ideal transformers, find the indicated phasors:

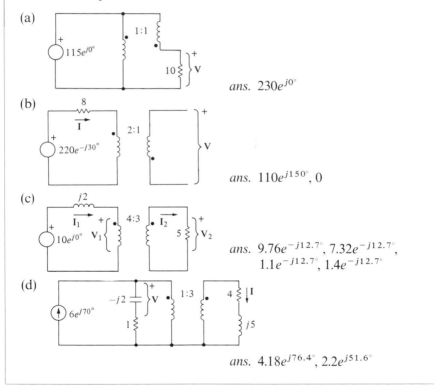

(a)

$$ans.\ 230e^{j0°}$$

(b)

$$ans.\ 110e^{j150°},\ 0$$

(c)

$$ans.\ 9.76e^{-j12.7°},\ 7.32e^{-j12.7°},$$
$$1.1e^{-j12.7°},\ 1.4e^{-j12.7°}$$

(d)

$$ans.\ 4.18e^{j76.4°},\ 2.2e^{j51.6°}$$

Transformers

Transformers are coupled inductors with large coupling coefficients. The controlled source and leakage inductance model, in terms of the coupling coefficient and turns ratio, is useful for such devices.

An ideal transformer is a model of coupled inductors for which the magnetic coupling coefficient is maximum, $k = 1$.

For an ideal transformer, the primary voltage is the turns ratio times the secondary voltage. The primary current is the inverse of the turns ratio times the secondary current.

The impedance as viewed from the primary of an ideal transformer is the secondary impedance times the square of the turns ratio.

10.10   Three-Phase Power Transmission

10.10.1   Balanced Three-Phase Power Flow

Consider the three networks of Figure 10-16 which are identical except for the phases of the source voltages. The source voltage phase angles are located symmetrically at $120°$ intervals, as indicated on the typical phasor diagram of Figure 10-16(b). In consequence of the symmetry, I_1, I_2, and I_3 differ only in that their phase angles are spaced $120°$ apart.

The instantaneous electrical power flow out of the first source is

$$p_1(t) = \frac{AB}{2} \cos(\theta - \phi) + \frac{AB}{2} \cos(2\omega t + \theta + \phi)$$

For the second and third sources, the power flows are

$$p_2(t) = \frac{AB}{2} \cos(\theta - \phi) + \frac{AB}{2} \cos(2\omega t + \theta + \phi + 240°)$$

$$p_3(t) = \frac{AB}{2} \cos(\theta - \phi) + \frac{AB}{2} \cos(2\omega t + \theta + \phi + 480°)$$

$$= \frac{AB}{2} \cos(\theta - \phi) + \frac{AB}{2} \cos(2\omega t + \theta + \phi + 120°)$$

Figure 10-16   Three symmetrical circuits

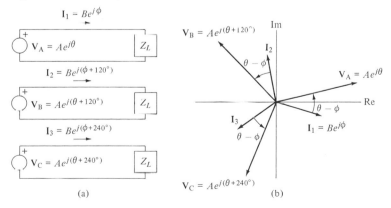

(a)   (b)

The total instantaneous power flow from the three sources is

$$p(t) = p_1(t) + p_2(t) + p_3(t)$$

$$= \frac{3AB}{2} \cos(\theta - \phi) + \frac{AB}{2} \{\cos(2\omega t + \theta + \phi)$$

$$+ \cos(2\omega t + \theta + \phi + 240°) + \cos(2\omega t + \theta + \phi + 120°)\}$$

$$= \frac{3AB}{2} \cos(\theta - \phi)$$

which is *constant*, the terms in brackets summing to zero.

Individually, the power flows vary with time, but the sum of the three powers is a constant. This *three-phase* arrangement is particularly advantageous for the generation and utilization of electrical power.

For single-phase power generation, the flow of mechanical power into a generator from, say, a turbine, varies with time as does $p_1(t)$. During part of the shaft rotation, a large amount of mechanical power must be converted to electrical power. During another part of the shaft rotation, there is less mechanical-electrical power conversion. In fact, electrical power is returned to the mechanical sources in any interval of time when $p_1(t)$ is negative. Not only is this time-varying power flow hard on the generator and turbine bearings, but it is inefficient since there are power losses in each conversion back and forth between electrical and mechanical energy.

If all three generators are connected to the same shaft (in practice, three symmetrical generator coils are placed in the same housing), the net power flow is constant. This result is true also of the load. The net power flow into a balanced three-phase load, which might consist of three motor coils on the same shaft, is constant.

10.10.2   Transmission Efficiency

A three-phase system, also offers advantages in the transmission of electrical power. To reduce the number of wires involved in the distribution of power, three main conductors and a common, *neutral*, conductor may be used, as indicated in Figure 10-17(a).

The current in the neutral wire is the sum of the three load currents that, from a phasor diagram such as that of Figure 10-16(b), are seen to add to zero. Only three wires are really needed then, as indicated in Figure 10-17(b). It is common to draw diagrams for three-phase networks in the manner of Figure 10-17(b), which is suggestive of phasor positions.

A three-phase system can transmit three times the power over three wires that a single-phase system can deliver over two wires, using the same generator voltage and line current. In practice, a neutral wire of relatively small

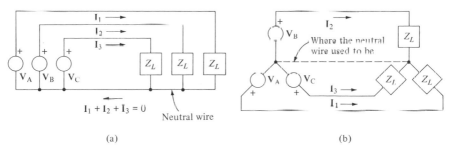

Figure 10-17   Balanced three phase wye system

diameter is often used to allow for small imbalances in the system. In any event, the neutral connection is wired to conductors buried in the earth (*grounded*), to prevent accumulation of static charge on the conductors.

Models with perfectly symmetrical three-phase sources and loads are referred to as *balanced* three-phase systems. If relatively small yet significant asymmetries are present, the system is called an *unbalanced* three-phase system. Unbalanced systems will not be considered further here, except to note that any network may be solved by conventional means; the symmetry of a balanced system allows simplified solutions.

10.10.3   Balanced Wye Systems

A simple model of a three-phase power system consists of wye connections of sources and loads such as the example of Figure 10-18(a), in which each generator is modeled by a Thévenin equivalent, each wire by a resistance, and each load by an impedance. Provided the three-phase system is balanced, the neutral wire may be added, and the three equivalent single-phase networks solved, as in Figure 10-18(b). Of course it is really only necessary to calculate a solution of one of the single-phase networks; the other network solutions differ only in phase, by 120° and 240°.

The solution for I_1 is

$$I_1 = \frac{500e^{j30°}}{5 + j5} = \frac{500e^{j30°}}{5\sqrt{2}e^{j45°}} = \frac{100}{\sqrt{2}}e^{-j15°}$$

Then

$$I_2 = \frac{100}{\sqrt{2}}e^{j105°}$$

$$I_3 = \frac{100}{\sqrt{2}}e^{j225°}$$

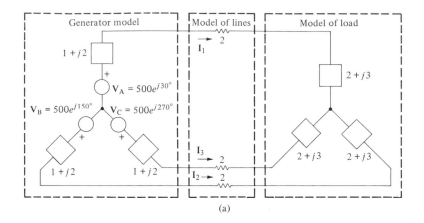

(a)

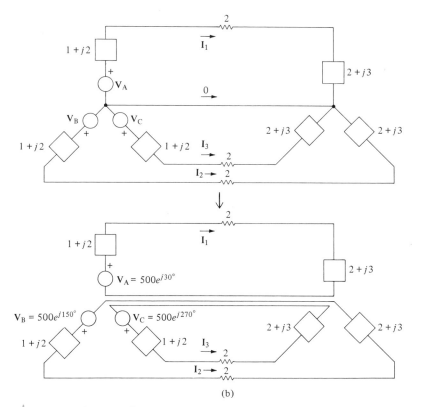

(b)

Figure 10-18   Solution of a wye system

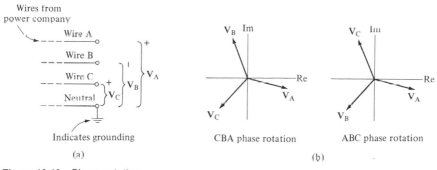

Figure 10-19   Phase rotation

10.10.4   Phase Rotation

Given the incoming wires of a three-phase power distribution system, it is often of the utmost important to know the *phase sequence*. Labeling the wires A, B, and C, the line-to-neutral voltages, Figure 10-19, can crest in the order A–B–C or in the order C–B–A.

The phase sequence makes quite a difference in many applications. For instance, interchanging connections will result in reversal of the direction of rotation of a three-phase induction motor.

D10-10

Draw a phasor diagram showing every voltage and current in the following system:

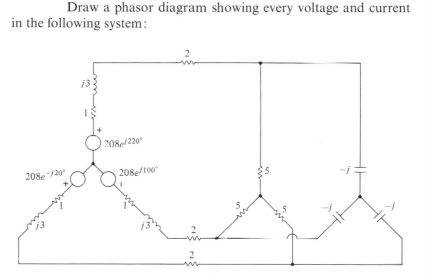

Three-Phase Power Transmission

Three-phase power transmission has the advantage of conveying more power per wire, using three wires, than a single-phase system using two wires of the same capacity. The total power transferred in a balanced three-phase system is constant with time.

   For analysis of a balanced wye-connected source and load, the neutral connection may be inserted if it is not already present, and the problem reduces to that of three symmetric single-phase systems. The neutral connection cannot be made if it is not already present in an *unbalanced* system.

   Phase rotation is the order in which the three line voltages in a three-phase system reach maximum, A–B–C or C–B–A.

10.11   Balanced Delta-Wye Transformation

10.11.1   Line Voltages

The line-to-line voltages corresponding to a three-phase wye connection of sources, Figure 10-20(a), may be easily found using a phasor diagram. The wye voltages are termed the *phase* voltages, to distinguish them.

   The voltage

$$\mathbf{V}_1 = \mathbf{V}_A - \mathbf{V}_C$$

is seen from the phasor diagram construction of Figure 10-20(b) to lead $\mathbf{V}_A$ in phase by $30°$ and to have magnitude $\sqrt{3}$ times the magnitude of $\mathbf{V}_A$.

Figure 10-20   Line-to-line voltage in terms of wye voltages

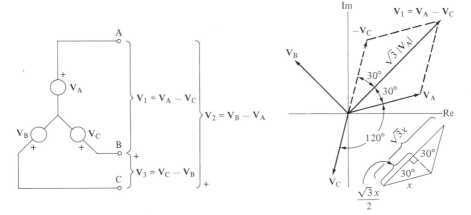

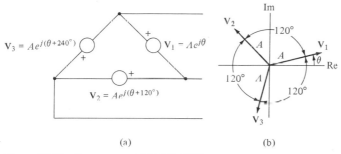

Figure 10-21 Sources connected in delta

The other two line voltages, $\mathbf{V}_2$ and $\mathbf{V}_3$, are of the same magnitude as $\mathbf{V}_1$ and are spaced at $120°$ intervals with $\mathbf{V}_1$, as shown on the phasor diagram of Figure 10-20(c).

10.11.2 Balanced Delta Systems

Another way to model a three-phase power system is with a delta connection of sources such as that shown in Figure 10-21(a). The voltage phasors, Figure 10-21(b), being equal in magnitude and at $120°$ intervals in phase, add to zero around the loop of sources, as they must. Any one of the three sources is redundant and could be removed so far as network solution is concerned, although this is seldom done in practice.

In this arrangement the delta-connected sources provide the line-to-line voltages. When the sources are wye connected, they provide the phase voltages.

If sources in delta are connected to a delta connection of impedances, the voltage across each impedance is a known source voltage, and the network solution is particularly easy, as in the example of Figure 10-22,

Figure 10-22 Solution for the load currents in a delta system

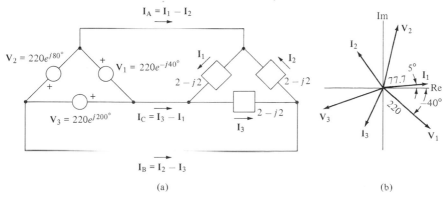

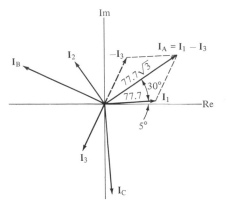

Figure 10-23 Phasor diagram of solution for line currents

for which

$$\mathbf{I}_1 = \frac{\mathbf{V}_1}{2 - j2} = \frac{220e^{-j40°}}{2.83e^{-j45°}}$$

$$= 77.7e^{j5°}$$

$$\mathbf{I}_2 = 77.7e^{j125°}$$

$$\mathbf{I}_3 = 77.7e^{j245°}$$

The line currents, $\mathbf{I}_A$, $\mathbf{I}_B$, and $\mathbf{I}_C$ may be found from the load currents by phasor addition similar to that used for finding line voltages from phase voltages. From the phasor diagram of Figure 10-23,

$$\mathbf{I}_A = 77.7\sqrt{3}\,e^{j35°}$$

$$\mathbf{I}_B = 77.7\sqrt{3}\,e^{j155°}$$

$$\mathbf{I}_C = 77.7\sqrt{3}\,e^{j275°}$$

10.11.3 Delta-Wye Source Transformation

Transformation between wye and equivalent delta sources may be used to simplify the solution of balanced three-phase systems. In the network of Figure 10-24, for example, the sources are wye connected and the load is delta connected. Converting the wye sources to equivalent delta ones gives a balanced delta system for which the solution is simple:

$$\mathbf{I}_1 = \frac{\mathbf{V}_1}{10 + j20} = \frac{100\sqrt{3}\,e^{j30°}}{22.36e^{j63.5°}}$$

$$= 7.75e^{-j33.5°}$$

Transformation from delta to wye sources is also useful. In the system of Figure 10-25, with delta-connected sources and wye-connected load, a delta-to-wye source transformation greatly simplifies the analysis.

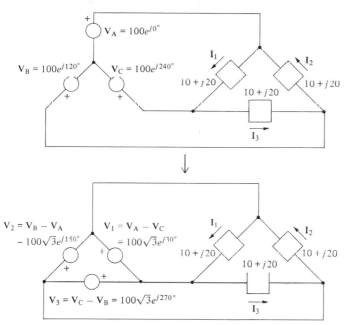

Figure 10-24 Wye to delta source transformation

Figure 10-25 Delta-to-wye source transformation

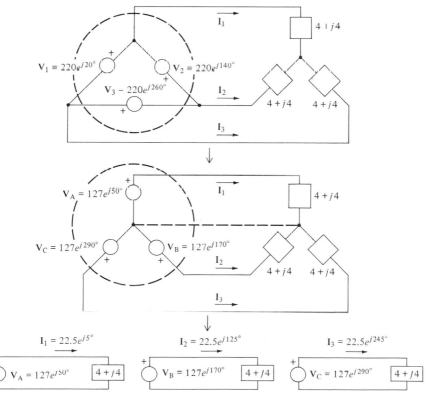

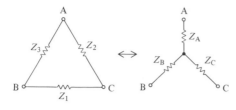

$$Z_1 = \frac{Z_A Z_B + Z_A Z_C + Z_B Z_C}{Z_A} \qquad Z_A = \frac{Z_2 Z_3}{Z_1 + Z_2 + Z_3}$$

$$Z_2 = \frac{Z_A Z_B + Z_A Z_C + Z_B Z_C}{Z_B} \qquad Z_B = \frac{Z_1 Z_3}{Z_1 + Z_2 + Z_3}$$

$$Z_3 = \frac{Z_A Z_B + Z_A Z_C + Z_B Z_C}{Z_C} \qquad Z_C = \frac{Z_1 Z_2}{Z_1 + Z_2 + Z_3}$$

Figure 10-26   General delta-wye impedance transformations

Figure 10-27   Delta-wye and wye-delta load transformations

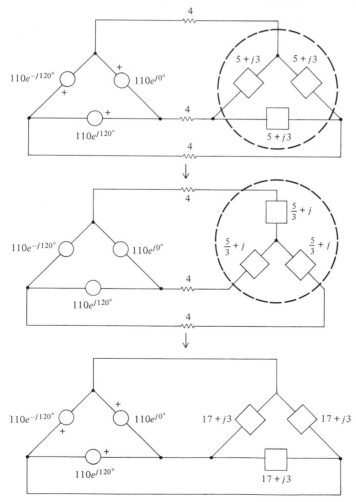

10.11.4 Delta-Wye Load Transformation

Transformation between equivalent delta and wye loads may also be used to simplify the solution of three-phase power networks. As developed for resistors in Section 4.9, conversion between equivalent delta and wye sets of impedances is summarized in Figure 10-26.

For *balanced* systems, where

$$Z_1 = Z_2 = Z_3, \qquad Z_A = Z_B = Z_C$$

the relation between the wye impedances Z_A and the delta impedances Z_1 reduces to

$$Z_A = \tfrac{1}{3}Z_1$$

Figure 10-27 shows an example of the use of delta-wye transformation to simplify the solution of a balanced three-phase system. First, the delta load is converted to an equivalent wye load, and the series impedances are combined. Then the combined wye impedances are converted to an equivalent delta load, giving a balanced delta system for solution. Alternatively, the load could have been left as a wye and the delta sources converted to an equivalent wye.

D10-11

Perform the indicated delta-wye or wye-delta source transformations, and draw a phasor diagram that shows the six phasors involved in each case.

(a)

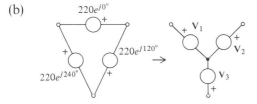

$100e^{j0°}$

$100e^{j240°}$ $100e^{j120°}$ $\rightarrow$ V_1 V_2 V_3

ans. $173e^{-j30°}, 173e^{j90°}, 173e^{j210°}$

(b)

$220e^{j0°}$

$220e^{j240°}$ $220e^{j120°}$ $\rightarrow$ V_1 V_2 V_3

ans. $127e^{j210°}, 127e^{j330°}, 127e^{j90°}$

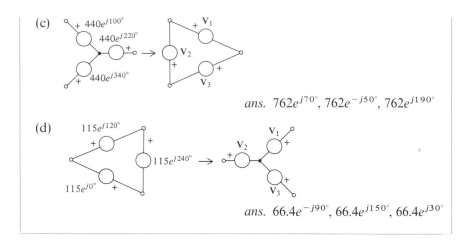

(c)

$ans.$ $762e^{j70°}$, $762e^{-j50°}$, $762e^{j190°}$

(d)

$ans.$ $66.4e^{-j90°}$, $66.4e^{j150°}$, $66.4e^{j30°}$

D10-12

Use delta-wye or wye-delta load transformations to find the indicated phasors:

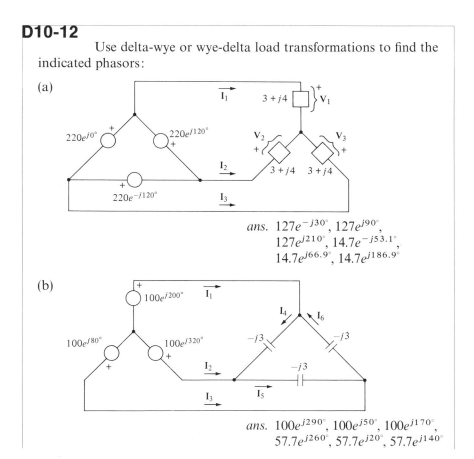

(a)

$ans.$ $127e^{-j30°}$, $127e^{j90°}$,
$127e^{j210°}$, $14.7e^{-j53.1°}$,
$14.7e^{j66.9°}$, $14.7e^{j186.9°}$

(b)

$ans.$ $100e^{j290°}$, $100e^{j50°}$, $100e^{j170°}$,
$57.7e^{j260°}$, $57.7e^{j20°}$, $57.7e^{j140°}$

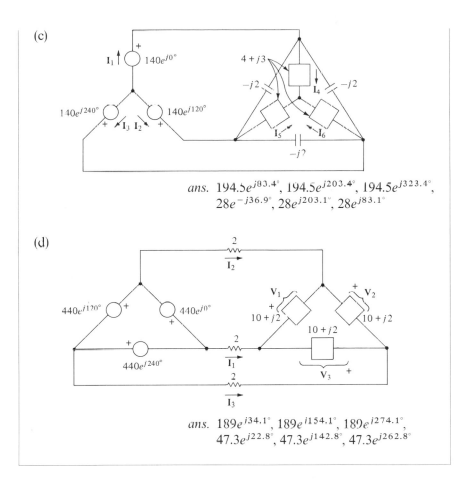

(c)

ans. $194.5e^{j83.4°}$, $194.5e^{j203.4°}$, $194.5e^{j323.4°}$,
$28e^{-j36.9°}$, $28e^{j203.1°}$, $28e^{j83.1°}$

(d)

ans. $189e^{j34.1°}$, $189e^{j154.1°}$, $189e^{j274.1°}$,
$47.3e^{j22.8°}$, $47.3e^{j142.8°}$, $47.3e^{j262.8°}$

D10-13

The following are typical of a class of EIT examination problems:

(a) A balanced, three-phase wye-connected 2400-W load has line-to-line voltages of 220 and a lagging power factor of 0.85. Find the line currents and the total volt-amperes of the load.

ans. 7.41, 2824

(b) A balanced, three-phase delta-connected load has phase voltages of 208 V and line currents of 15 A. Total power delivered to the load is 2200 W. Find the load impedance per phase and the power factor.

ans. 41.6, 0.705

(c)   A balanced, three-phase system has phase voltages of 95 V at the generator and 90 V at the load. Each of the three elements of a delta-connected load require 600 VA at a power factor of 0.7. What are the loses (per wire) in the lines?

ans. 33.3

(d)   A three-phase generator supplies 3800 VA at a power factor of 0.95, lagging. Line loses are 100 W per wire, and 3000 W are delivered to the load. What are the loses in the generator?

ans. 310

Balanced Delta-Wye Transformation

For three-phase systems involving a mixture of delta- and wye-connected sources and loads, it is often useful to convert a set of wye sources to an equivalent set of delta sources. For balanced systems, the delta source voltages have amplitude $\sqrt{3}$ as large as the wye voltages and lag them by 30°.

Delta-wye transformation may be used to convert a wye load to an equivalent delta load, or vice versa. For balanced systems, the wye impedances are one-third the delta impedances.

Chapter Ten Problems

Basic Problems

Instantaneous, Average, and Reactive Power

1.   For the following networks, find the instantaneous electrical power flow into each element as a function of time.

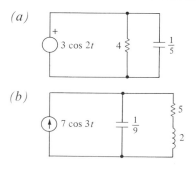

(a)

(b)

2. For the following networks, find the average electrical power flow into each element and verify that the net average power flow out of sources equals the net average power flow into the other elements:

(a)

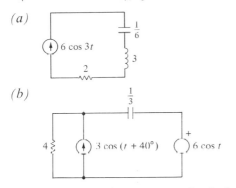

(b)

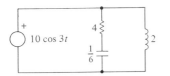

3. For the following network, find the reactive power flow into each element and verify that the net reactive power flow out of sources equals the net reactive power flow into the other elements:

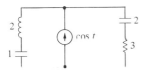

4. For the following network, find the complex power flow into each element and verify that the net complex power flow out of sources equals the net complex power flow into the other elements:

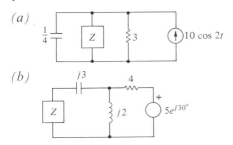

Maximum Power Transfer

5. For the networks below, find the value of Z for which the average power flow into Z is maximized:

(a)

(b)

Single-Phase Power

6.   Draw a phasor diagram showing the indicated voltages and currents in the following system for $C = 0$. Then show the same quantities on a second phasor diagram for C chosen to give unity power factor at the load.

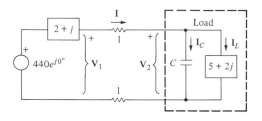

Transformers

7.   Find the indicated phasors. The transformers are ideal, and there are two phasors to find in each problem:

(a)

(b)

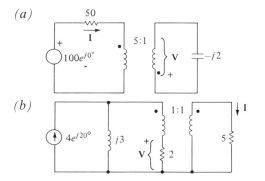

Three-Phase Power

8.   Draw phasor diagrams showing each indicated voltage and current in the following balanced three-phase system:

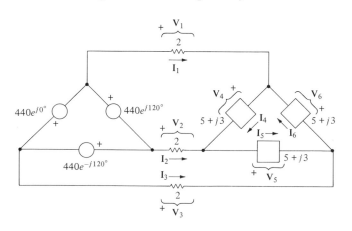

Practical Problems

The following problems are representative of a class found on many EIT examinations:

1. A certain load consists of $10 + j4$ in parallel with $-j3$. What is the overall power factor?

2. A generator supplies two parallel 115-V loads at a 0.86 power factor, lagging. One load is 4800 VA at a power factor of 0.9, lagging. Is the power factor of the other load leading or lagging? If the other load consumes 4200 W, what are its volt-amperes and its power factor?

3. An 1100-VA, 60-Hz, 220-V load has a power factor of 0.85, lagging. Find the value of parallel capacitance that will correct the load power factor to unity.

A high-power radio transmitter installation. The amplitude or frequency of a sinusoidal "carrier" signal is varied to represent the sound, picture, or data to be transmitted. The power generated, which is delivered to an antenna, is typically thousands of watts.   (*Photo courtesy of the Radio Corporation of America*.)

Power transmission lines leading from a switching yard to a distant metropolis. The switching yard contains transformers for altering the line voltages, circuit breakers to protect the lines and equipment from faults such as lightning strikes, and huge switches to change the routing of electrical power to bypass damage or for maintenance. (*Photo courtesy of Southern California Edison Co.*)

4. A 3000-W, 125-V load has a power factor of 0.8, lagging. What value of parallel reactance will correct the power factor to 0.95, lagging?

5. A balanced three-phase system has a line-to-line voltage of 440. One load, which is wye connected, delivers 2000 W at a power factor of 0.9, leading; and a second, delta-connected load, consumes 3500 watts at 0.8 power factor, lagging. Find the line current and the power factor of the system.

Advanced Problems

Average Power

1. An element is said to be *passive* if for sinusoidal voltage and current of any frequency, the average electrical power flow is always *into* the element. Show that an element with impedance $Z(s)$ is passive if and only if

$$\text{Re}\left[Z(s = j\omega)\right] \geq 0$$

for all ω.

A large power transformer. This unit is three-phase and is designed to operate at 220 kV. (*Photo courtesy of Southern California Edison Co.*)

A wattmeter is a meter that indicates average power. Mechanical watt meters involve two coils, one for voltage, the other for current. A force between the coils is developed that is proportional to the voltage-current product. The average force, indicated by the meter scale, is the average of the voltage-current product, which is the average power.

Wattmeter symbols are shown in Figure 10-28. The voltage-sensing terminals are labeled V and are connected as one would connect a voltmeter,

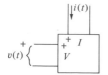

Figure 10-28 Wattmeter symbol

across the voltage to be sensed. The current sensing terminals I are connected as an ammeter is connected, in series with the current to be sensed.

The plus signs beside a voltage and a current terminal indicate the senses
of positive voltage and current that produce an upward deflection of the
meter pointer. For voltage and current

$$v(t) = A \cos(\omega t + \theta)$$
$$i(t) = B \cos(\omega t + \phi)$$

with the reference senses shown in Figure 10-28, the wattmeter will indicate
the quantity

$$P = \frac{AB}{2} \cos(\theta - \phi)$$

2. What will be the reading of the wattmeter in the network of Figure 10-29?

3. A wattmeter may be used to measure the power flow from a source in
a three-phase system, as shown in Figure 10-30(a) for sources in wye. For
a perfectly balanced power system, the total power flow is three times this
reading.

Show that in a balanced three-phase system, the wattmeter connection
of Figure 10-30(b) will produce a wattmeter reading of $(3/2)P \pm (\sqrt{3}/2)Q$,
where P and Q are the average and the reactive power of one source, and
where the algebraic sign depends on the phase rotation.

4. In an unbalanced three-phase power system, the total power flow could
be measured using three wattmeters, each responding to the power flow
out of one of the sources, as indicated in Figure 10-31(a). In terms of rms
phasors, the total power is

$$P = |\mathbf{V}_1||\mathbf{I}_1| \cos \theta_1 + |\mathbf{V}_2||\mathbf{I}_2| \cos \theta_2 + |\mathbf{V}_3||\mathbf{I}_3| \cos \theta_3$$

Figure 10-29 Advanced Problem 2

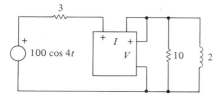

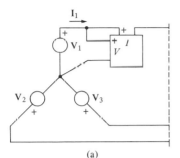 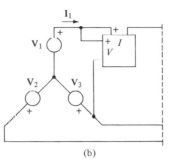

(a)   (b)

Figure 10-30 Advanced Problem 3

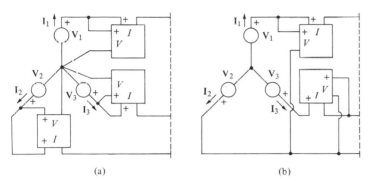

(a)   (b)

Figure 10-31 Advanced Problem 4

where

$$\theta_1 = \angle V_1 - \angle I_1$$
$$\theta_2 = \angle V_2 - \angle I_2$$
$$\theta_3 = \angle V_3 - \angle I_3$$

Show that the commonly used arrangement of *two* wattmeters in Figure 10-31(b) produces readings the sum of which are the total power flow out of the sources.

In this two-wattmeter method of power measurement, one of the wattmeter readings may be negative, meaning that the magnitude of its reading should be subtracted from that of the other wattmeter to give the total power flow.

5. There are other numbers of balanced circuits besides three that have the property of constant net power flow. Find a two-phase system for which the total instantaneous power flow is constant. How many wires are necessary for power transmission in this system?

Index